AF478545

Advances in Applied Nanotechnology for Agriculture

ACS SYMPOSIUM SERIES **1143**

Advances in Applied Nanotechnology for Agriculture

Bosoon Park, Editor

U.S. Department of Agriculture, Agricultural Research Service
Richard B. Russell Research Center
Athens, Georgia

Michael Appell, Editor

U.S. Department of Agriculture, Agricultural Research Service
National Center for Agricultural Utilization Research
Peoria, Illinois

Sponsored by the
ACS Division of Agricultural and Food Chemistry, Inc.

American Chemical Society, Washington, DC

Distributed in print by Oxford University Press

Library of Congress Cataloging-in-Publication Data

Advances in applied nanotechnology for agriculture / Bosoon Park, editor, U.S. Department of Agriculture, Agricultural Research Service, Richard B. Russell Research Center, Athens, Georgia, Michael Appell, editor, U.S. Department of Agriculture, Agricultural Research Service, National Center for Agricultural Utilization Research, Peoria, Illinois ; sponsored by the ACS Division of Agricultural and Food Chemistry, Inc.

pages cm. -- (ACS symposium series ; 1143)
Includes bibliographical references and index.
ISBN 978-0-8412-2802-3
1. Agricultural innovations. 2. Nanotechnology. I. Park, Bosoon. II. Appell, Michael. III. American Chemical Society. Division of Agricultural and Food Chemistry.
S494.5.I5A285 2013
620'.5--dc23

2013035409

The paper used in this publication meets the minimum requirements of American National Standard for Information Sciences—Permanence of Paper for Printed Library Materials, ANSI Z39.48n1984.

PRINTED IN THE UNITED STATES OF AMERICA

Foreword

The ACS Symposium Series was first published in 1974 to provide a mechanism for publishing symposia quickly in book form. The purpose of the series is to publish timely, comprehensive books developed from the ACS sponsored symposia based on current scientific research. Occasionally, books are developed from symposia sponsored by other organizations when the topic is of keen interest to the chemistry audience.

Before agreeing to publish a book, the proposed table of contents is reviewed for appropriate and comprehensive coverage and for interest to the audience. Some papers may be excluded to better focus the book; others may be added to provide comprehensiveness. When appropriate, overview or introductory chapters are added. Drafts of chapters are peer-reviewed prior to final acceptance or rejection, and manuscripts are prepared in camera-ready format.

As a rule, only original research papers and original review papers are included in the volumes. Verbatim reproductions of previous published papers are not accepted.

ACS Books Department

Contents

Indexes

Preface

Agricultural products influence most aspects of life, including food and feed, feedstocks for bio-based products, and everyday materials, such as fuels, textiles, and furniture. Advances in technology are necessary to address the future global needs from agriculture. Nanotechnology is a promising field focused on the unique chemical properties of materials with a dimension of 1–100 nm. Nanotechnology has the potential to revolutionize agricultural and food systems with various applications including food safety, quality, product traceability, better nutrient delivery systems, enhancing packaging performance, and improving agricultural and food processing. This volume of the ACS Symposium Series focuses on the unique challenges of applying nanotechnology to benefit the agriculture sector and the food industry.

This book is based on a series of nanotechnology-related symposia sponsored by the Division of Agricultural and Food Chemistry at spring ACS national meetings between 2009 and 2013. During this period tremendous progress was made on practical applications of nano-based technologies to address agricultural problems, including those in the areas of food safety, development of new value-added biomaterials, and nutraceutical/flavor delivery. This book initiated from the interest of symposium speakers to share their work with wider audiences. Fourteen chapters were selected and written for this special issue. The chapters were developed independently and are arranged according to topic. The authors are from a wide range of disciplines and include food scientists, chemists, engineers, biologists, medical researchers, and physicists.

The goal of this book is to provide the perspectives of scientists working with nanotechnology to address agricultural problems. The research presented within this book was conducted in proper laboratory environments meeting facility, local, and national safety standards. Applied nanotechnology research benefits from the proactive and concurrent efforts by the safety community to provide guidance and assess risks associated with nanomaterials as the technology develops. Challenges remain for more broad utilization of nanotechnology in the field, including more thorough assessment of occupational risks of exposure to nanomaterials, regulation, and public support of nanotechnology in daily lives.

We are extremely grateful to the authors and peer-reviewers for contributing their expertise to this book. The editors would like to thank the Division of Agricultural and Food Chemistry for providing venues for the symposia and financial support. We are also thankful to ACS publications (Tim Marney, Arlene Furman, Bob Hauserman, and Ashlie Carlson), James Oxley (previous symposium co-organizer), and the contributors and participants of the symposia. It has been a pleasure to work with dedicated people possessing the talent and strong desire to advance this field.

Bosoon Park
Quality and Safety Assessment Research
USDA-ARS, Richard B. Russell Research Center
950 College Station Rd.
Athens, GA 30605 U.S.A.

Michael Appell
Bacterial Foodborne Pathogens & Mycology Research
USDA-ARS, National Center for Agricultural Utilization Research
1815 N. University St.
Peoria, IL 61604 U.S.A.

Chapter 1

Gold Nanotechnology for Targeted Detection and Killing of Multiple Drug Resistant Bacteria from Food Samples

Paresh Chandra Ray,* Sadia Afrin Khan, Zhen Fan,
and Dulal Senapati

Department of Chemistry, Jackson State University,
Jackson, Mississippi 39217, U.S.A.
*E-mail: paresh.c.ray@jsums.edu. Fax: 601-979-3674.

Contamination of food like lettuce, egg, and peanut butter by pathogenic bacteria is very common in this world. Since last decade, food technology is facing another challenge due to the infection of multiple drug resistant bacteria (MDRB). This demands an ultrasensitive sensing technology for the rapid detection from food and new approaches for the killing of bacterial pathogens without using the currently available antibiotics. This chapter deals with the basic concepts and critical properties of the gold nanostructures that are useful for the MDRB sensing and killing. Herein, we discuss our recent reports on bio-conjugated gold nanomaterial based strategies for the MDRB detection from food samples. We also discuss the photothermal applications of gold nanotechnology for MDRB killing.

Introduction

Despite modern treatment processes, contamination of food by pathogenic bacteria like *Salmonella, E. coli* is very common in this world (*1–5*). *Salmonella* outbreak from eggs is one of the biggest food contaminations in 2010 (*1–5*). On the other hand, *E. coli* outbreak in ground beef, lettuce and vegetables is common for the food industry (*1*). The worldwide food production industry is worth about 578 billion US dollars and often food recalls due to the presence of pathogenic bacteria contamination are becoming an economical problem for the food industry.

As a result, there is a high demand for the development of rapid, sensitive, and reliable assay to identify harmful pathogens from food samples (*6–10*). Though the currently existing conventioonal methods are selective and sensitive, since they rely on a series of enrichment steps, they are too slow from the perspective of industrial needs (*3–8*). As a result, scientists have been searching for the alternative methods for decades (*5–12*). In the last five years, we have reported the development of gold nanomaterial based highly sensitive assay for the detection of different pathogens. Although it is currently in the initial stage, current gold nanomaterial-enabled detection strategies from food samples show promise, but the technology needs further development to meet industrial demands (*13–20*). In terms of detection of bacteria from water, the use of nanotechnology has led to the production of selective, rapid & sensitive multi-analytes assays (*21–30*). Due to the presence of large surface area, a large number of target-specific recognition elements such as antibody, aptamers and peptides can be attached to the gold nanomaterial surface, which allows selective and sensitive sensing of different pathogens (*31–47*). Since gold nanostructures possess unique shape and size-dependent optical properties, gold nanosystem became very attractive in their use in technological system for diagnostic applications (*38–55*). Due to the lack of toxicity, scientists have shown great interest to use gold nanosystems for sensing and imaging (*50–60*). As a result, bio-conjugated gold nanoparticles can be used for several pathogens simultaneously as well as selectively even at the single bacterium level (*60–65*). The purpose of this overview is to provide the current status of the field with a focus of work from our laboratory and include a vision for future applications of gold-nanoparticles for food safety.

Since the development of penicillin in the 1940s, antibiotics are responsible for saving countless human lives (*2–10*). There are around 42 billion U.S. dollars sales of antibiotics every year in the USA alone. However, due to the intensive use of the antibiotics, human pathogens are becoming resistant to many available antibiotics (*2–10*). According to the World Health Organization (WHO) (*2*). there may be another 1-2 decades left for people to use the existing antibiotics. All the above facts clearly indicate that the new approaches for the treatment of infectious bacterial pathogens that do not rely on traditional therapeutic regimes, is very urgent for our society's health care. One promising method, still in its initial stage, is to use nanomaterial-based photothermal killing of harmful bacteria selectively (*17–23*). In the last few years, we have shown that unique size and shape dependent optical properties of gold nanomaterials can be used for the selective non-invasive photothermal lysis applications for MDRB and other pathogens (*17–23*). In this book chapter, we summarize the recent promising reports mainly from our group and suggest future strategies to identify, target or destroy pathogens.

Label-Free Colorimetric Sensing

Since last one decade, scientists are working on the development of simple colorimetric assay for bacteria. As we know, solid material exhibits vibrational modes which are associated with the collective oscillations of their atoms (*10–18*). In the case of noble metals like gold, as the size is reduced to a few nanometers scale, which is comparable to the wavelength of light, a new very strong absorption is observed (*18–25*). It is due to the collective oscillations of the electrons in the conduction band from one surface of the particle to the other and known as localized surface Plasmon resonances (LSPR) (*25–32*).

In the presence of LSPR, electric fields near the particle's surface are greatly enhanced and the particle's optical extinction reaches maximum at the plasmon resonant frequency at visible and NIR wavelength depending on the size and shape of gold nanoparticles (*32–40*). Due to the huge surface plasmon enhancement, absorption cross-sections of noble gold nanoparticles are 5-7 orders of magnitude more than any available organic dye molecules (*22–25, 58–61*). As a result, each metal nanoparticle can be considered as an optical probe equivalent to several million dye molecules (*40–47*). Due to the above fact, very high detection sensitivity can be achieved using gold nanoparticle based assays. Since the colors of gold nanoparticle solutions are highly dependent on the interparticle distance of nanoparticles, the working principle of the gold nanoparticle based colorimetric sensor is based on the fact that when individual gold nanoparticles come into close proximity, due to the interparticle plasmon coupling, a distinct color change is usually observed (*10–20*).

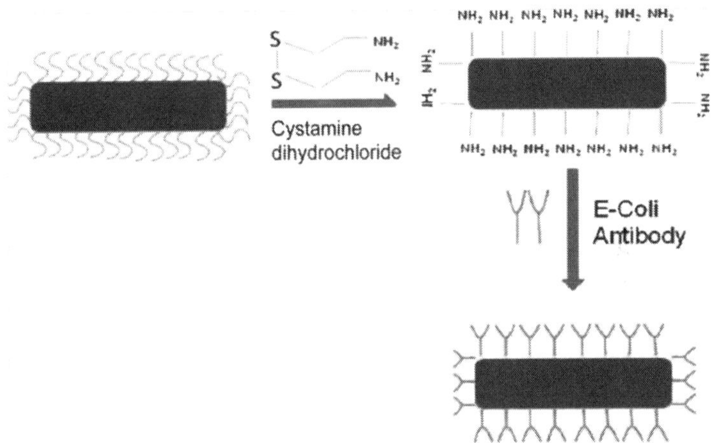

Scheme 1. Schematic representation shows the synthesis of anti-E. coli antibody-conjugated nanorod. (Reprinted with permission from Ref. (20), Copyright 2009, American Chemical Society.)

In 2005, Berry et. al. (*16*) reported that due to the strong electrostatic interactions, cetyltrimethylammonium bromide (CTAB)-functionalized gold nanorods can conformally deposit to form a monolayer on Bacillus cereus. We have reported (*20*) gold nanorod based colorimetric detection assay for *E. coli* bacteria. At first we have modified the gold nanorod with anti-*E. coli* antibody, as shown in Scheme 1. The working principle for our colorimetric assay is based on the fact that *E. coli* bacteria are more than an order of magnitude larger in size (1−3 µm) than the anti-*E. coli* antibody-conjugated gold nanorods. In the presence of *E. coli* bacteria, several gold nanorods conjugate with one *E. coli* bacterium, as shown in Figure 1. As a result, anti-*E. coli* antibody-conjugated gold nanorods undergo aggregation in the presence of bacteria (*20*). Due to the aggregation in the presence of bacteria, the interparticle distances of nanoparticles decrease. As a result, the color change takes place (as shown in Figure 1). This bioassay is rapid, takes less than 15 min from bacterium binding to detection and analysis, and is convenient and highly selective. This color change is also due to the change of the refractive index near the nanoparticle surface in the presence of bacteria. As the refractive index near the nanoparticle surface increases, the nanoparticle extinction spectrum shifts to longer wavelengths. In our case, when antibody attached nanoparticles conjugate with one bacteria through antigen-antibody interaction, local refractive index on the nanoparticle surface increases and as a result, a red-shift in the plasmon mode is expected.

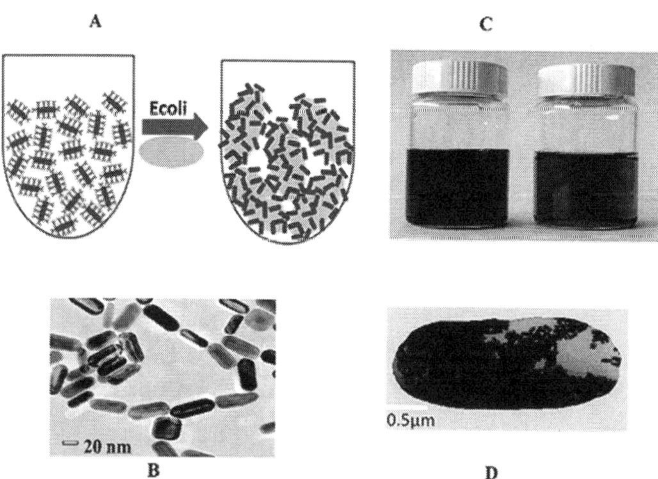

Figure 1. (A) Schematic representation of anti-E. coli antibody-conjugated nanorod-based sensing of E. coli bacteria. (B) TEM image of anti-E. coli antibody-conjugated nanorods before the addition of E. coli bacteria. (C) Photograph showing colorimetric change upon the addition of E. coli bacteria (10⁴ cfu/mL), and (D) TEM image demonstrating aggregation of the gold nanorods after the addition of E. coli bacteria (10³ cfu/mL). (Reprinted with permission from Ref. (20), Copyright 2009, American Chemical Society.)

Next, we have reported (*21*) bio-conjugated egg shape gold nanomaterial assay for the selective sensing of *salmonella* bacteria. In this case, at first we have modified the nanoparticles with anti-*salmonella* antibody (as shown in Scheme 2), for selective *salmonella* bacteria detection. Our reported experimental data have shown that the sensitivity of colorimetric assay for *Salmonella* detection is 10^4 bacteria cells/mL (*21*), which is comparable to the detection limit (10^4–10^5cfu/mL) of ELISA. Our reported result shows that oval shape gold nano particle based bioassay is rapid, takes less than 5 min from bacterium binding to detection and analysis. It is also convenient and highly selective. Next, to understand whether our oval shape gold nanoparticle based colorimetric assay is highly selective, we have also found out how our colorimetric assay responded to the addition of *E. coli* bacteria to anti *salmonella* antibody conjugated oval shape gold nanoparticle. As shown in Figure 2C, no color change has been observed even after the addition of 10^6 *E. coli* bacteria. Similarly, we have not observed any color change when we added 10^6 *Salmonella typhimurium* bacteria to anti *E. coli* antibody conjugated oval shape gold nanoparticles. On the other hand, a distinct color change was observed when we added 10^6 *E. coli* bacteria to anti-*E. coli* antibody conjugated oval shape gold nanoparticles.

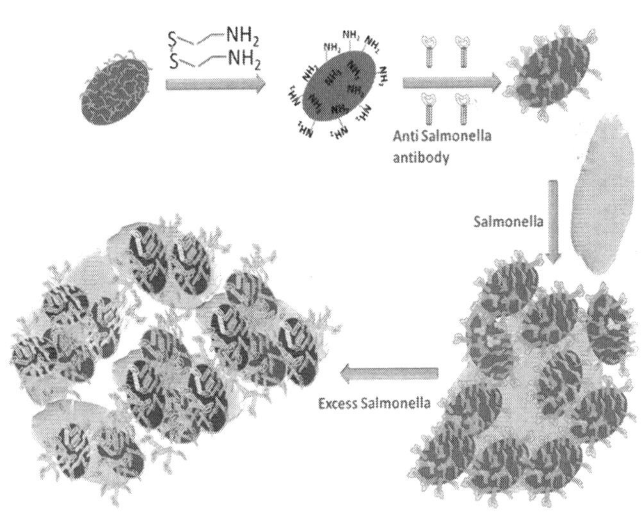

Scheme 2. Schematic representation showing antibody-conjugated nanotechnology-driven approach using oval shape gold nanoparticle to selectively target Salmonella bacteria. (Reprinted copyright permission from from Ref. (21), Copyright 2010, Wiley- VCH.)

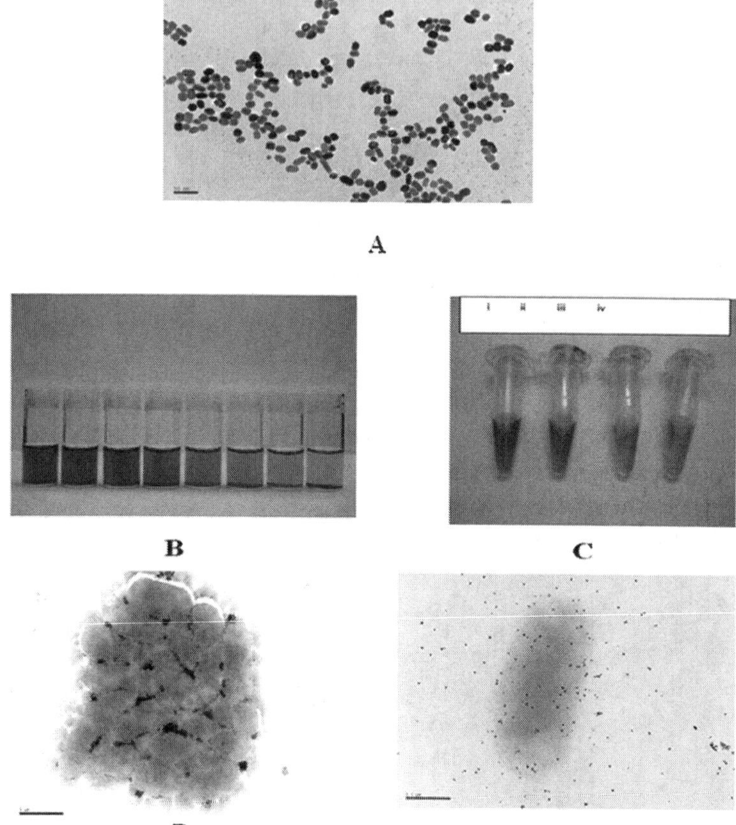

Figure 2. A) TEM picture of the freshly prepared oval shape gold nanoparticle. B) Photograph showing colorimetric change upon the addition of (from left to right) 10, 50, 100, 500, 1000, 5000, 10000, 50000, 100000, 500000 CFU/mL of Salmonella bacteria. C) Photograph showing colorimetric change upon the addition of i) 10^6 E. coli bacteria to anti Salmonella antibody conjugated oval shape gold nanoparticles, ii) 10^6 salmonella bacteria to anti E. coli antibody conjugated oval shape gold nanoparticles, iii) 10^3 E. coli bacteria to anti E. coli antibody conjugated oval shape gold nanoparticles, iv) 10^6 E. coli bacteria to anti E. coli antibody conjugated oval shape gold nanoparticles. D) TEM image demonstrating the formation of bigger microbial clusters in the presence 10^6 cfu/mL Salmonella bacteria. E) TEM image of Salmonella bacteria in the presence of anti E.coli antibody conjugated oval shape gold nanoparticles. (Reprinted with permission from Ref. (21), Copyright 2010, Wiley- VCH.)

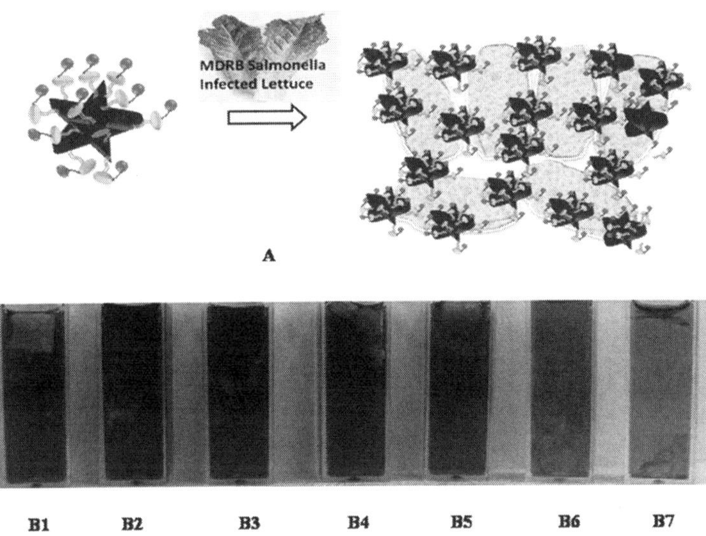

Figure 3. A) Schematic representation showing our gold nanotechnology-driven colorimetric approach for the rapid selective screening of MDRB Salmonella bacteria from infected lettuce. B) Photograph showing colorimetric change upon the addition of infected lettuce on antibody-conjugated gold nanoparticle. Lettuce was infected by B1) 10^2CFU/gm MDRB Salmonella, B2) 10^6 CFU/gm Salmonella ser. Agona bacteria, B3) 10^6CFU/gm E. coli bacteria, B4) the mixture of 10^6CFU/gm Salmonella ser. Agona and 10^6CFU/gm E.coli, B5) the mixture of 10^6CFU/gm Salmonella ser. Agona and 10^2CFU/gm MDRB Salmonella, B6) the mixture of 10^6CFU/gm Salmonella ser. Agona and 10^3CFU/gm MDRB Salmonella, B7) 10^3CFU/gm MDRB Salmonella. (Reprinted with permission from Ref. (22), Copyright 2011, Royal Society of Chemistry.)

Next, to understand whether our gold nanotechnology based assay can detect MDRB from food samples, we have tested the selectivity and sensitivity of our assay using MDRB *Salmonella DT104* infected romaine lettuce. We have reported (*22*) the selectivity and sensitivity of our colorimetric assay using ampicillin, chloramphenicol, streptomycin, sulfonamides, and tetracycline antibiotics drug resistant (MDRB) *S. typhimurium DT104* (ATCC 700408) infected romaine lettuce. For this purpose, we purchased lettuce from a local vender (Jackson, Mississippi, USA) and then chopped them into small pieces. After that, the chopped lettuces were infected with Salmonella DT104. This MDRB *Salmonella* DT104 infected lettuce samples were used for MDRB detection. Figure 3 shows the selectivity and sensitivity of our colorimetric assay result, when antibody-conjugated gold nanoparticles were mixed with various concentrations of MDRB *S. typhimurium DT104* infected lettuce samples. The color change is mainly due to the aggregation of antibody conjugated star shape gold nanoparticle in the presence of *S. typhimurium DT104*, as shown in Figure 4. Next, to understand whether our gold nanoparticle based colorimetric assay is highly selective, we have also tested if our colorimetric assay responded to the addition of *Salmonella ser. Agona* or *E. coli O157:H7* bacteria infected lettuce. As shown in Figure 3, no color change has been observed even after the addition of lettuce sample infected by10^6 CFU/gm *Salmonella ser. Agona* or *E. coli* bacteria to M3038 antibody conjugated popcorn shape gold nanoparticle. Since M3038 antibody conjugated star shape nanoparticle does not bind with *Salmonella ser. Agona* or *E. coli* bacteria, as shown in Figure 4, no color change has been observed. We have also tested for the mixture of bacteria, as shown in Figure 3. Our result clearly shows that the color change is observed only when MDRB 10^3 CFU/gram or more of *Salmonella* DT104 are present. The sensitivity of this colorimetric assay was 10^3 CFU/gram.

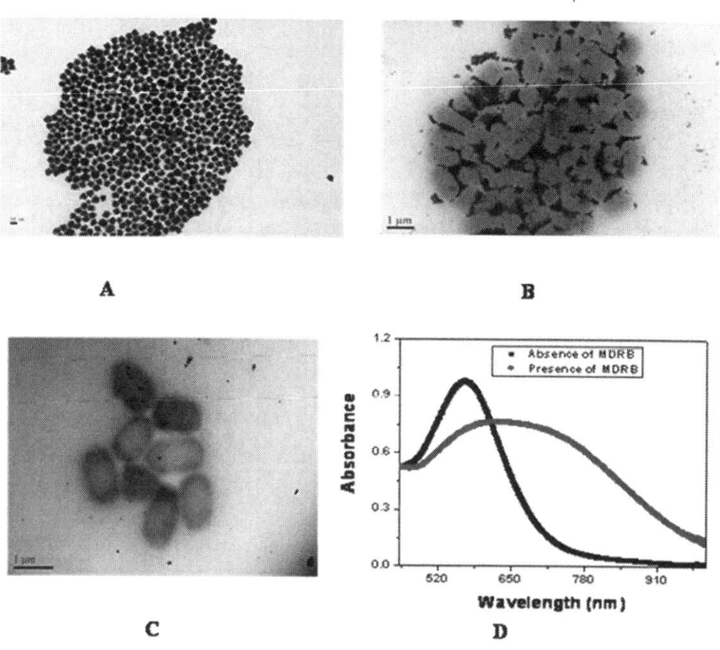

Figure 4. TEM image showing A) monoclonal M3038 antibody-conjugated popcorn shape gold nanoparticles before the addition of MDRB. B) Aggregation of gold nanoparticles as well as the formation of microbial clusters in presence of 10^5 CFU/mL MDRB Salmonella. C) Very little binding is observed in the presence of 10^6 CFU/mL Salmonella ser. Agona bacteria. D) Plot demonstrating the absorption spectral change in the presence of 10^5CFU/mL MDRB Salmonella DT104. (Reprinted with permission from Ref. (22), Copyright 2011, Royal Society of Chemistry.)

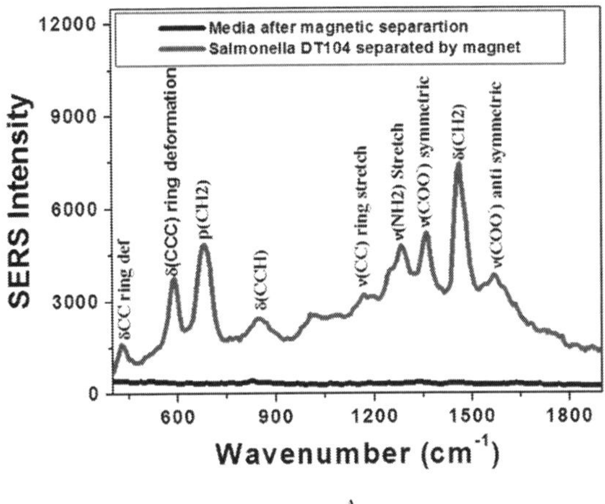

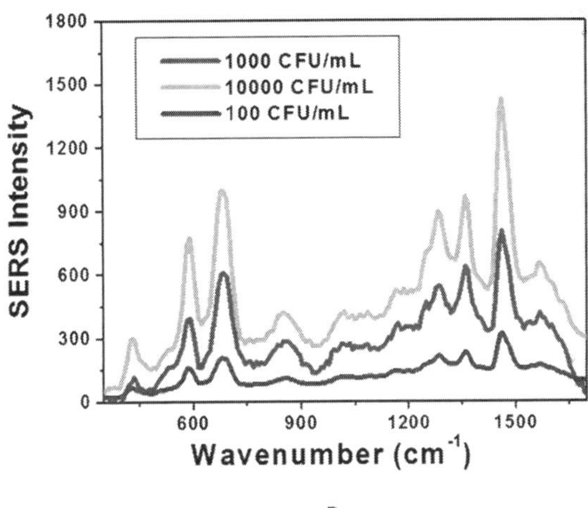

Figure 5. A) Plot shows SERS intensity from bacteria conjugated nanoparticle after magnetic separation and re-suspension with PBS. All the observed Raman signals are directly from Salmonella DT104. No SERS signal has been observed from supernatant, which is mainly PBS. (B) Plot shows SERS intensity variation from different concentrations of MDR Salmonella DT104. All the reported concentrations are before the magenic separation. (Reprinted with permission from Ref. (65), copyright 2012, Wiley-VCH.)

Label-Free SERS Sensing

The possibility of observing Raman signals, which are normally very weak, with enhancements on the order of 10^8–10^{14} allow surface enhanced Raman spectroscopy (SERS) to be unique for ultrasensitive pathogen sensing (*51–58*). In addition to sensitivity, one of the other important features in case of SERS is the specificity that usually can be achieved by controlling the chemistry around the metal surface (*59–65*). As a result, SERS based sensor for the bacteria is highly valuable. Premasiri et. al. (*41*) reported SERS probe based on gold nano particle aggregate covered glass chips for the detection of microorganisms through their Raman spectra analysis. They have shown that the enhancement of intensity factors can be greater than 10^4 per bacterium.

Very recently, we have reported (*65*) the detection of MDRB *Salmonella* bacteria using SERS, without any tagged dye. The largest Raman scattering enhancements are observed when molecules are residing in the fractal space between aggregated colloidal nanoparticles (*51–55*). As shown in Figure 4, our data clearly show that the MDR *Salmonella* DT104 helps to generate "hot spots" through aggregation of the multifunctional M3038 antibody nanoparticles via antigen-antibody interaction. It is the perfect condition to use SERS spectra as "fingerprint" for whole MDRB organism, as shown in Figure 5. Also, in nano-popcorn, like nano-star, the central sphere acts as an electron reservoir while the tips are capable of focusing the field at their apexes, which provides huge field enhancement of scattering signal.

Figure 5 shows the SERS spectra from the suspensions of the nanoparticle-*Salmonella* conjugates after magnetic separation. Since the bacteria cell wall consists of proteins, lipids, carbohydrates, one can expect to see the SERS spectra from the vibrational mode of the above compositions. Vibrational assignments of the observed SERS peak are shown in Figure 5. Our observed SERS bands are in good agreement with the Raman bands reported for different microorganisms in the literature (*41, 65*).

The Raman enhancement, G, was measured experimentally by direct comparison with normal & SERS Spectra as shown below (*51–56*),

$$G = [I_{\text{SERS}}]/[I_{\text{Raman}}] \times [M_{\text{bulk}}]/[M_{\text{ads}}]$$

where I_{SERS} is the intensity of a 1460 cm^{-1} vibrational mode in the surface-enhanced spectrum in the presence of *Salmonella* DT104, and I_{Raman} is the intensity of the same mode in the bulk Raman spectrum from only bacteria. M_{bulk} is the number of bacteria used in the bulk, M_{ads} is the number of bacteria adsorbed and sampled on the SERS-active substrate. M_{bulk} was calculated using colony counting. For M_{ads} calculation, after magnetic separation, we have also performed colony counting. All spectra are normalized for the integration time.

An enhancement factor estimated from the SERS signal and normal Raman signal ratio for 1460 cm^{-1} band is approximately 8.5×10^7. No significant changes in Raman frequencies were observed in comparison to the corresponding SERS and Raman bands. To evaluate the sensitivity of our SERS probe, different concentrations of MDR *Salmonella* DT104 from one stock solution were evaluated. As shown in Figure 5B, the SERS intensity is quite sensitive to the concentration of MDR *Salmonella* DT104. Our experimental results clearly demonstrate that the sensitivity of our popcorn shape nanoparticle based label-free SERS probe is as low as 100 CFU/mL MDR. Our data also indicate that SERS can be used as a finger print for MDRB.

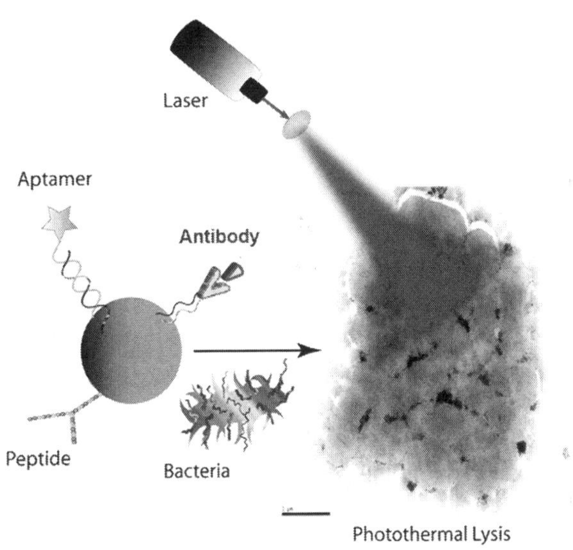

Scheme 3. Schematic representation showing bacteria conjugated nanoparticle based photothermal hyperthermic destruction of MDRB. (Reprinted with permission from Ref. (22), Copyright 2011, Royal Society of Chemistry.)

MDRB Killing Using Photothermal Process

Gold nanoparticles of different sizes and shapes with optical properties tunable in the visible to near-infrared can be used for the hyperthermic destruction of MDRB using NIR light (*30–50*), as shown in Scheme 3. When the light of appropriate wavelength is absorbed by bacteria conjugated desired gold nanoparticles, absorbed light will be converted into heat by rapid electron-phonon relaxation followed by phonon-phonon relaxation. This highly localized heat generated by the gold nanoparticle attached with bacteria can be used to destroy MDRB selectively by disrupting the cell membrane. For photothermal destruction, gold nanoparticles serve as "light-directed nano heaters", which is very useful for the selective laser photothermolysis of pathogens (*15–23, 30–33, 65*). Gold nanoparticles are known to absorb light several millions of times stronger than the organic dye molecule and, as a result, in the presence of light, temperature rises on the order of a few tens of degrees (*15–23, 30–33, 38*). As shown in Figure 6, this photothermal hyperthermia process destroys MDRB via cell damage using different thermal effects such as denaturation of proteins/enzymes, induction of heat-shock proteins, metabolic signaling disruption, endothelial swelling, microthrombosis, etc. (*15–23, 30–33, 38*).

When the incident laser frequency overlapped with the plasmon absorption maximum of the gold nanoparticles conjugated pathogens, the selective heating and destruction of MDRB can be achieved at a much lower laser powers than that required to destroy healthy bacteria to which bio-conjugated nanoparticles do not bind specifically. Zharov et. al. (*19*) reported localized killing of *S. aureus* in vitro by combining laser and photo-thermal technique. Later, Norman et. al. (*17*) reported photothermal killing using gold nanorods that have been covalently linked with antibodies to selectively destroy *Pseudomonas aeruginosa,* which was obtained from the upper respiratory tract of sinusitis patients.

Table 1. Laser irradiation time to kill 100% MDRB in lettuce samples

Concentration of MDRB (CFU/gm) in lettuce samples	*Laser Irradiation Time (minutes) to Kill 100% Bacteria*
10^2	6
10^3	9
10^4	14
10^5	20
10^6	25
10^7	30

(Reprinted with permission from Ref. (*23*), Copyright 2011, Royal Society of Chemistry.)

All initial studies did not demonstrate the selectivity of their assay with respect to other pathogens, which is very important before this technique can be used for real life sample. In 2010, we have reported selective photothermal killing of salmonella bacteria using antibody conjugated popcorn shape gold nanoparticle in the presence of 670 nm light exposure (*21*). As shown in Figure 6, our result shows that when oval shape gold nanoparticles are attached to bacterial cells, the localized heating that occurs during 670 nm irradiation is able to cause irreparable cellular damage within 15 minutes light exposure.

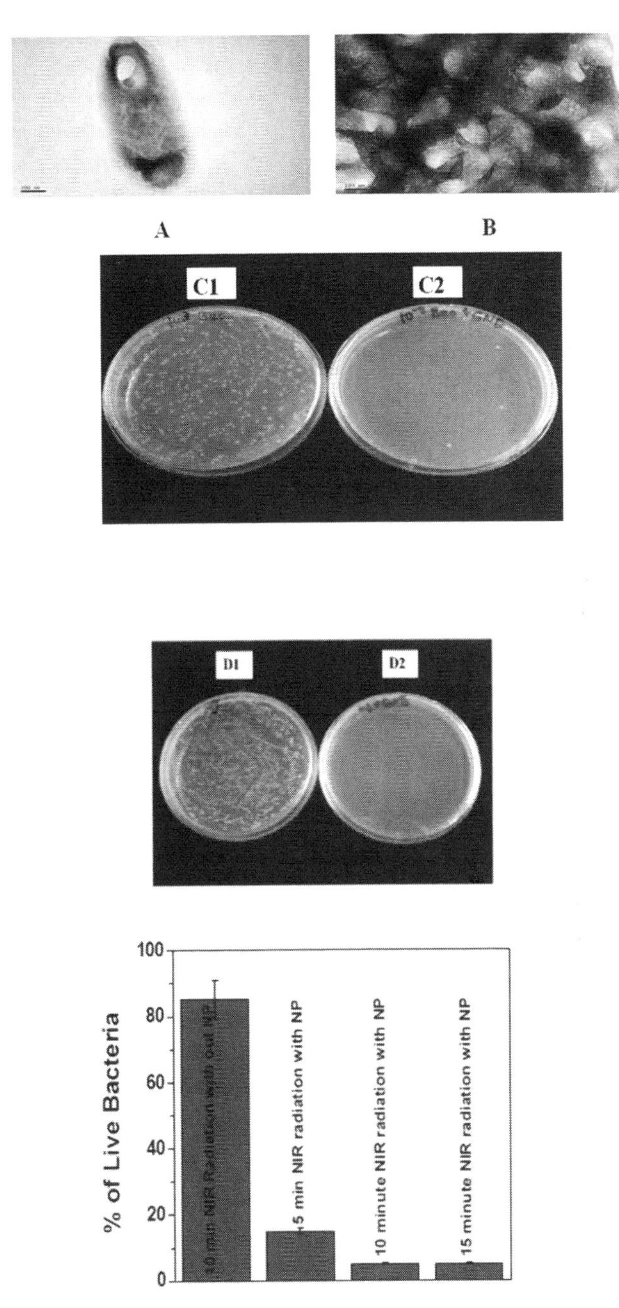

Figure 6. TEM image demonstrating irreparable damage of single bacterial cell surface when anti–Salmonella antibody coated NP conjugated S. typhimurium bacteria were exposed to 676 nm NIR radiation for 10 minutes. B) TEM image

To understand whether our oval shape gold nanoparticle based photo thermal lysis assay is highly selective, we have also performed experiments to find out how our assay responds to the addition of 10^6/ml *E. coli* bacteria to anti salmonella antibody conjugated oval shape gold nanoparticle. As shown in Figure 6, most of the *E. coli* bacteria remain alive even after 15 minutes of irradiation using 670 nm light.

On the other hand, 97% bacteria were killed when we used salmonella bacteria instead of *E. coli* bacteria. So our experimental results clearly show that our bioconjugated oval shape gold nanoparticle based photo-thermal lysis process is highly selective (*21*). On the basis our results, it is clear that a nanotechnology-driven approach for rapid detection and treating antibiotic resistant bacteria is feasible.

After that, to demonstrate that our gold nanotechnology based assay can kill MDRB from the food sample, we have reported (*23*) a photothermal assay to kill MDRB using MDRB *Salmonella DT104* infected romaine lettuce, as shown in Figure 7. Our selective photothermal lysis of MDRB *Salmonella* typhimurium DT104 is based on the fact that monoclonal M3038 antibody- conjugated popcorn shaped gold nanoparticles can readily and specifically bind with *Salmonella typhimurium* DT104 bacterium O-antigen, through antibody–antigen recognition. Figure 7C shows the time dependent photothermal killing of MDRB *Salmonella typhimurium* DT104, which clearly shows that 100% bacteria can be killed with 20 minute of exposure to 670 nm light. Table 1 shows how the laser irradiation time varies with the concentration of bacteria (CFU/gm) to kill 100% of MDRB from MDRB infected lettuce sample.

As shown in Table 1, 100 % of bacteria can be destroyed by 6 to 30 minutes of laser irradiation treatment, depending on the MDRB concentration (10^2 -10^7 CFU/gm). Though we have shown bio-conjugated gold nanotechnology based photothermal killing of MDRB in food sample, it is fair to admit that at this relatively early stage of development, we have to do much more before it can be used for daily life food sample.

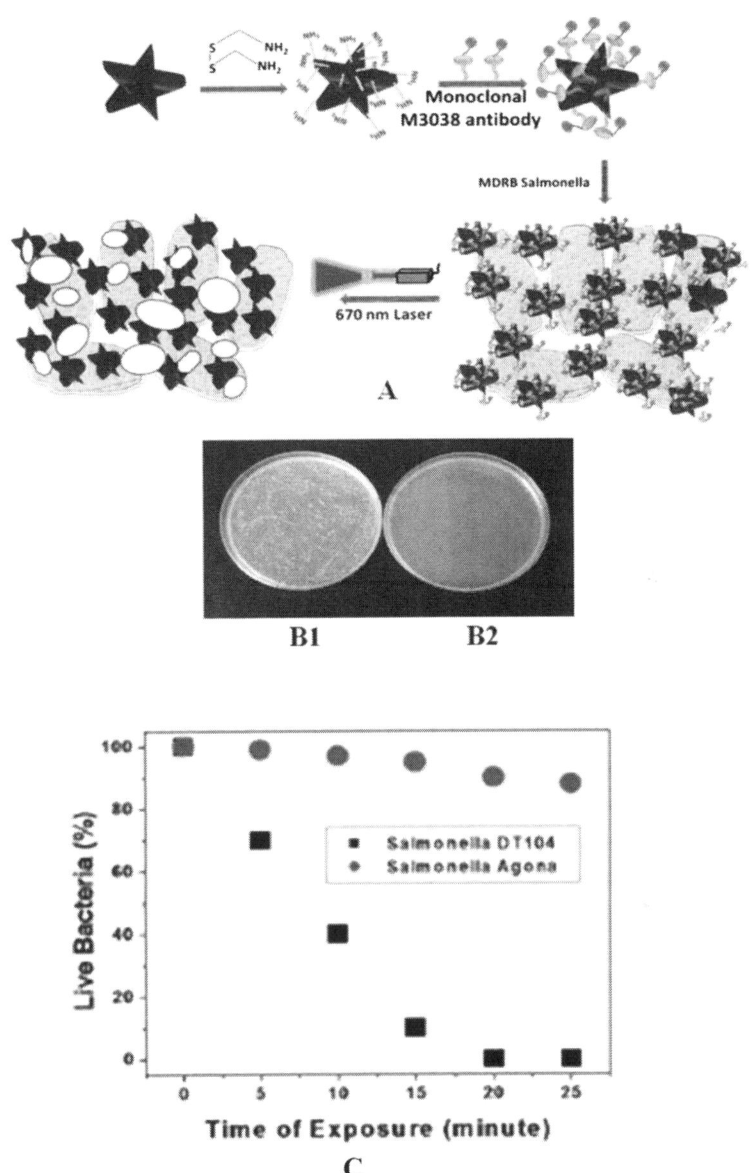

Figure 7. (A) Schematic representation showing the monoclonal M3038 antibody conjugated popcorn shaped gold nanotechnology-driven approach to selective photothermal killing of MDRB Salmonella. (B) Colonies of MDRB Salmonella DT104 (B1) before and (B2) after exposure to 670 nm light with 200 mW cm² power for 20 minutes in the presence of monoclonal M3038 antibody-conjugated popcorn shaped gold nanoparticles . (C) Plot showing time dependent effect of photothermal lysis when antibody attached popcorn shaped gold nanoparticle

conjugated Salmonella DT104, Salmonella Agona bacteria were treated using 200 mW/cm² 670 nm light. (Reprinted with permission from Ref. (23), Copyright 2011, Royal Society of Chemistry.)

Conclusion and Outlooks

In this chapter, we have discussed the great potential of bio-conjugated gold nanomaterials for the possible application of MDRB diagnostics from food sample and photothermal killing of MDRB. Our results demonstrate that beautiful plasmonic properties of metal nanoparticles are useful for bacteria labeling and photothermal killing. We have reported that after conjugation with bacteria specific aptamer/antibody, gold nanoparticle can be used for label-free sensing of MDRB very selectively. We have shown that bio-conjugated gold nanomaterials also can serve as "nanoscopic heaters" in the presence of suitable wavelength light, which can be very useful for the selective killing of MDRB without antibiotics. Though the use of bio-conjugated gold nanomaterial for MDRB sensing from food sample and selective photothermal killings are only just the beginning, it represents one of the highly promising areas of scientific inquiry into the food technology field. We believe that it will likely lead to the development of exciting techniques or powerful combinations of existing ones for MDRB detection from food sample & killing.

While there are several advantages of using gold nanomaterial based assays, there still remain a number of challenges which needs to be solved before it will be useful for MDRB detection and killing from food sample. Development of reproducible synthetic procedures will help to produce nanomaterials with systematically precise size, shapes, compositions and with specific absorption and light-scattering properties, which will make these materials ideal for multiplexed bacterial detection and killing. Additionally, techniques for surface modification and patterning will need to be advanced. Even after 15 years of extensive nanomaterial research, it is quite hard to precisely control the number of functional molecules on the surface of nanoparticles, which clearly indicates that we need to develop better strategies for uniform surface modification as well as reproducible functionalization. Problems of nonspecific binding, nanoparticle aggregation and environmental stability need to be addressed, before it can be used for sensing from food sample in the presence of competing targets. In parallel for each study, toxicity and side effects need to be addressed in a serious and systematic way as a function of nanoparticle size, shape, and surface coating. As a result, an understanding of biological response and environmental remediation is necessary before they can be useful in food processing technology.

Acknowledgments

Dr. Ray thanks NSF-PREM grant # DMR-0611539, NSF-CREST grant # HRD-0833178 and NSF-RISE grant # HRD-1137763 for their generous funding.

References

1.   Centers for Disease Control and Prevention. *E. coli (Escherichia coli)*; http://www.cdc.gov/ecoli/outbreaks.html, accessed at 20/08/2012.

2.   World Health Organization. *Tuberculosis (TB)*; http://www.who.int/tb/challenges/mdr/en/index.html, accessed at 10/12/2010

3.   Nikaido, H. *Annu. Rev. Biochem.* **2009**, *78*, 119–146.

4.   Andersson, D. I.; Hughes, D *Nat. Rev.* **2010**, *8*, 260–271.

5.   Ronholm, J.; Zhang, Z.; Cao, X.; Lin, M. *Hybridoma* **2011**, *30*, 43–52.

6.   Allen, H. K.; Donato, J; Wang, H. H.; Cloud-Hansen, K. A.; Davies, J.; Handelsman, J. *Nat. Rev. Microbiol.* **2010**, *8*, 251–259.

7.   Chiu, C. H.; Su, L. H.; Chu, C. H.; Wang, M. H.; Yeh, C. M.; Weill, F. X.; Chu, C. *J. Clin. Microbiol.* **2006**, *44*, 2354–2358.

8.   Pignato, S.; Coniglio, M. A.; Faro, G.; Lefevre, M.; Weill, F.-X.; Giammanco, G. *Foodborne Pathog. Dis.* **2010**, *7*, 945–951.

9.   Perron, G. G.; Quessy, S.; Letellier, A.; Bell, G. *Infec., Genet. Evol.* **2007**, *7*, 223–228.

10. Ram, S.; Vajpayee, P.; Shanker, R. *Environ. Sci. Technol.* **2007**, *41*, 7383–7388.

11. Smith, P. *Emerging Infect. Dis.* **2008**, *14*, 1327–8.

12. Patel, J. R.; Bhagwat, A. A.; Sanglay, G. C.; Solomon, M. B. *Food Microbiol.* **2006**, *23*, 39–46.

13. Andersson, D. I. *Curr. Opin. Microbiol.* **2006**, *9*, 461–465.

14. Vikesland, J.; Wigginton, K. R. *Environ. Sci. Technol.* **2010**, *44*, 3656–3669.

15. Khan, S. A.; Singh, A. K.; Senapati, D.; Fan, Z.; Ray, P. C. *Chem. Soc. Rev.* **2012**, *41*, 3193–3209.

16. Berry, V.; Gole, A.; Kundu, S.; Murphy, C. J.; Saraf, R. F. *J. Am. Chem. Soc.* **2005**, *127*, 17600–17602.

17. Norman, R. S.; Stone, J. W.; Gole, A.; Murphy, C. J.; Sabo-Attwood, T. L. *Nano Lett.* **2008**, *8*, 302–306.

18. Wang, C.; Irudayaraj, J. *Small* **2008**, *4*, 2204–2208.

19. Zharov, V. P.; Mercer, K. E.; Galitovskaya, E. N.; Smeltzer, M. S. *Biophys J.* **2006**, *90*, 619–627.

20. Singh, A. K.; Senapati, D.; Wang, S.; Griffin, J.; Neely, A.; Candice, P.; Naylor, K. M.; Varisli, B.; Kalluri, J. R.; Ray, P. C. *ACS Nano* **2009**, *3*, 1906–1912.

21. Wang, S.; Singh, A. K.; Senapati, D.; Neely, A.; Yu, H.; Ray, P. C. *Chem. A Eur. J.* **2010**, *16*, 5600–5606.

22. Khan, S. A.; Singh, A. K.; Senapati, D.; Fan, Z.; Ray, P. C. *Chem. Commun.* **2011**, *47*, 9444–9446.

23. Khan, S. A.; Singh, A. K.; Senapati, D.; Fan, Z.; Ray, P. C. *J. Mater. Chem.* **2011**, *21*, 17705–17709.

24. Jarvis, R. M.; Brooker, A.; Goodacre, R. *Anal. Chem.* **2004**, *76*, 5198–5202.

25. Ravindranath, S. O.; Mauer, J. L.; Deb-Roy, C.; Irudayaraj, J. *Anal. Chem.* **2009**, *81*, 2840–2846.

26. Xiao, Z.; Farokhzad, O. C. *ACS Nano* **2012**, *6*, 3670–3676.

27. Golightly, R. S.; Doering, W. E.; Natan, M. J. *ACS Nano* **2009**, *3*, 2859–2869.

28. Kell, A. J.; Stewart, G.; Ryan, S.; Peytavi, R.; Boissinot, B.; Huletsky, A.; Bergeron, M. G.; Simard, B. *ACS Nano* **2008**, *2*, 1777–1788.

29. Chen, T.; Wang, H.; Chen, G.; Wang, Y.; Feng, Y.; Teo, W. S.; Wu, T.; Chen, H. *ACS Nano* **2010**, *4*, 3087–3094.

30. Ray, P C *Chem. Rev.* **2010**, *110*, 5332–5365.

31. Huang, X.; El-Sayed, I. H.; Qian, W.; El-Sayed, M. A. *J. Am. Chem. Soc.* **2006**, *128*, 2115–2120.

32. Singh, A K.; Lu, W.; Senapati, D.; Khan, S. A.; Fan, Z.; Senapati, T.; Demeritte, T.; Beqa, L.; Ray, P. C. *Small* **2011**, *7*, 2517–2525.

33. Lu, W.; Singh, A. K.; Khan, S. A.; Senapati, D.; Yu, H.; Ray, P. C. *J. Am. Chem. Soc.* **2010**, *132*, 18103–18114.

34. Kaittanis, C.; Santra, S.; Perez, J. M. *J. Am. Chem. Soc.* **2009**, *131*, 12780–12791.

35. Xie, J.; Zhang, Q.; Lee, J. Y.; Wang, D. I. *ACS Nano* **2008**, *2*, 2472–2480.

36. Rule, K. L.; Vikesland, P. J. *Environ. Sci. Technol.* **2009**, *43*, 1147–1152.

37. Li, F.; Zhao, Q.; Wang, C.; Lu, X.; Li, X. F.; Le, X. C. *Anal. Chem.* **2010**, *82*, 3399–3403.

38. Jain, P. K.; Huang, X.; El-Sayed, I. H.; El-Sayed, M. A. *Acc. Chem. Res.* **2008**, *41*, 1578–1586.

39. Lu, W.; Arumugam, S. R.; Senapati, D.; Singh, A. K.; Arbneshi, T.; Khan, S. A.; Yu, H.; Ray, P. C. *ACS Nano* **2010**, *4*, 1739–1749.

40. Dasary, S. S. R.; Singh, A. K.; Senapati, D.; Yu, H.; Ray, P. C. *J. Am. Chem. Soc.* **2009**, *131*, 13806–13812.

41. Premasiri, W. R.; Moir, D. T.; Klempner, M. S.; Kriege, N.; Jones, G.; Ziegler, L. D. *J. Phys. Chem. B* **2005**, *109*, 312–320.

42. Wang, Y.; Lee, K.; Irudayaraj, J. *J. Phys. Chem. C.* **2010**, *114*, 16122–16128.

43. Griffin, J.; Singh, A. K.; Senapati, D.; Rhodes, P.; Mitchell, K.; Robinson, B.; Yu, E.; Ray, P. C. *Chem.—Eur. J.* **2009**, *15*, 342–351.

44. Darbha, G. K.; Rai, U. S.; Singh, A. K.; Ray, P. C. *J. Am. Chem. Soc.* **2008**, *130*, 8038–8042.

45. Griffin, J.; Ray, P. C. *J. Phys. Chem. B* **2008**, *112*, 11198–11201.

46. Schmucker, A. L.; Harris, N.; Banholzer, M. J.; Blaber, M. G.; Osberg, K. D.; Schatz, G. C.; Mirkin, C. A. *ACS Nano* **2010**, *4*, 5453–5463.

47. Darbha, G. K.; Ray, S.; Ray, P. C. *ACS Nano* **2007**, *3*, 208–214.

48. Ray, P. C. *Angew. Chem., Int. Ed.* **2006**, *45*, 1151–1154.

49. Wijaya, A.; Schaffer, S. B.; Pallares, I. G.; Kimberly, H.-S. *ACS Nano* **2009**, *3*, 80–86.

50. Cao, Y. W. C.; Jin, R. C.; Mirkin, C. A. *Science (Washington, DC, U. S.)* **2002**, *297*, 1536–1540.

51. Nie, S.; Emory, S. R. *Science* **1997**, *275*, 1102–1104.

52. Giljohann, D. A.; Seferos, D. S.; Prigodich, A. E.; Patel, P. C.; Mirkin, C. A. *J. Am. Chem. Soc.* **2009**, *131*, 2072–2073.

53. Porter, M. D.; Lipert, R. J.; Siperko, L. M.; Wang, G.; Narayanana, R. *Chem. Soc. Rev.* **2008**, *37*, 1001–1011.

54. Tiwari, V. S.; Oleg, T.; Darbha, G. K.; Hardy, W.; Singh, J. P.; Ray, P. C. *Chem. Phys. Lett.* **2007**, *446*, 77–82.

55. Casadi, F.; Leona, M.; Lombardi, J. R.; Van Duyne, R. *Acc. Chem. Res.* **2010**, *43*, 782–791.

56. Brus, L. *Acc. Chem. Res.* **2008**, *41*, 1742–1749.

57. Pallaoro, A.; Braun, G. B.; Reich, N. O.; Moskovits, M. *Small* **2010**, *6*, 618–622.

58. Baik, J. M.; Lee, S. J.; Moskovits, M. *Nano Lett.* **2009**, *9*, 672–676.

59. Yan, B.; Thubagere, A.; Premasiri, R. W.; Ziegler, L. D.; Negro, L. D.; Reinhard, B. M. *ACS Nano* **2009**, *3*, 1190–1202.

60. Lu, W.; Singh, A. K.; Khan, S. A.; Senapati, D.; Yu, H.; Ray, P. C. *J. Am. Chem. Soc.* **2010**, *132*, 18103–18114.

61. Lorenzo, L. R.; Javier, F.; Abajo, G.; Liz-Marzn, L. M. *J. Phys. Chem. C* **2010**, *114*, 7336–7340.

62. Khoury, C. G.; Vo-Dinh, T. *J. Phys. Chem.C* **2008**, *112*, 18849–18859.

63. Hu, Y. S.; Jeon, J.; Seok, T. J.; Lee, S.; Hafner, J. H.; Drezek, R. A.; Choo, H. *ACS Nano* **2010**, *4*, 5721–5730.

64. Zelada-Guillén, G. A.; Riu, J.; Düzgün, A.; Rius, F. X. *Angew. Chem., Int. Ed.* **2009**, *48*, 7334–7337.

65. Fan, Z.; Senapati, D.; Khan, S. A.; Singh, A. K.; Hamme, A.; Yust, B.; Sardar, D.; Ray, P. C. *Chem. A Eur. J.* **2012** ASAP Article.

Chapter 2

Nanocolloid Substrates for Surface-Enhanced Raman Scattering (SERS) Sensor for Biological Applications

Jaya Sundaram,[1] Bosoon Park,[*,1] and Yongkuk Kwon[2]

[1]United State Department of Agriculture, Agricultural Research Service, Russell Research Center, Athens, Georgia 30605, U.S.A.
[2]Animal and Plant Quarantine Agency, Anyang, Korea
[*]E-mail: bosoon.park@ars.usda.gov.

Raman spectroscopy gives light scattering to detect and characterize various properties of biological materials; however, since Raman scattering signals are very weak, methods for fabricating surfaces with patterned, roughened substrates are primarily needed to enhance the signals. Preparation of stable substrates using simple methods is a current priority in SERS research. Biopolymer encapsulated with silver nanoparticle (BeSN) substrate was prepared by reducing silver nitrate into silver using trisodium citrate with addition of polyvinyl alcohol as a stablizer. Optical properties of biopolymer based silver nanoparticles were analyzed with UV/Vis spectroscopy as well as hyperspectral imaging microscopy. UV/Vis spectra showed plasmon resonance absorption at 460 nm; distinct plasmon resonance peak from the hyperspectral imaging spectral profile explained homogeneous nature of SERS substrates. SERS signals of *trans*-1,2-bis(4-pyridyl)ethylene (BPE) and Rhodamine 6G were measured to check substrate repeatability. Good signal repeatability was observed with less noise and spot-to-spot variation from substrates prepared at various times. The SERS substrates were used to detect and differentiate two different serotypes (Typhimurium and Enteritidis) of *Salmonella spp.* In addition to *Salmonella spp.*, other foodborne pathogens such as *E. coli, Listeria* innocua, *Staphylococcus* aureus and also *Salmonella* Infantis were detected using the

BeSN SERS substrates. SERS spectral data were obtained with a HeNe laser of 785 nm excitation, between 400 and 1800cm^{-1} from 15 different spots of a target sample on the substrate at three different times. Principal component analysis (PCA) was carried out to differentiate *Salmonella* serotypes, resulting in a total of 98% variation among the serotypes. SERS spectral comparison of different serotypes indicated that they have similar cell wall and cell membrane structures which were identified by spectral regions distinctively between 520 and 1050 cm^{-1}. Major differences between serotypes were observed between 1200 and 1700 cm^{-1} which represents biochcemical characteristics of DNA and RNA components of the cells.

Introduction

Salmonella are Gram-negative, facultative bacteria of the family Enterobacteriaceae (*1*). Improper cooking, reheating, and handling of food may lead to *Salmonella* outbreaks (*2*). *Salmonella* Typhimurium is responsible for most cases of salmonellosis in the United States. During past several years, a *Salmonella* outbreak was linked to peanut butter contamination that caused 329 illnesses, and *Salmonella* tainted eggs caused 1,200 illnesses. Also, it was reported that poultry and poultry products are the major sources of *Salmonella* contamination (*3*). Since *Salmonella* can grow and survive even in adverse environmental conditions (low nutrient concentrations and temperatures between 5.9 °C and 54 °C), control of *Salmonella* in food processing industries is a major concern in terms of food safety (*4*). Food safety depends on ability to detect and differentiatc foodborne pathogens that cause severe outbreaks. Conventional microbiological identification methods are time consuming and labor intensive (*5*). Since food is a complex matrix, it is difficult to directly detect the presence of bacterial pathogens in the food matrices. Thus, development of methods for rapid detection of foodborne pathogens on the food continues and being high priority for food industries. Furthermore, rapid pathogen detection would be very beneficial in quality control during large scale food processing and production operations.

Various technologies such as Fourier transform infrared spectroscopy (FTIR) as well as Raman spectroscopy have been studied for pathogen detection based on their optical characteristics. Raman spectroscopy techniques provide spectral information furnished by molecular vibrations that are provided by both mid- and near-infrared spectral regions. Molecular bonds directly affect the vibrational spectra and give specific fingerprints that can be used to characterize the target samples (*6–10*). The advantage of Raman spectroscopy is that Raman spectrum for water molecules is very weak. Since water is the major component in most biological samples such as foodborne bacteria, use of optical methods with FTIR are limited in identification and characterization of biological materials. While, Raman spectroscopic method can be applied to detect foodborne pathogens (*11*,

12) without interuption of water. Raman spectra also produce sensitive peaks formed by carbon-carbon bond than other infrared spectroscopic methods (*13, 14*), which helps in biological sample identification.

Raman spectroscopy gives inelastic light scattering to characterize various properties of analytes molecules. In Raman spectroscopy, the inherent weakness of Raman signals can be found with an increase of 6 or more order of magnitude through surface enhaced Raman scattering (SERS). SERS gives large signal enhancement of Raman scattering of the sample molecules that are placed on certain rough metal surfaces (*15–17*). Surface plasmon on the substrate of roughened surface allows a part of plasma energy applied through excitation wavelength to radiate and enhance the scattered signals; whereas, smooth surfaces lose the excitation energy as heat and do not provide scattered signal enhancement. Plasmon is the electron excitation of a metal conductor. If this excitation is restricted on the conductive metal surface, it is called surface plasmon. This surface plasmon can either propagate on grating/rough surface or localize on spherical particle surface. Thus, it is necessary to have surface roughness/grating or curvature for surface plasmonic excitation under the light (*18–22*). SERS effects are characterized by two different mechanisms. One is as excitation of surface plasmon which creates local electromagnetic field enhancement, and the other one is adsorption of analyte molecules on the substrate surface. Based on electromagnetic theory, SERS signal enhancement depends on many factors such as excitation wavelength, optical properties of the substrate as well as their surface morphology (*22–26*).

Metals such as silver, gold and copper have been identified as suitable SERS substrate metals. Many different methods such as vacuum evaporation, lithography to make nanoparticle array, laser ablation, electrode deposition, and colloid preparation (*27–30*) have been developed and proposed to prepare metal substrates with nanostructures for SERS. Each method has advantages and disadvantages to fullfill the requirements as substrate, such as optical properties, reproducibility, surface morphology, stability, and signal enhancement factors (*31*). Among various methods, colloidal nanostructures prepared with gold and silver metals have been most frequently used. Although gold, silver and copper are used for SERS nanosubstrates, silver has been identified as universal substrate metal due to its broad plasmon resonance in visible-near infrared region and its higher stability than gold.

Also it was found through many experiments that the most intense Raman scattering signals were observed with silver metal (*31–34*). Several forms of silver nanostructures such as silver nanorod, dendrite, nanowire, silver nanoparticle colloid (*35–42*) are being used for SERS research. Several different methods including chemical reduction, photo-induced reduction, and laser ablation are used for silver nanoparticle colloid preparation. Among those methods, chemical reductions of silver nitrate solutions to synthesize silver nanoparticle colloids are the commonly used methods for development of SERS substrates. Nanostructures used as SERS substrates to analyze biological samples, such as foodborne pathogens and toxins, mainly depend on their plasmonic resonance scattering. Such plasmonic resonance depends on not only shape but also size of the nanoscale metal structures. Thus, spherically shaped metal nanoparticles provide high energy level plasmonic resonance, which gives enhanced scattering signals (*43*). Spherical nanoparticles provide signal enhancement in the order of approximately 10^6. For certain samples, it is possible to get signal enhancement as high as from 10^{10} to 10^{11}.

Research on silver colloidal nanoparticles has been extensively conducted to characterize analytes for SERS applications. Since it is relatively easy to prepare spherical silver nanoparticles from metal salt through the chemical reduction process, many researchers have used this process to fabricate nanosubstrates to detect different biological samples (*44*). Silver colloids contain nano size (10-100 nm) spherical or quasi spherical shaped nanoparticles. Most common agents for reducing silver nitrate salt are tri sodium citrate (called Lee-Meisels method) (*45*) or sodium borohydride (Creighton method) (*46*). Sometimes fructose is also used as reducing agent (*47*). However silver nanoparticles in colloidal suspension have instability, resulting in formation of nanoparticle aggregation and gives short shelf-life for substrates. Particle aggregation also develops hot spots on the substrate surface and leads to poor signal enhancement (*48*). Aggregation of metal nanoparticles can be controlled by introducing stabilizers such as polymers during the synthesis of metal nanoparticles. Polymers such as poly (vinyl pyrrolidone), gelatin, sodium poly (D-L glutamic acid), and polyvinyl alcohol are good stabilizers because of their thermal stability and chemical resistance. These chemicals are water soluble and biodegradable (*49*). Therefore, adding them to silver colloidal suspensions protects nanoparticle aggregation and increases the substrate stability and shelf-life. Since research efforts for affordable SERS substrate fabrication are being conducted as biosensing tools to detect pathogens and biochemical contaminants, development and characterization of potential SERS substrates are extremely important.

Therefore, in this paper we report fabrication of novel biopolymers encapsulated with silver nanoparticles (BeSN), optical and morphological characteritics of nanoparticles using UV-Visible spectroscopy, hyperspectral microscopic imaging, and Transmission Electron Microscopy (TEM). Also, standard chemicals, namely *trans*-1,2-bis(4-pyridyl)ethylene (BPE) and Rhodamine 6G dye were used to measure SERS signal enhancement efficiency of BeSN. Another major focus of this paper is to prove the concept of a biopolymer encapsulated with silver nanoparticles as a SERS substrate to detect and differentiate two serotypes (Typhimurium and Enteritidis) of *Salmonella*. Futher confirmation of BeSN SERS substrate efficiency in detection of other food borne pathogens has been investigated with *Salmonella* Infantis, *E,Coli, Listeria innocua* and *Staphylococcus aureus*.

Fabrication of Biopolymer Nanocolloidal SERS Substrate

In reference with authors previously published work (*45*), 2% solution of polyvinyl alcohol (PVA) with molecular weight of 98,000 (Sigma Aldrich Chemicals, St. Louis, MO, USA) was prepared in deionized distilled water by stirring and heating simultaneously until temperature of solution to be 60 °C. PVA was used to increase the stability of substrate. One- hundred mg of silver nitrate (AgNO$_3$, Sigma Aldrich Chemicals, USA) was added to the PVA solution, followed by adding 20 ml of 1% trisodium citrate (Sigma Aldrich Chemicals, USA) solution (*46*) to reduce silver nitrate to silver. The reduced silver nanoparticles encapsulated the polyvinyl alcohol nanoparticles to make biopolymer encapsulated with silver nanoparticles (BeSN), which composed of core of biopolymer nanoparticle surrounded by shell of silver nanoparticles. Figure 1 shows the flow diagram for preparation of BeSN. Figure 2 shows the schematic of single BeSN. It shows that inner core is PVA polymer nanospheres and the outer shell layer is silver nanoparticle obtained from citrus reduction (*45, 46*).

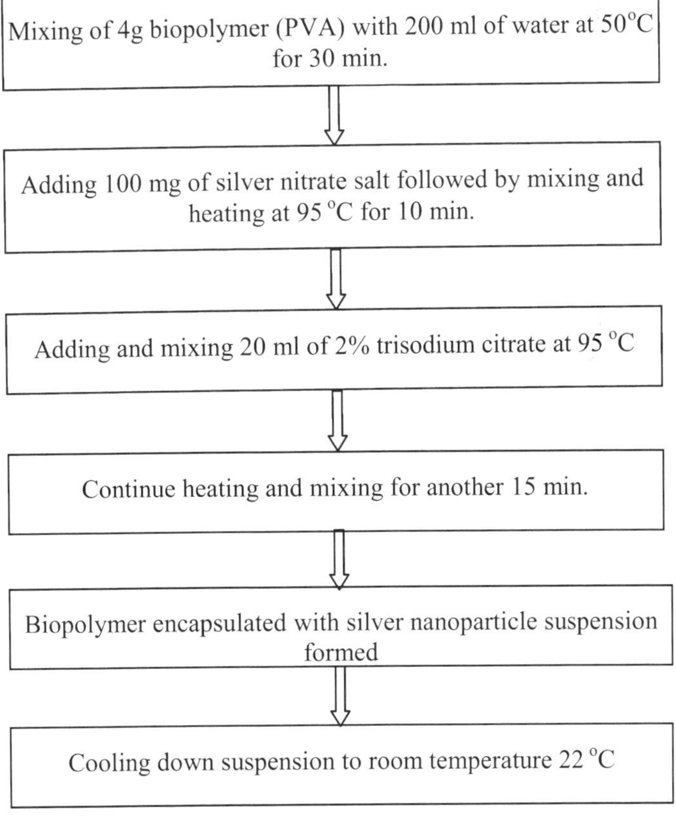

Figure 1. Flow diagram of biopolymer encapsulated with silver nanoparticle (BeSN) preparation.

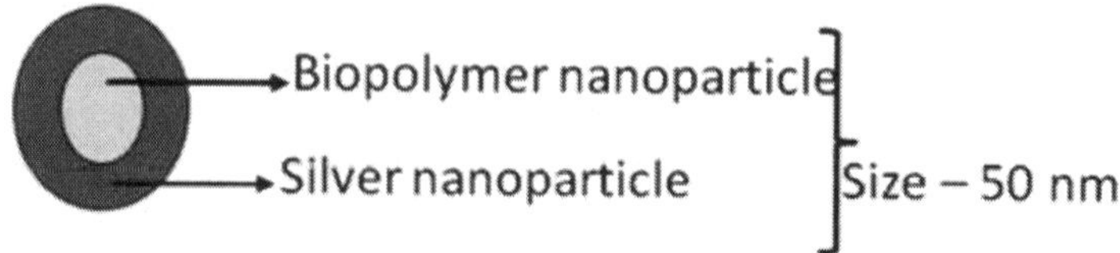

Figure 2. A schematic of biopolymer encapsulated with silver nanoparticles (BeSN).

Relatively uniform biopolymer encapculated with silver nanoparticles are shown in Figure 3a. TEM micrographs (Figure 3a and 3b) show the silver encapsulated nanoparticles (BeSN) and PVA particles without silver encapsulation. Both figures illustrate that nanoparticles are monodispersed without particle aggregation. The role of PVA has stabilized the nanoparticles without aggregation. Negatively charged silver ions obtained through the citrate reduction process, adsorbed onto the positively charged PVA particles and created silver encapsulated biopolymer hybrid nanoparticles (BeSN). By comparing the Figure 3a and 3b, it is clear that silver nanoparticles covered the biopolymer nanoparticles to make a shell on the core of biopolymer. During this process, PVA was expected to prevent the aggregation of silver nanoparticles and to stabilize the particles in colloidal solution. The final particle size and stability of formed nanoparticles are influenced by the molecular weight of PVA, the amount of PVA in solution to encapsulate, and the amount of silver ions produced by reduction reaction. It could be noted that the role of PVA is very important to keep the nanocolloidal solution stable without any particle aggregation for several months at room temperature. This is confirmed by the Figure 3c, which shows pictures of BeSN prepared at different times, with and without biopolymer. Stability of these BeSN prepared at different times was confirmed through their plasmonic absorption in the electromagnetic spectrum at the UV/Vis range. It gave an absorption peak very close to the value obtained from a freshly prepared sample of the some composition. If there is nanoparticle aggregation after the storage period, the interparticle distances become smaller than the diameter of the individual nanoparticles and it gives plasmon resonance to red shift and absorption peak at a longer wavelength region.

Optical Characteristics of Biopolymer Nanocolloidal SERS Substrates

Optical Characteristics with UV-Visible Spectroscopy

Optical properties of BeSN was analysed with UV-Visible (UV/Vis) absorption spectroscopy and hyperspectral microscopic imaging. UV/Vis spectroscopy is one of the widely used techniques for structural and optical characterization of silver nanoparticles (*48–51*). UV/Vis absorption spectra are very sensitive to particle size and its aggregation. It measures silver nanoparticle plasmonic resonance intensity. Silver nanoparticles have strong plasmonic absorption in the visible region due to the surface plasmon resonance. Silver biopolymer nanoparticles were diluted with deionized distilled water to 1:20 ratio (1 part of nanoparticle suspension and 20 parts of water). Diluted sample was exposed to UV/Vis light and the absorption spectra were recorded from 350 nm to 600 nm (figure 4). Free electrons in silver metal nanoparticles give surface plasmonic resonance absorption band when a light wave hits the metal nanoparticle (*52, 53*). Plasmon resonance is shifted into visible region of electromagnetic spectrum of silver metal. When the frequency of electromagnetic field is resonant with the free electron motion, a strong absorption takes place. This absorption strongly depends on the size and shape of silver nanoparticle (*52, 53*). The UV/Vis absorption spectra of BeSN is shown in Figure 4. It shows the peak of plasmon resonance absorption band at 460 nm. The stability of silver biopolymer nanoparticles is observed by measuring the absorption peak after 9 months of storage. Absorption peak position was observed at 457 nm. Absorption peaks between 400 – 480 nm is the normal characteristics of silver nanoparticle in UV/Vis region. Biopolymer encapsulated silver nanoparticles fabricated in this study also gave strong absorption peak in the visible region, which also confirms that silver nanoparticles were on the outer core of biopolymer particle as a shell to respond to the exposure of UV/Vis light.

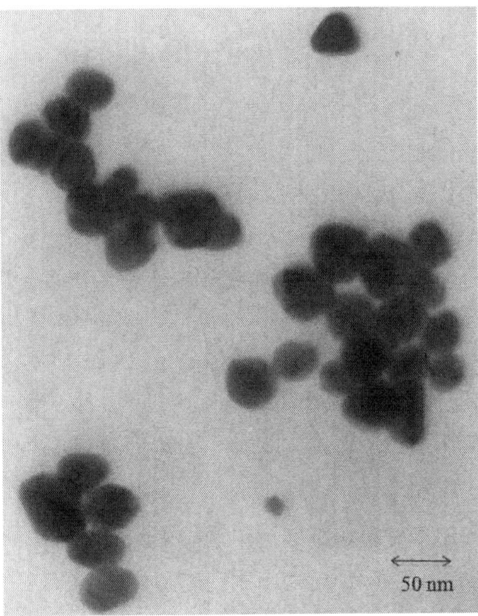

Figure 3a. TEM image of biopolymer encapsulated with silver nanoparticle (BeSN).

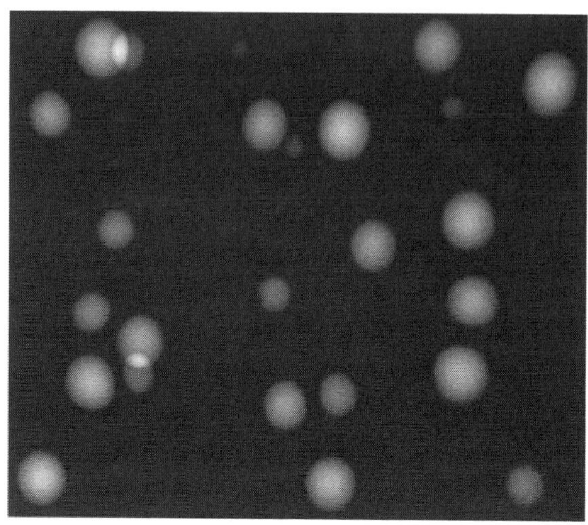

Figure 3b. TEM image of biopolymer without silver nanoparticle encapsulation.

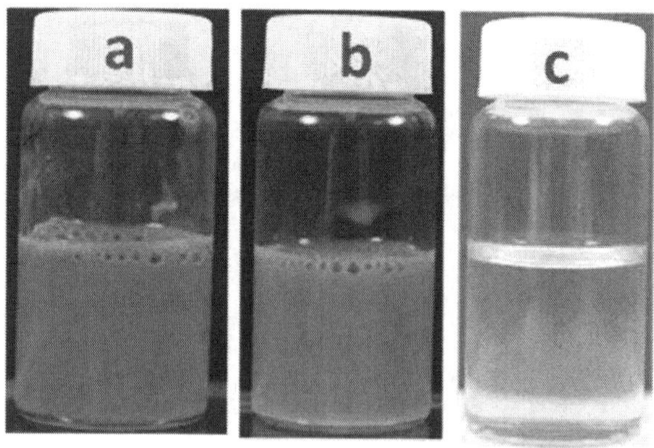

Figure 3c. Silver nanoparticle prepared at different period of time: (a) BeSN prepared on January 2012; (b)BeSN prepared on May 2011; (c)Silver nanoparticle prepared on May 2011 without biopolymer.

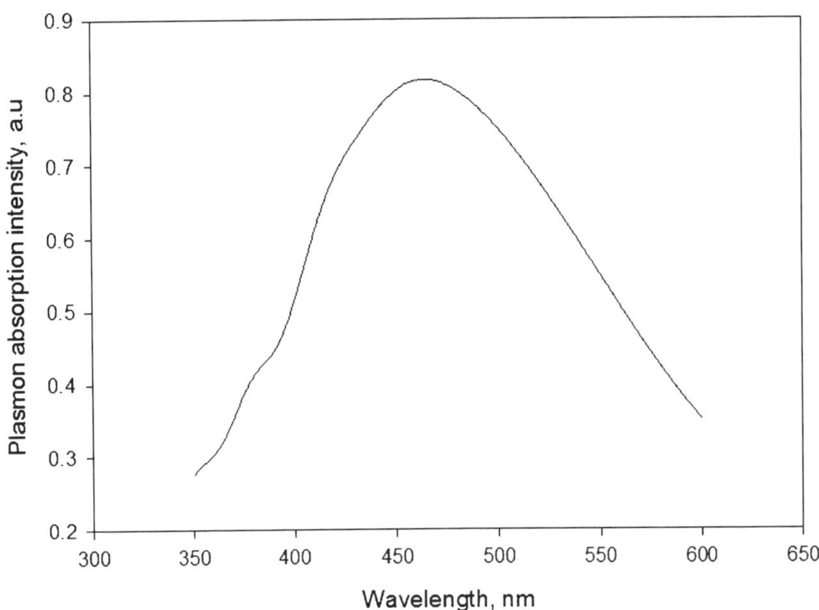

Figure 4. Plasmonic absorption of BeSN in electromagnetic spectrum in UV/Vis range.

Substrate Characterization with Hyperspectral Microscopic Imaging

The hyperspectral microscopic imaging (HMI) method was used to measure scattering characterization of BeSN. HSI with a metal-halide light source was used for measurement of light scattering from BeSN. HSI analysis gives plasmon excitation map of the silver surrounding the biopolymer nanoparticle and its spectral distribution. Approximately 10 µl of diluted sample (1 part of nanoparticle suspension with 20 parts of water) was coated on glass microscopic slides after a prewash with methanol followed by air drying. After complete drying, one micro liter of PBS buffer (7.4 pH) was dropped on the dried sample and covered by a cover slip with firm pressure with fingers. The sample slide was scanned with 100x objective lens under the darkfield illuminating metal halide light source. Hyperspectral images were focused well to obtain the sharp and clear image of photon emission as bright light spots comes from the BeSN, which was scattered by the light source. Hyperspectral images were captured with a acousto-optic tunable filter (AOTF)-based HMI system with 4 nm spectral resolution between 450 and 800 nm. The gain and exposure time for image acquisition were 4 and 250 ms, respectively. The hyperspectral image gives details about the BeSN plasmon resonance effect and plasmon resonance profile to evaluate the scattering intensity of substrate for SERS use.

HMI allowed photon excitation and did mapping of plasmonic excitation with spectral resolution of 4 nm. In theory, the light emission from metal nanoparticles is due to the excitation of localized surface plasmon (*54, 55*). Figure 5 shows an image of hundreds of light induced photon emission due to the plasmonic excitation of nanoparticles which are visible as blue dots in Figure 5. This figure also shows some green dots, which might be from some other particles other than BeSN, however there are significantly less of these particles than BeSN. Surface palsmon resonance (SPR) from metal nanoparticles is very intense and a sensitive imaging technique under darkfield illumination (*56–58*). Thus, the frequency of light corresponding to plasmonic resonance is scattered by BeSN and examples of images of SPR scattering of BeSN with darkfield background are shown in Figure 5.

When the nanoparticles are scanned at different wavelength ranges, the interaction between light and nanoparticle varies with wavelength as well as electric field (*53*). Changes in electric field causes change in the electromagnetic field surrounding the particles and generates thousands of electromagnetic bands. The incident light illuminates these bands and HMI captures these electromagnetic bands as images with corresponding intensities (*59*). The light emissions called plasmonic excitation obtained by the HMI technique is displayed in Figure 5. Photon induced emission spectral profiles from BeSN nanoparticle at specific wavelength and PVA biopolymer nanoparticle are shown in Figure 6. The sharp Plasmon resonance peak indicates the homogeneous nature of nanoparticles. Multiple profiles were colleted from various parts of the HMI image and it shows that all were overlapped. This is evidence for homogeneity of BeSN. The intensity profile of HMI from BeSN is much higher than biopolymer alone.

Figure 5. Hyperspectral microscope imaging of plasmonic excitation from BeSN.

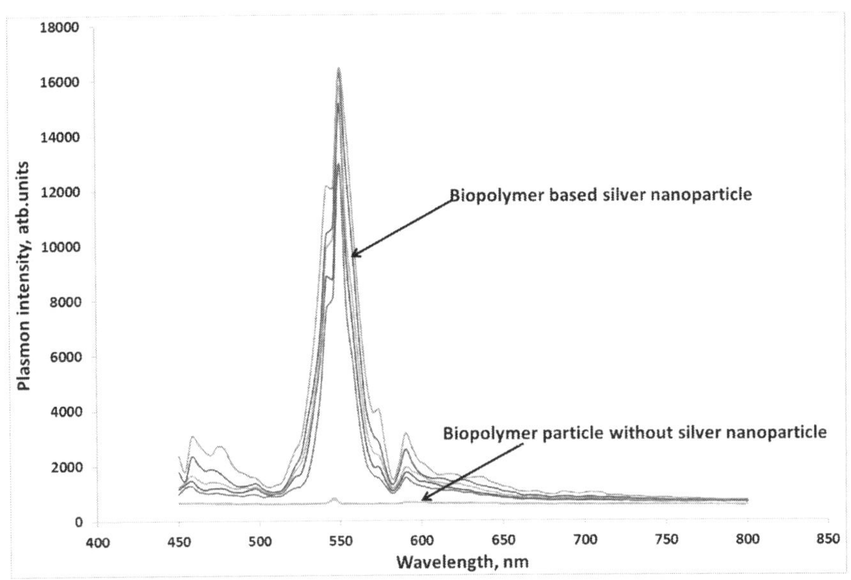

Figure 6. Surface plasmon resonance (SPR) spectral profile of biopolymer encapsulated with and without silver nanoparticles.

Reproducibility and Repeatability of BeSN Substrates

The repeatability and reproducibility of BeSN were evaluated by collecting SERS signals from Rhodamine 6G dye samples. Substrates used in this experiment were prepared at different time using same protocol. The surface uniformity of BeSN substrates was evaluated by SERS singal mapping of 5x5 μm area on a empty substrate, coated on a stainless steel plate.

Biopolymer encapsulated with silver nanoparticles were coated on a clean stainless steel plate of 1.75 cm diameter by thermal evaporation (60-70°C) under the chemical hood. After complete evaporation of water, these particles were used as SERS substrates. SERS signals were collected with confocal Raman spectrometer (HR800 LabRam, Jobin-Yvon-Horiba, Edison, New Jersey, USA). A laser excitation wavelength of 785 nm was produced by HeNe 20 mW laser source with a CCD detector (1024x256 pixels of 26 μm). Samples were scanned with a confocal hole size of 100 μm with a 50x magnification. Confocal was used to adjust the depth and corresponding focal plane at the top of surface. The substrates which were made at different times was scanned with mapping mode over an area of 5x5 μm to assess the uniformity of substrates. Likewise, SERS spectral data were collected using Rhodamine 6G dye (Sigma Aldrich, USA), adsorbed on substartes. Three substrates from two different batches were used for collecting SERS signals of Rhodamine 6G from the five different spots on each substrate. Since dyes produced strong signals, it was possible to verify SERS signal enhancement as well as detection limit of SERS. The concentration of Rhodamine 6G for analyzing repeatability and reproducibility of BeSN substrates was 10^{-6} M.

Figure 7 shows the Raman mapping spectral profiles from empty substrates without any reference analytes over an area of 5x5 µm. The gap between two spectral profiles represents the variation of scanned area on substrates. Usually, more gap bewteen two spectra represents larger variation on the surface uniformity of substrates. The gap obtained from BeSN is minimum that indicates overall surface uniformity of the substrate. In general, surface uniformity generates less noise, resulted in higher SERS signal enhancement with less spot to spot variation. Figure 8 shows mean and standard deviation of SERS signals from Rhodamine 6G sample at 10^{-6}M concentration, for repeatability test of susbtares. In this case, SERS mean intensity was obtained from 10 replicates with two substrates prepared at different time and 5 different spots on each substrate. The standard deviation of spectra in this figure demonstrated that the substrates have reliable signal repeatability. This confirmed that there are little spot-to-spot and batch-to-batch variation on the substrates.

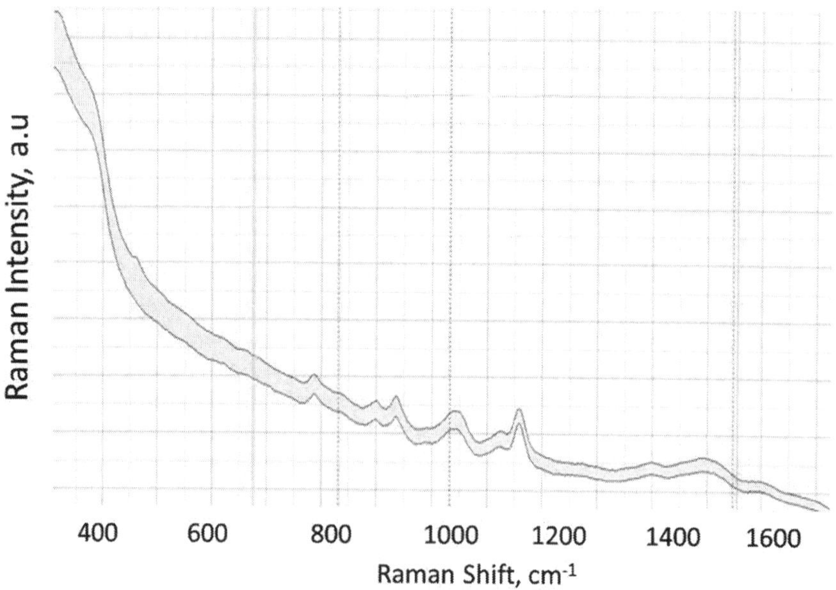

Figure 7. Mapping spectral profile of BeSN substrate without any analytes.

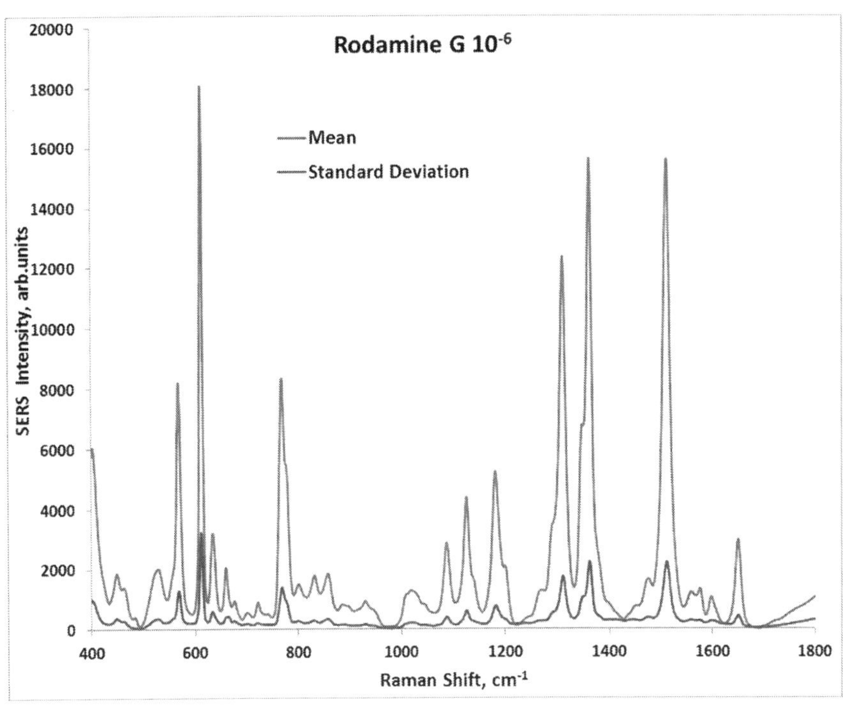

Figure 8. Mean and standard deviation of Rhodamine G at 10^{-6} M concentration (n=10; 5 spots in each on two substrates).

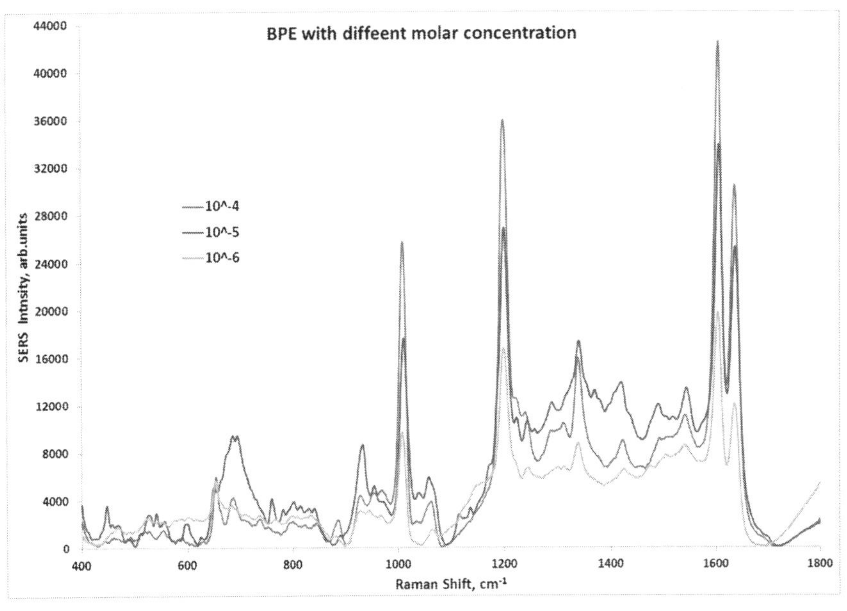

Figure 9. SERS spectra of BPE with various concentrations (n=10; 5 spots on each substrate with two different batch).

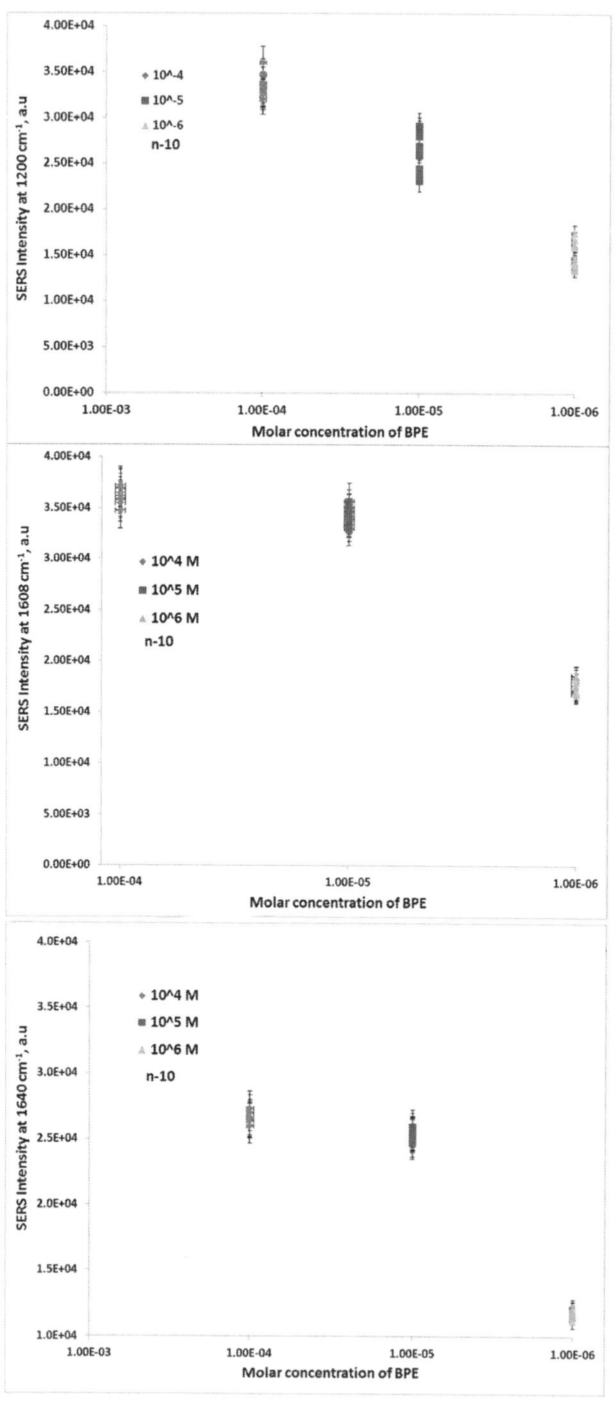

Figure 10. SERS intensity obtained from BeSN substrates at C=C band of BPE with different molar concentrations.

SERS with BeSN Substrates for Chemical Constituents

Figure 9 shows the average SERS signal obtained from *trans*-1,2-bis(4-pyridyl)ethylene (BPE) with various concentrations using same procedure as Rhodamine 6G dye samples. In figure 9, SERS signals obtained from the characteristics of chemical confirmation bands (e.g., 1200 cm^{-1}, for C=C for BPE) were highly enhanced.

The major phenomena that confine SERS analytes within the electromagnetic field are physisorption and chemisorptions through a self-assembled monolayer of the respective analytes. These two phenomena depend on the natural adhesive force between analytes and nano metal surface (*60*). Spectroscopic properties of analytes such as electronic absorption bands with frequencies that are close to the laser excitation is important; because they have influence on SERS signals. When laser excitation wavelength overlaps with the electronic absorption band of an analyte, Raman scattering intensity is amplified, resulting in specific chemical moieties of that analyte (*61*).

Figure 10 shows the reproducibility of SERS intensity at 1200, 1608 and 1640 cm^{-1} bands for BPE with different molar concentrations. Integrated intensities of these bands obtained from various spots of the substrates were plotted individually in figure 10. Error bar in each point represents the spot-to-spot variation of the Raman intensity at their respective Raman shift (frequency) collected from 10 different locations from two substrates at 5 different spots from each substrate. As seen in figure 10, Raman intensity at 1200 cm^{-1} decreased by lowering molar concentration of BPE solution. In SERS spectra, bond at 1200 cm^{-1} represents C=C in-plane ring mode. The peaks at both 1608 and 1640 cm^{-1} frequency also represent C=C bond in aromatic ring stretching mode and stretching mode, respectively (*62*).

SERS with BeSN Substrates for Foodborne Pathogen Detection

Bacteria Sample Preparation

Bacterial strains for this study were isolated by the Poultry Processing and Swine Physiology Unit, Russell Research Center in Athens, GA. Isolates of *Salmonella* Typhimurium, strain MH 68123 (from chicken rinse), and *Salmonella* Enteritidis strain MH 42841 (from ground chicken) were tested. Stock cultures of bacteria were stored at 4°C. Fresh cultures of two serotypes (Typhimurium and Enteritidis) of *Salmonella* were prepared by transferring aliquots of stock cultures to Tryptic Soy Broth (TSB) (Becton, Dickinson, and Co., Sparks, MD, USA) and incubating for 18 hours at 37°C. After incubation, broth of each culture was transferred to 15 mL sterile centrifuge tubes, and centrifuged at 20-22°C with 2800 rcf (5000 rpm) for 10 minutes (Labnet, Model Hermle Z300, Hermle LaborTechnik, Germany). Bacterial pellets were washed by suspending in 10 mL of sterile deionized water and centrifuging as previously described for a total of 3 times. Since cell membrane damage may occur if bacteria cells were holding in water over 12 hours, SERS measurements were taken as soon as cell suspensions was made. The experiments were replicated 3 times, and samples were prepared in duplicates during each replicate. Approximately 5 µL of cell suspension from each duplicate was used for SERS data acquisition.

Bacterial SERS Data Acquisition

As described in SERS signal acquisition of dye samples, SERS signals of *Salmonella* serotypes were collected with BeSN substrates. A total of 15 spots were scanned for each serotype from all replicate samples. Similarly, three different data sets were collected from each serotype that were enriched at three different times. Spot size and corresponding laser power of laser beam to scan the samples were approximately 1.74 µm and 15.55 mW, respectively. Bacteria were focused through top objective lens. Confocal was used to adjust the depth at its focal plane on top surface of samples, deposited on BeSN substrates. With a size of 1.74 µm laser beam, approximately two bacteria cells could be exposed to the lase spot. Scattering intensity of SERS spectra between 400 cm^{-1} and 1800 cm^{-1} were collected and recorded from each serotype samples for the analysis.

Data Analysis

The SERS spectral data were analyzed with multivariate data analysis software (Unscrambler, version 9.8, CAMO Software, Inc., Woodbridge, NJ). Spectral data were subjected to baseline correction pretreatment to avoid shifts of any variable below the baseline, so that all SERS spectra of each serotype could have the same base line. This preprocessing helps determine any difference among the spectral data quickly. Multivariate statistical analysis technique of Principal Component Analysis (PCA) was performed to develop classification model using calibration data set. PCA model was used to show the natural clusters in the data set and also to describe differences between sample clusters (*63, 64*). Score value obtained for each principal component (PC) provided the best fit value for each sample. The Soft independent modeling of class analogy (SIMCA) analysis was also performed to validate the PCA classification model. The different data set were used for validation in SIMCA analysis. To test models, spectral data of each serotype in validation data set were classified using a PCA calibration model developed.

PCA classification model gave 98 % variation between two serotypes (Typhimurium and Enteritidis) of *Salmonella*. Figure 11 (a) shows the score plot obtained from PCA classification analysis on two different *Salmonella* serotypes. Score plot shows that PC1 explained a maximum of 92 % of the variation between *Salmonella* Typhimurium and Enteritidis. PC2 explained 6% of variation, resulting in a total of 98 % variation between two serotypes. Figure 11(a) also illustrates that samples in each serotype of *Salmonella* are close to each other on a score plot, which indicates similarity in cell composition. This finding can be further confirmed by their minimum difference in leverage values which are the distance of each sample point from the center of cluster in the model. *Salmonella* Typhimurium cells have positive score values; whereas, *Salmonella* Enteritidis cells have negative score values, which indicates that *Salmonella* Typhimurium has more absorbance values over the selected spectral region than

Salmonella Enteritidis. Negative score values are a result of mean centering data, i.e., it is a geographic center of the *Salmonella* Enteritidis cells based on the absorbance values obtained in the given spectral region. Score plot also indicates the PC, which distinguishes variation among the samples (*45*). Figure 11 (a) shows that PC1 is able to distinguish the samples of *Salmonella* Typhimurium and *Salmonella* Enteritidis cells with higher accuracy. PCA model developed was tested with spectral data obtained from the samples for validation with a SIMCA classification method. SIMCA classification for two serotypes of *Salmonella* for validation data set performed with 89 % classification accuracy (PC1 of 71% and PC2 of 18%).

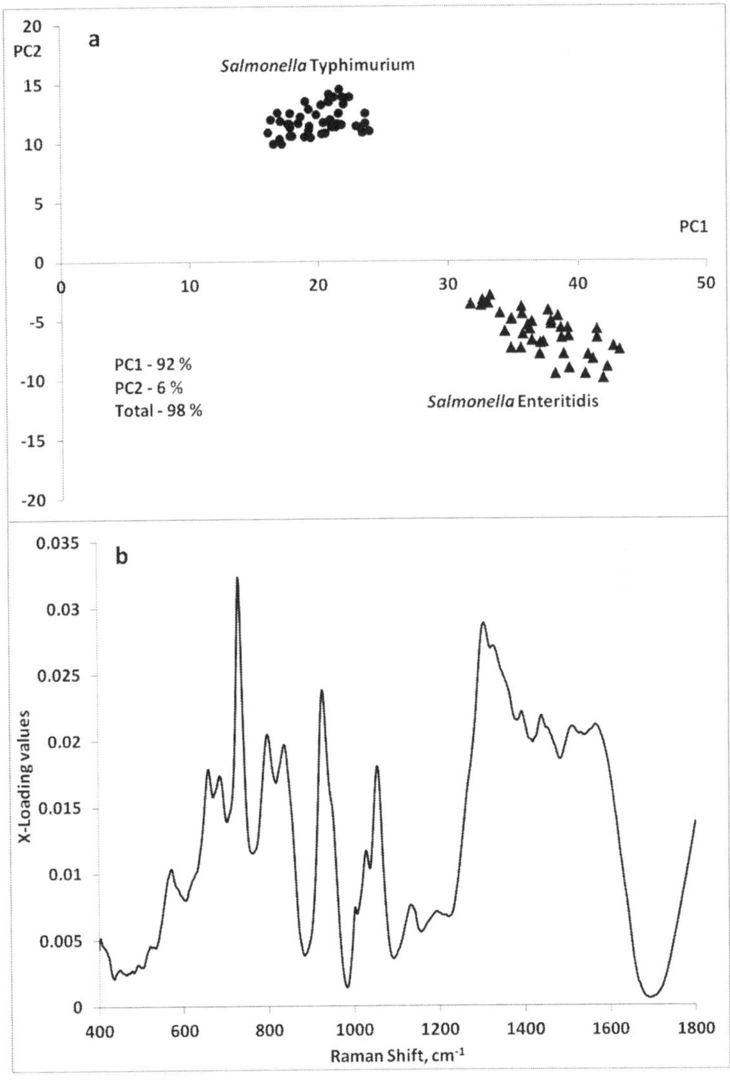

Figure 11. PCA classification model a) Score plot; b) X-loading plot.

Loading plot in Figure 11 (b) shows the frequencies that have significant contribution to the variation described by PC1, which can be identified by higher loading values (absolute values). Frequencies with higher loading value carry most of information about the difference between two serotypes of *Salmonella* (*45*). Spectral bands at 730 cm⁻¹, 1050cm⁻¹, and 1304 cm⁻¹ have higher loading values that demonstrate significant contributions for classification of two *Salmonella* serotypes of Typhimurium and Enteritidis (*65–67*). These regions are possibly due to the peptidoglycan present in outer layer of bacteria cell walls, carbohydrates, proteins, lipid, and amino acids side chains of DNA/RNA nucleic acid structures. It indicates that information about bacteria cell protein and their amino acid structures of DNA/RNA could be major factors that responsible for differentiating two different serotypes of *Salmonella*. Frequency range between 1300 and 1600 cm⁻¹ in the loading plot shows many small peaks with high loading values. These peaks also indicate difference between two serotypes which are responsible for classification accuracy (*45*).

Similarly, detection method of SERS with BeSN was examined with different foodborne pathogens including *Salmonella* Infantis, *E. coli*, *Listeria innocua* and *Staphylococcus aureus*. Figure 12 shows the SERS signals of these foodborne pathogens collected with BeSN substrates along with SERS signals of empty substrates. These signals were average of 30 spectra collected at three different times with 10 different spots from bacterial samples immobilized on the substrates. Raman signal without BeSN shows no significant bands that affect the SERS signals from pathogens for detection and identification.

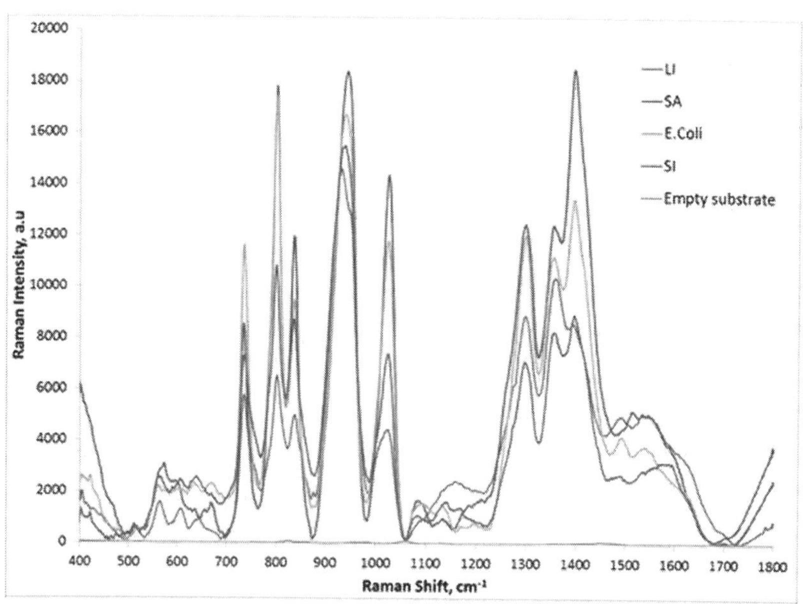

Figure 12. SERS signals of various foodborne pathogens and empty substrate.

Conclusions

Present study describes the preparation, characterization and efficiency of SERS substrates, for the detection and classification of foodborne pathogens. Prepared SERS substrates showed overall surface uniformity, which is important to obtain better signal enhancement from the analytes. The uniformity of surface also determines signal to noise ratio as well as spot to spot variation in SERS signals. Results obtained from various foodborne pathogens showed that BeSN can be used as promising SERS substrates for rapid detection and classification of foodborne pathogens. The method to develop BeSN substrates for SERS is very efficient, inexpensive, reproducible, stable, and repeatable for better SERS signals. With BeSN substrates, further resaerch needs to be conducted for developing spectral libraries from different foodborne pathogens to develop rapid detection method with high sensitivity and selectivity using SERS technology in conjunction with conventional microbiological validation methods.

Acknowledgments

This work was partly supported by USDA/NIFA and Animal and Plant Quarantine Agency, South Korea. The authors would like to thank Dr. Nasreen Bano for taking hyperspectral images for this work.

Mention of trade names or commercial products in this article is solely for the purpose of providing specific information and does not imply recommendation or endorsement by the USDA. The USDA is an equal opportunity provider and employer.

References

1. Garbutt, J. *Essentials of Food Microbiology*, 1st ed; CRC Press: London, U.K., 1997.
2. Gorman, R.; Adley, C. C. *J. Clinic Microbiol.* **2004**, *42*, 2314–2316.
3. Zhang, G.; Ma, L.; Patel, N.; Swaminathan, B.; Wedel, S.; Doyle, M. P. *J. Food Prot.* **2007**, *70*, 997–1001.
4. Tietjen, M.; Fung, D. Y. *Crit. Rev. Microbiol.* **1995**, *21*, 53–83.
5. Cotton, T. M.; Kim, J. M.; Chumanov, G. D. *J. Raman Spectrosc.* **1991**, *22*, 729–742.
6. Herne, T. M.; Ahern, A. M.; Garrrell, R. L. *J. Am. Chem. Soc.* **1991**, *113*, 846–854.
7. Campion, A.; Kambhampati, P. *Chem. Soc. Rev.* **1998**, *27*, 241–250.
8. Cavalu, S.; Cinta-Pinzaru, S.; Leopold, N.; Kieffer, W. C. *Biopolym. Biospectrosc.* **2001**, *62*, 341–348.
9. Kneipp, K.; Kneipp, H.; Manoharan, H.; Hanlon, E. B.; Itzkan, I.; Desari, R. R.; Feld, M. S. *Phys. Rev.* **1998**, *E57*, R6281.
10. Weldon, M. K.; Morris, M. D. *Appl. Spectrosc.* **2000**, *54*, 124–130.
11. Premasiri, W. R.; Moir, D. T.; Klempner, M. S.; Krieger, N.; Jones, G.; Ziegler, L. D. *J. Phys. Chem. B* **2005**, *109*, 312–320.

12. Atanu, S.; Mujacic, M.; Davis, E. J. *Anal. Bioanal. Chem.* **2006**, *386*, 379–1386.

13. Chumanov, G. D.; Efremov, R. G.; Nabiev, I. R. *J. Raman Spectrosc.* **1990**, *21*, 43–48.

14. Vo-Dinh, T.; Houck, K.; Stokes, D. L. *Anal. Chem.* **1994**, *66*, 3379–3383.

15. Cotton, T. M.; Kim, J. M.; Chumanov, G. D. *J. Raman Spectrosc.* **1991**, *22*, 729.

16. Herne, T. M.; Ahern, A. M.; Garrrell, R. L. *J. Am. Chem. Soc.* **1991**, *113*, 846.

17. Grabb, E. S.; Buck, R. P. *J. Am. Chem. Soc.* **1989**, *111*, 8362.

18. Wantanabe, T.; Maeda, H. *J. Phys. Chem.* **1989**, *93*, 3258.

19. Chumanov, G. D.; Efremov, R. G.; Nabiev, I. R. *J. Raman Spectrosc.* **1990**, *21*, 43.

20. Haes, A. J.; Van Duyne, R. P. *J. Am. Chem. Soc.* **2002**, *124*, 10596–10604.

21. Liu, G.; Zheng, H.; Liu, M.; Zhang, Z.; Dong, J.; Yan, X.; Li, X. *J. Nanosci. Nanotechnol.* **2011**, *11*, 9523–9527.

22. Wang, H.; Zhenghua, A. N.; Ren, Q.; Wang, H.; Mao, F.; Chen, Z.; Shen, X. *J. Nanosci. Nanotechnol.* **2011**, *11*, 10886–10890.

23. Haes, A. J.; Van Duyne, R. P. *Anal. Bioanal.Chem.* **2004**, *379*, 920–930.

24. Weldon, M. K.; Morris, M. D. *Appl. Spectrosc.* **2000**, *54*, 20.

25. Cavalu, S.; Cinta-Pinzaru, S.; Leopold, N.; Kieffer, W. C. *Biopolymers.* **2001**, *62*, 341.

26. Campion, A.; Kambhampati, P. *Chem. Soc. Rev.* **1998**, *27*, 241–250.

27. Sengupta, A.; Mujacic, M.; Davis, E. J. *Anal. Bioanal Chem.* **2006**, *386*, 1379–1386.

28. Premasiri, W. R.; Moir, D. T.; Klempner, M. S.; Kricger, N.; Jones, G.; Ziegler, L. D. *J. Phys. Chem B* **2005**, *109*, 312–320.

29. Kahraman, M; Yazıcı, M. M.; Sahin, F; Culha, M. *Langmuir.* **2008**, *24*, 894–901.

30. Jarvis, R. M.; Goodacre, R. *Anal. Chem.* **2004**, *76*, 40–47.

31. Ales, P; Kvitek, L; Prucek, R.; Kolar, M.; Vecerova, R.; Pizurova, N.; Sharma, V. K.; Nevecna, T.; Zboril, R. *J. Phys. Chem. B* **2006**, *110*, 16248–16253.

32. Guzman, G. M.; Dille, J.; Godet, S. *Int. J. Chem. Biol. Eng.* **2009**, *2*, 104–111.

33. Yang, Y.; Huang, Z.; Nogami, M.; Tanemura, M.; Yamaguchi, K.; Li, Z.; Zhou, F.; Huang, Y. *J. Nanosci. Nanotechnol.* **2011**, *11*, 10930–10934.

34. Chang, J.; Lee, J.; Najeeb, C. K.; Kim, J. *J. Nanosci. Nanotechnol.* **2011**, *11*, 6253–6257.

35. Jarvis, R. M.; Brooker, A.; Goodacre, R. *Anal. Chem.* **2004**, *76*, 5198–5202.

36. Chou, K. S.; Lu, Y. C.; Lee, H. H. *Mater. Chem. Phys.* **2005**, *94*, 429.

37. Zeiri, L.; Bronk, B. V.; Shabtai, Y.; Eichler, J.; Efrima, S. *Appl. Spectrosc.* **2004**, *58* (1), 33–40.

38. Zeiri, L.; Efrima, S. *J. Raman Spectrosc.* **2005**, *36*, 667–675.

39. *Encyclopedia of Nanoscience and Nanotechnology*; Nalwa, H. S., Ed.; American Scientific Publishers: Los Angeles, 2011; Vols. 1–25.

40. Tu, C. H.; Lo, S. S. *J. Nanosci. Nanotechnol.* **2011**, *11*, 10575–10578.

41. Kim, M. H.; Kim, D. H.; Kim, J. H.; Woo, H. G.; Lee, B. G.; Yang, K. S.; Kim, B. H.; Park, Y. J.; Sohn, H. *J. Nanosci. Nanotechnol.* **2011**, *11*, 7374–7377.

42. Çınar, S.; Gündüz, G.; Mavis, B.; Çolak, Ü. *J. Nanosci. Nanotechnol.* **2011**, *11*, 3669–3679.

43. Song, K. C.; Lee, S. M.; Park, T. S.; Lee, B. S. *Korean J. Chem. Eng.* **2009**, *26*, 153–155.

44. Zeman, E. J.; Schatz, G. C. *J. Phys. Chem.* **1987**, *91*, 634.

45. Sundaram, J.; Park, B.; Hinton, A.; Lawrence, K. C.; Kwon, Y. *J. Food Measur. Charact.* **2012** DOI:10.1007/s11694-012-9133-0.

46. Lee, P. C.; Meisel, D. *J. Phys. Chem.* **1982**, *86*, 3391.

47. Creighton, J. A.; Blatchford, C. G.; Albrecht, M. G. *J. Chem. Soc., Faraday Trans.* **1979**, *75*, 790.

48. Darroudi, M.; Ahmad, M. B.; Abdullah, A. H.; Ibrahim, N. A. *Int. J. Nanomed.* **2011**, *6*, 569–574.

49. Kim, K. D.; Han, D. N.; Kim, H. T. *Chem. Eng. J.* **2004**, *104*, 55.

50. Yu, D. G.; Lin, W. C.; Lin, C. H.; Chang, L. M.; Yang, M. C. *Mater. Chem. Phys.* **2007**, *101*, 93–98.

51. Kelly, K. L.; Coronado, E.; Zhao, L. L; Schatz, G. C. *J. Phys. Chem. B* **2003**, *107*, 668–677.

52. Kapoor, S. *Langmuir.* **1998**, *14*, 1021.

53. Ozin, G. A. *Adv. Mater.* **1992**, *4*, 612.

54. Evanoff, D. D.; Chumanov, G. *ChemPhysChem* **2005**, *6*, 1221–1231.

55. Kreibig, U. *Z. Phys.* **1970**, *234*, 307.

56. Liz-Marza'n, L. M. *Langmuir.* **2006**, *22*, 32–41.

57. Murray, W. A.; Auguie, B.; Barnes, W. L. *J. Phys. Chem. C* **2009**, *113*, 5120–5125.

58. Jensen, T. R.; Malinsky, M. D.; Haynes, C. L.; Van Duyne, R. P. *J. Phys. Chem. B* **2000**, *104*, 10549–10556.

59. Bouillard, J. S.; Vilain, S.; Dickson, W.; Zayats, A. V. *Opt. Express* **2010**, *18*, 16513–16519.

60. Homola, J. *J. Chem. Rev.* **2008**, *108*, 462–493.

61. Herne, T. M.; Ahern, A. M.; Garrrell, R. L. *J. Am. Chem. Soc.* **1991**, *113*, 846–854.

62. Chaney, S. B.; Shanmukh, S.; Dluhy, R. A.; Zhao, Y. P. *Appl. Phys. Lett.* **2005**, *87*, 31908.

63. Krzanowski, W. J. *Principles of Multivariate Analysis: A User's Perspective*; Oxford University Press: Oxford, U.K., 1988.

64. Martens, H.; Naes, T. *Multivariate Calibration*; Wiley: New York, 1989.

65. Jarvis, R. M.; Brooker, A.; Goodacre, R. *Anal. Chem.* **2004**, *76*, 5198–5202.

66. Sengupta, A.; Laucks, M. L.; Davis, E. J. *Appl. Spectrosc.* **2005**, *59*, 1016–1023.

67. Jarvis, R. M.; Goodacre, R. *Anal. Chem.* **2004**, *76*, 40–47.

Chapter 3

AuNP-DNA Biosensor for Rapid Detection of *Salmonella enterica* Serovar Enteritidis

Evangelyn C. Alocilja,* Deng Zhang, and Connie Shi

[1]Department of Biosystems and Agricultural Engineering, Michigan State University, East Lansing, Michigan 48824
*E-mail: alocilja@msu.edu.

Salmonella enterica serovar Enteritidis is one of the most frequently reported causes of foodborne illness. A rapid and sensitive detection of the insertion element (*Iel*) gene of *Salmonella* Enteritidis on screen-printed carbon electrodes (SPCE) is reported in this paper. The biosensor system includes two nanoparticles: gold nanoparticles (AuNPs) and magnetic nanoparticles (MNPs). The AuNPs are coated with the 1st target-specific DNA probe (pDNA) which can recognize the target DNA (tDNA) and the MNPs are coated with the 2nd target-specific pDNA. After mixing the nanoparticles with the tDNA, the sandwich complex (MNP-2ndpDNA/tDNA/1stpDNA-AuNP) is formed. The whole complex is magnetically separated and applied on a screen-printed carbon electrode chip after washing away the unreacted AuNPs. Following oxidative gold metal dissolution in an acidic solution, the released Au^{3+} ions are directly quantified by differential pulse voltammetry (DPV) at a potential of +0.4 V (vs. Ag/AgCl). Using this technique, the detection limit of this SPCE biosensor for the insertion element (*Iel*) gene of *Salmonella* Enteritidis is as low as 0.7 ng/mL.

Introduction

The Centers for Disease Control and Prevention (CDC) estimates that 1.4 million people in the US are infected with salmonellosis and that 1,000 people die each year (*1*). Environmental sources of the organism include water, soil, insects, factory surfaces, kitchen surfaces, animal feces, raw meats, raw poultry, and raw seafoods. One of the most common serotypes is *Salmonella enterica* serovar Enteritidis *(Salmonella* Enteritidis*)*. Persons infected with *Salmonella* Enteritidis usually develop diarrhea, fever, and abdominal cramps 12 to 72 hours after infection. In some patients, especially infants and young children, pregnant women, and elderly, *Salmonella* infection may spread from the intestines to the blood stream, and then to other body sites and can be life-threatening unless the person is treated promptly with antibiotics.

Common detection methods for *Salmonella* Enteritidis include microbiological, immunological and molecular biological techniques. Although microbiological detection is accurate, it often relies on time-consuming growth in culture media, followed by isolation, biochemical identification, and sometimes serology, and need special reagents and facilities. Immunological detection systems are specific but their sensitivity is low. Molecular PCR-based detection method is sensitive however, PCR is complex, expensive, time-consuming, and labor-intensive. So a rapid, sensitive detection and valid identification of *Salmonella* Enteritidis is vital within the overall context of food safety and public health.

A biosensor is an analytical device that integrates a biological sensing element with a transducer to quantify a biological event into an electrical output. It is a promising device for rapid detection of pathogenic microorganisms. Based on different sensing elements, the biosensor can be divided into immuno-sensor (*2–4*), DNA sensor (*5*), cell-based biosensor (*6*), aptasensor (*7*), and enzymatic sensor (*8*). In recent years, interest for DNA-based diagnostic tests has been growing. The development of systems allowing DNA detection is motivated by applications in many fields: DNA diagnostics, gene analysis, fast detection of biological warfare agents, and forensic applications (*9*). DNA sensor usually relies on the immobilization of a single-stranded DNA probe (pDNA) onto a surface to recognize its complementary DNA target sequence by hybridization. Transduction of hybridization of DNA can be measured electronically (*10, 11*), optically (*9, 12, 13*) electrochemically (*5, 14–16*), or by using mass-sensitive devices (*17*).

In this paper, a biosensor based on gold nanoparticles (AuNPs) conjugated to DNA probes for the detection of the insertion element (*Iel*) gene of *Salmonella* Enteritidis is described. The biosensor utilizes oligonucleotide-modified gold nanoparticles (AuNPs) as electrochemical indicator and magnetic nanoparticles (MNPs) for easy and clean separation from the sample. After hybridization with the target DNA (tDNA), a magnetic field is used to separate the sandwich complex consisting of MNP-2ndpDNA/tDNA/1st pDNA-AuNP. Following oxidative gold metal dissolution in an acidic solution, the released Au^{3+} ions are measured by differential pulse voltammetry (DPV) on a disposable screen-printed carbon electrode (SPCE).

Experimental

Materials and Reagents

Hydrogen tetrachloroaurate (III) trihydrate and sodium citrate dihydrate were used for the synthesis of gold nanoparticles. 1,4-Dithio-DL-threitol (DTT) was used for the cleavage of oxidized thiolated oligonucleotides before conjugation. Anhydrous sodium acetate, 1,6-hexanediamine, $FeCl_3 \cdot 6H_2O$, and ethylene glycol were used for the synthesis of amine-coated magnetic nanoparticles (MNPs). All the above reagents were purchased from Sigma (St. Louis, MO). Nap-5 column was purchased from GE Healthcare (Piscataway, NJ), which was used to purify the DNA product from the DTT solution. Sulfosuccinimidyl 4-N-maleimidomethyl cyclohexane-1-carboxylate (sulfo-SMCC; Pierce, WI) was used as a cross-linker between thiolated DNA probes (pDNA) and amine-coated MNPs. Sulfo-NHS acetate (Pierce, WI) was used to block unreacted sulfo-SMCC. All solutions were prepared in double distilled water.

Apparatus

The target DNA (tDNA) was amplified by a thermocycler (Mastercycler Personal, Eppendorf). Gel electrophoresis (Runone System) was used for confirmation of PCR product. After purification of the PCR product, a spectrophotometer (SmartSpec 3000, Bio-Rad Laboratories) was used to measure the concentration of tDNA samples, as well as pDNA. In the characterization experiments of AuNPs, a UV-Vis-NIR Scanning Spectrophotometer (UV-3101PC, Shimadzu) was used to determine the absorbance of AuNPs, and transmission electron microscopy (TEM, JEOL100 CXII) was used to characterize the nanoparticles. The dimension of MNPs was characterized by TEM and their magnetic profile was characterized by Quantum Design MPMS SQUID (Superconducting quantum interference device) magnetometer. All magnetic separation was done using a magnetic separator (FlexiMag, SpheroTech). A centrifuge (Micro12, Fisher Scientific) was used for separation and purification of AuNPs. An incubator (HS-101, Amerex Instrument Inc.) was used to enrich the bacteria and hybridization reaction. Screen-printed carbon electrode (Gwent, England) and a potentiostat/galvanostat (263A, Princeton Applied Research, TN) were used for electrochemical measurement.

Genomic DNA Preparation

A culture of *Salmonella* Enteritidis was grown by the standard microbiological culture. DNA isolation was performed from 1 mL *Salmonella* Enteritidis culture using the QiaAmp DNA Mini Kit (Qiagen Inc., Valencia). The forward primer used was 5′- CTAACAGGCGCATACGATCTGACA -3′; the reverse primer was 5′- TACGCATAGC GATCTCCTTCGTTG -3′ (*18*). The DNA probes (pDNAs) of *Iel* gene were as follows: 1st pDNA on AuNPs: 5′-AATATGCTGCCTACTGCCCTACGCTT-SH-3′; 2nd pDNA on MNPs: 5′-SH - TTATGTAGTCCTGTATCTTCGCCGT -3′. All oligonucleotides were purchased from IDTDNA Inc. (Coralville, IA).

Fabrication and Functionalization of Nanoparticles

Gold nanoparticles (AuNPs) were synthesized using a published procedure (*19*). Hydrogen tetrachloroaurate (III) trihydrate aqueous solution (1 mM, 50 mL) was heated with stirring on a hotplate. Once it refluxed vigorously, the solution was slowly titrated with 5 mL of 38.8 mM sodium citrate. The solution turned from yellow to clear, to black, to purple and finally deep red. AuNP dimension and spectroscopic properties were characterized by a transmission electron microscopy (TEM) and UV-vis-NIR scanning spectrophotometer. Amine-functionalized MNPs were synthesized by following a facile one-pot template-free method (*20*). A solution of 1,6-hexanediamine (6.5 g), anhydrous sodium acetate (2.0 g) and $FeCl_3 \cdot 6H_2O$ (1.0 g) in ethylene glycol (30 mL) was stirred vigorously at 50 °C to give a transparent solution. This solution was then transferred into an oven and reacted at 198 °C for 6 h. The MNPs were characterized by TEM and Quantum Design MPMS SQUID magnetometer.

To ensure full reactivity, thiol-modified oligonucleotides should be reduced by DTT immediately before use. Otherwise, the thiol group on the oligonucleotides could not form a self-assembled monolayer on the surface of AuNPs due to loss of active thiol group. So, the AuNPs synthesized previously (1mL), and the purified thiolated 1st pDNA (0.05 nmol) were then mixed together resulting in the formation of 1stpDNA-AuNPs. A self-assembled monolayer of thiolated pDNA formed on the surface of AuNPs. After a serial salt addition (*19*), the particles were stabilized for long-term storage at room temperature.

For the MNPs, the polyamine-functionalized iron oxide particles (1 mg) were reacted with 300 µg of sulfo-SMCC bifunctional linker for 2 h in 1 mL coupling buffer (0.1M PBS buffer, 0.2M NaCl, pH 7.2). The supernatant was removed after magnetic separation and the MNP-cross-linker conjugate was rinsed with the coupling buffer 3 times. The reduced thiolated 2nd pDNA (1 nmol) was added into 1 mL coupling buffer containing sulfo-SMCC-modified MNPs and reacted for 8 h. The functionalized MNPs were then suspended in 35 mL of 10 mM sulfo-NHS acetate. The solution was incubated and shaken at room temperature to block the unreacted sulfo-SMCC on the surface of MNPs. After passivation, the particles were centrifuged at 4000 rpm for 1 min and washed with passivation buffer (0.2M Tris, pH 8.5) and then with a storage buffer (10 mM PBS buffer, 0.2M NaCl, pH 7.4).

DNA Hybridization

A solution containing tDNA in PCR tube was put in the thermocycler at 95 °C for 10 min to separate the double-stranded DNA (dsDNA) into ssDNA. The serially diluted tDNA (40 µL) were mixed with 1stpDNA-AuNPs (40 µL) and 0.8 mg of MNP-2ndpDNA in 200 µL assay buffer (10mM PBS buffer, 0.15M NaCl, 0.1%SDS, pH 7.4). The hybridization reaction was maintained at a temperature of 45 °C for 45 min in an incubator. After the sandwich complex (MNP-2ndpDNA/ tDNA/1stpDNA-AuNP) was formed, the assay was put on the magnetic separator for 3 min and then the supernatant was removed. Unreacted solution components (DNA and AuNPs) were washed away 5 times with 500 µL assay buffer in order to effectively remove AuNPs that were not specifically bound to the MNPs through hybridization. Finally, the sandwich complex was resuspended in 100 μL of 0.1 M HCl solution. Figure 1 shows a schematic of the DNA hybridization and sandwich complex formation.

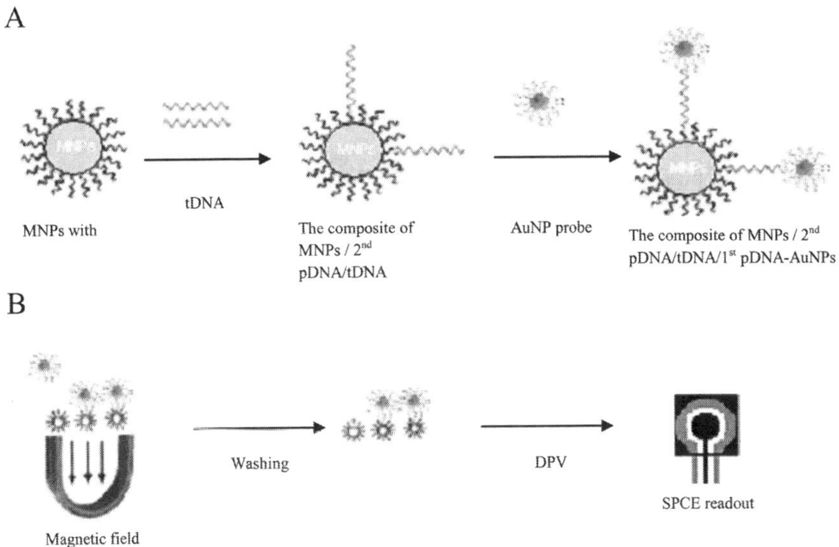

Figure 1. Schematic of the biosensor detection. (A) formation of MNP-2nd pDNA/tDNA/1stpDNA-AuNP sandwich complex; (B) Target DNA separation and electrochemical detection of AuNPs.

Electrochemical Detection on SPCE

The SPCE is composed of two electrodes: carbon electrode (working electrode) and silver/silver chloride electrode (counter and reference electrode). The working area is limited by a meshed well. Cyclic voltammetry (CV) was used to characterize the AuNPs on SPCE. One hundred microliters of the sandwich complex was allowed to accumulate on the SPCE for 10 min. A constant 1.25V was applied to the electrode for 2 min to facilitate the oxidation of the AuNPs (21). Cyclic voltammetry (CV) was performed from 1.4V to 0.0V at a scan rate of 100 mV/s. The same procedure was also performed on the control group, a solution containing 0.1M HCl.

The sandwich complex in 100 μL of 0.1 M HCl was deposited onto the SPCE. The accumulation time of the AuNPs was varied for 0 min, 10 min and 20 min. The same CV procedure was performed to optimize the accumulation time of the electrochemical detection.

Sandwich complexes from different tDNA were measured by differential pulse voltammetry (DPV) on SPCE under the optimum accumulation time. The potential was scanned from 1.25V to 0.0V with a step potential of 10 mV, modulation amplitude of 50 mV, and scan rate of 33.5 mV/s (*21*). Figure 2 shows the schematic of the electrochemical detection of AuNPs. Following oxidative gold metal dissolution in an acidic solution, the released Au^{3+} ions were reduced on SPCE and directly quantified by differential pulse voltammetry (DPV). A potentiostat/galvanostat with PowerSuite software (Princeton Applied Research, TN) was used for electrochemical measurement and data analysis.

Results and Discussion

Characterization and Functionalization of Nanoparticles

The AuNP dimension and spectroscopic properties were characterized by using a transmission electron microscope (TEM) and a UV-Vis-NIR scanning spectrophotometer, respectively. Figure 3A shows a transmission electron microscopy (TEM) image of our synthesized AuNPs with an average diameter of 15 nm and an absorption peak at 519 nm wavelength. The dimension of AuNPs is homogenous. After one month storage in room temperature, the AuNPs did not aggregate and their spectroscopic absorbance property was stable.

The MNPs were characterized by TEM and Quantum Design MPMS SQUID magnetometer. Figure 3B shows The TEM image of our synthesized MNPs, and their magnetic properties (inset). The diameter of MNPs is around 100 nm and the magnetic hysteresis has a maximum of 74.6 EMU/g at room temperature. Our previous research showed that the conjugation between nanoparticles and thiolated oligonucleotides is efficient (*22*).

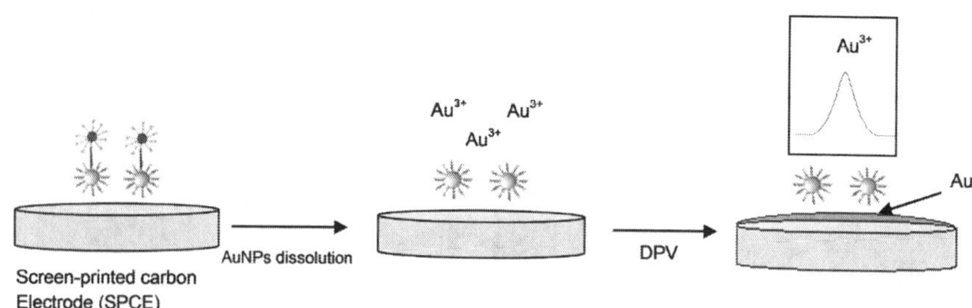

Figure 2. Schematics of the electrochemical measurement of AuNPs.

CV of Unmodified AuNPs on SPCE

Compared with conventional electrodes, screen-printed carbon electrodes (SPCEs) have several advantages, such as simplicity, convenience, and low cost (*23, 24*). Biosensors based on SPCEs have been extensively used for the detection of glucose (*25, 26*), cholesterol (*27*), antigens (*28*), and DNA (*29, 30*).

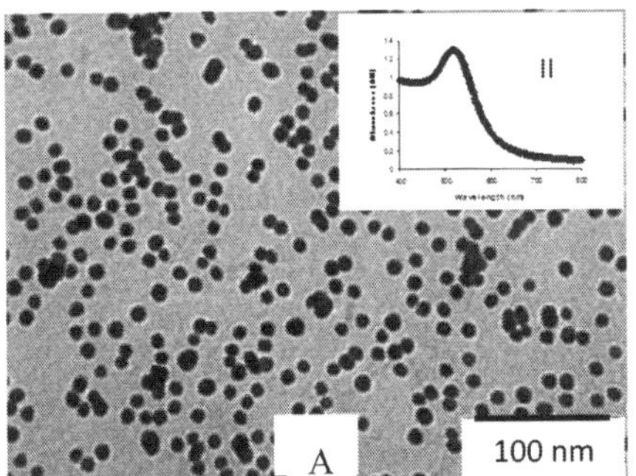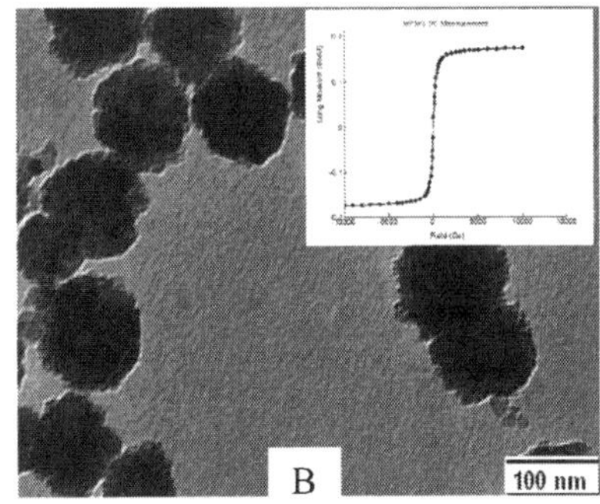

Figure 3. (A) A) TEM image of AuNPs showing ~15 nm in diameter. Inset: Absorbance spectrum of AuNPs showing a peak at 520 nm. (B) TEM image of MNPs showing ~100 nm in diameter, Inset: Magnetic hysteresis of MNPs.

Oligonucleotide-functionalized AuNPs were used as signal indicator in this study because of their ease of fabrication, greater oligonucleotide binding capabilities, stability under a variety of conditions, and good electrochemical properties (*31, 32*).

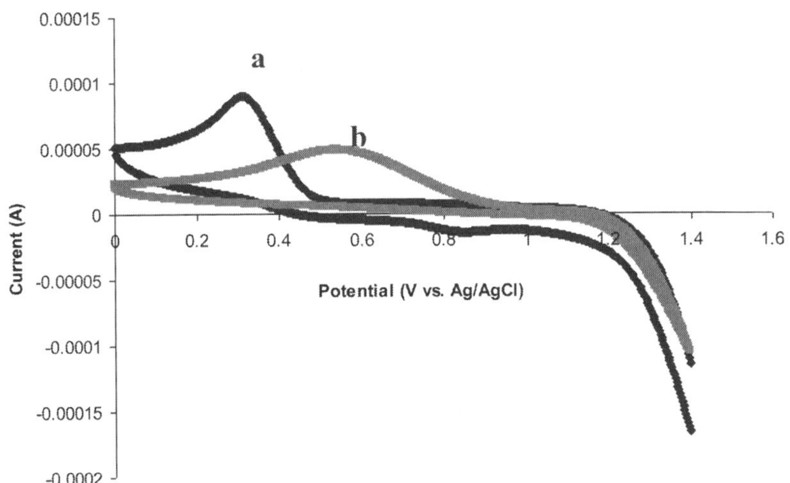

Figure 4. Cyclic voltammogram of unmodified AuNPs (1.17 x 10^{-8} M) on SPCE: (a) AuNPs in 0.1M HCl; (b) 0.1 M HCl only. CV scan from 1.4V to 0.0, scan rate 100 mV/s after 1.25 V electrooxidation for 2 min.

The cyclic voltammograms of AuNPs and reagents are depicted in Figure 4. Figure 4b shows that when there are no AuNPs, the carbon electrode is oxidized when performing oxidative procedure at 1.25V for 2 min. The carbon electrode has a reduction peak at 0.58 V. When AuNPs are present in the solution, the AuNPs are oxidized instead of the carbon electrode under 1.25V for 2 min. The curve (Figure 4a) shows that there is a reduction peak of Au^{3+} at 0.35 V. Meanwhile, the background signal at 0.58 V from the carbon electrode is suppressed greatly by AuNPs. The possible reason is that the surface of SPCE is covered by a self-assembled layer of AuNPs due to physical adsorption. The SPCE is protected by AuNPs and this prevents the electrooxidation process on the carbon electrode surface.

Optimization of Detection of Probe Modified-AuNPs

The accumulation time is the time from dropping the sample on the SPCE to the beginning of electrochemical detection. It is an important factor for the electrochemical measurement of AuNPs because more AuNPs would be adsorbed on the SPCE surface when the accumulation time is longer. The electrooxidation process would generate more Au^{3+} ions in the solution, which could be translated to a higher DPV signal and thus improve the biosensor sensitivity.

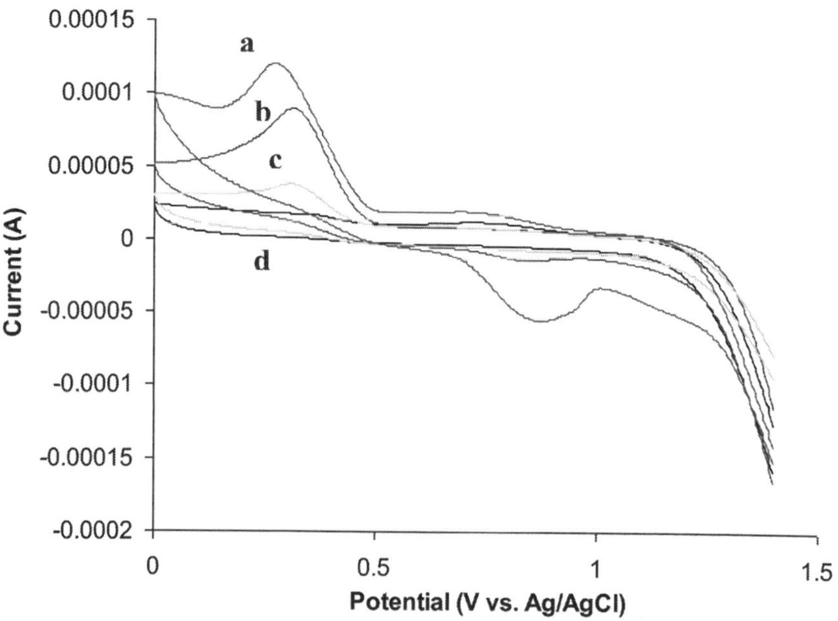

Figure 5. Cyclic voltammogram of probe-modified AuNPs in 0.1 M HCl on SPCE: (a) 20 min accumulation; (b) 10 accumulation; (c) no accumulation; (d) control, no AuNPs and no accumulation. Conditions: electrooxidation potential, +1.25 V; electrooxidation time, 2 min, CV scan from +1.4 V to 0 V, scan rate 100 mV/s, unstirred solution.

The sample solution was dried after 20 min at room temperature. The drying allowed most of the AuNPs to adsorb physically on the SPCE surface. Then 100 µL of 0.1M HCl was added again for DPV measurement. Figure 5 shows the DPV current for 0, 10, and 20 min of accumulation times. Figure 5a has 20 min of accumulation time which allows for more AuNPs to be reduced or oxidized, leading to a comparatively high signal. With 10 min and 0 min accumulation times, peak current are at 0.86 mA and 0.36 mA, respectively. They are only 72% and 30% of the reduction peak current compared to 1.21 mA for the 20 min accumulation time. Drying led to more AuNPs available on the surface of SPCEs, resulting in the higher sensitivity of the SPCE biosensor. Accumulation time of 20 min was used for the succeeding DPV detection of the DNA sandwich complex.

DPV Detection of DNA Sandwich Complex

Figure 6 shows the DPV response of the sandwich complex (MNP-2ndpDNA/ tDNA/1st pDNA-AuNP) after hybridization for various tDNA concentrations (0.7-700 ng/mL). Following oxidative gold metal dissolution in acidic solution, the Au^{3+} was reduced at the potential around 0.4V (vs. Ag/AgCl) on the SPCE. Inset shows that the peak current has a log-linear relationship with increasing concentrations of tDNA. The detection limit is 0.7 ng/mL of target DNA.

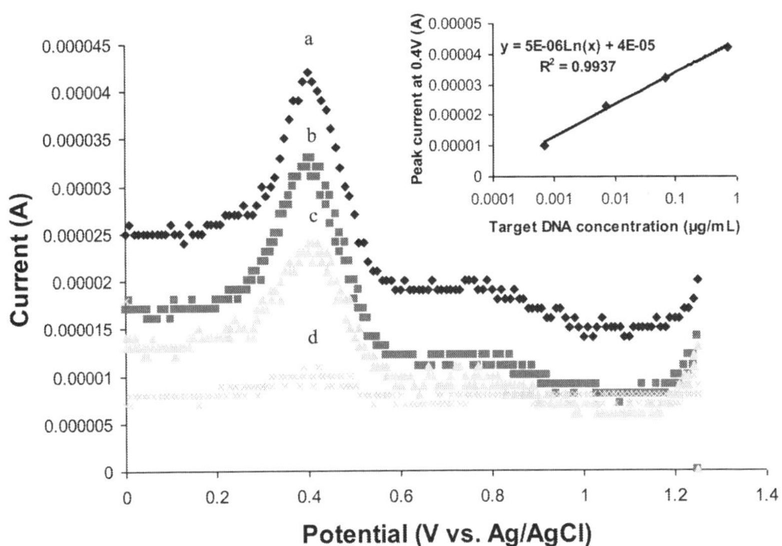

Figure 6. DPV hybridization response of different concentrations of DNA targets on SPCE (a: 700 ng/mL; b: 70 ng/mL; c: 7 ng/mL; d: 0.7 ng/mL) and the calibration plot between peak current and the tDNA concentration (inset). DPV scan from +1.25 V to 0 V, step potential 10 mV, modulation amplitude 50 mV, scan rate 50 mV/s, and nonstirred solution.

Conclusion

This paper has described a AuNP-based electrochemical biosensor for the detection of the insertion element gene of *Salmonella* Enteritidis. The biosensor takes advantage of the favorable electrochemical properties of AuNPs for a rapid and reliable detection on a disposable SPCE, and MNPs for easy and clean separation. The synthesized AuNPs were homogenous and stable, making them excellent materials for biosensing. The synthesized MNPs showed a good magnetic profile. The conjugation reaction between the two nanoparticles and thiolated oligonucleotides was efficient. After the pDNA on nanoparticles hybridized with tDNA, the sandwich complex formed, consisting of MNP-2ndpDNA/tDNA/1stpDNA-AuNP. This complex was applied directly onto the SPCE after removing the unreacted AuNPs. The SPCE did not need surface modification. During the oxidation process, the existence of AuNPs suppressed carbon oxidation. Deposition of AuNPs on the SPCE for 20 min allowed more AuNPs to be oxidized and increased the sensitivity of the biosensor. The current peak of DPV for Au^{3+} reduction had a log-linear relationship with increasing tDNA concentration. The total detection time was about 1 hour. The results showed that the detection limit of the insertion element (*Iel*) gene of *Salmonella* Enteritidis was 0.7 ng/mL (or 1.5× 10^{-16} mol). This disposable biosensor has great potential for quickly detecting foodborne and bio-threat agents.

Acknowledgments

This paper was developed under DHS Science and Technology Assistance Agreement No. 2007-ST-061-0000 03 awarded by the U.S. Department of Homeland Security. It has not been formally reviewed by DHS. The views and conclusions contained in this document are those of the authors and should not be interpreted as necessarily representing the official policies, either expressed or implied, of the U.S. Department of Homeland Security. The Department of Homeland Security does not endorse any products or commercial services mentioned in this publication. Connie Shi was a high school student conducting summer research under the High School Honors Science Program.

References

1. CDC (Centers for Disease Control and Prevention). *Salmonella surveillance summary, 2004*; U.S. Department of Health and Human Services, 2005.
2. Muhammad-Tahir, Z.; Alocilja, E. C.; Grooms, D. L. *IEEE Sens J.* **2005**, *5* (4), 757–762.
3. Pal, S.; Alocilja, E. C.; Downes, F. P. *Biosens. Bioelectron.* **2007**, *22* (9-10), 2329–2336.
4. Pal, S.; Ying, W.; Alocilja, E. C.; Downes, F. P. *Biosyst. Eng.* **2008**, *99* (4), 461–468.
5. Zhang, D.; Alocilja, E. C. *IEEE Sens. J.*. **2008**, *8* (6), 775–780.
6. Wanzenboeck, H. D.; Dominizi, K.; Hagl, P.; Almeder, C.; Bertagnolli, E.; Bogner, E.; Wirth, M.; Gabor, F. *Electrochem. Soc.* 2004, 8, 453–458.

7. Torres-Chavolla, E.; Alocilja, E. C. *Biosens. Bioelectron.* **2011**, *26*, 4614–4618.

8. Mathew, F. P.; Alocilja, E. C. *Biosens. Bioelectron.* **2005**, *20* (8) (SPEC ISS), 1656–1661.

9. Sassolas, A.; Leca-Bouvier, B. D.; Blum, L. J. *Chem. Rev.* **2008**, *108* (1), 109–139.

10. Lee, T. Y.; Shim, Y. B. *Anal. Chem.* **2001**, *73* (22), 5629–5632.

11. Yuk, J. S.; Jin, J.-H.; Alocilja, E. C.; Rose, J. B. *Biosens. Bioelectron.* **2009**, *24* (5), 1348–1352.

12. Dubertret, B. *Nat. Mater.* **2005**, *4* (11), 797–798.

13. Pu, K.-Y.; Pan, S. Y.-H.; Liu, B. *J. Phys. Chem. B* **2008**, *112* (31), 9295–9300.

14. Drummond, T. G.; Hill, M. G.; Barton, J. K. *Nat. Biotechnol.* **2003**, *21* (10), 1192–1199.

15. Jin, J.-H.; Zhang, D.; Alocilja, E. C.; Grooms, D. L. *IEEE Sens. J.* **2008**, *8* (6), 891–895.

16. Weng, J.; Zhang, J.; Li, H.; Sun, L.; Lin, C.; Zhang, Q. *Anal. Chem.* **2008**, *80* (18), 7075–7083.

17. Huang, G. S.; Wang, M. T.; Hong, M. Y. *Analyst* **2006**, *131* (3), 382–387.

18. Wang, S. J.; Yeh, D. B. *Lett. Appl. Microbiol.* **2002**, *34* (6), 422–427.

19. Hill, H. D.; Mirkin, C. A. *Nat. Protoc.* **2006**, *1* (1), 324–336.

20. Wang, L.; Bao, J.; Wang, L.; Zhang, F.; Li, Y. *Chem.—Eur. J.* **2006**, *12* (24), 6341–6347.

21. Pumera, M.; Aldavert, M.; Mills, C.; Merkoci, A.; Alegret, S. *Electrochim. Acta* **2005**, *50* (18), 3702–3707.

22. Zhang, D.; Carr, D. J.; Alocilja, E. C. *Biosens. Bioelectron.* **2009**, *24* (5), 1377–1381.

23. Miscoria, S. A.; Desbrieres, J.; Barrera, G. D.; Labbe, P.; Rivas, G. A. *Anal. Chim. Acta* **2006**, *578* (2), 137–144.

24. Susmel, S.; Guilbault, G. G.; O'Sullivan, C. K. *Biosens. Bioelectron.* **2003**, *18* (7), 881–889.

25. Guan, W. J.; Li, Y.; Chen, Y. Q.; Zhang, X. B.; Hu, G. Q. *Biosens. Bioelectron.* **2005**, *21* (3), 508–512.

26. Xu, H.; Li, G.; Wu, J.; Wang, Y.; Liu, J. *Conf. Proc. IEEE Eng. Med. Biol. Soc* **2005**, *2*, 1917–1920.

27. Carrara, S.; Shumyantseva, V. V.; Archakov, A. I.; Samori, B. *Biosens. Bioelectron.* **2008**, *24* (1), 148–15028.

28. Kim, J.-H.; Seo, K.-S.; Wang, J. *IEEE Sens. J.* **2006**, *6* (2), 248–252.

29. Hernandez-Santos, D.; Diaz-Gonzalez, M.; Gonzalez-Garcia, M. B.; Costa-Garcia, A. *Anal. Chem.* **2004**, *76* (23), 6887–6893.

30. Ruffien, A.; Dequaire, M.; Brossier, P. *Chem. Commun.* **2003**, *7*, 912–91332Cambridge, England.

31. Demers, L. M.; Ostblom, M.; Zhang, H.; Jang, N. H.; Liedberg, B.; Mirkin, C. A. *J. Am. Chem. Soc.* **2002**, *124* (38), 11248–11249.

32. Lytton-Jean, A. K. R.; Mirkin, C. A. *J. Am. Chem. Soc.* **2005**, *127* (37), 12754–12755.

Chapter 4

Nanoparticles and Biophotonics as Efficient Tools in Resonance Energy Transfer-Based Biosensing for Monitoring Food Toxins and Pesticides

Munna S. Thakur,* Rajeev Ranjan, Aaydha C. Vinayaka, Kunhitlu S. Abhijith, and Richa Sharma

Fermentation Technology & Bioengineering Department CSIR-Central Food Technological Research Institute Mysore-570 020, India
*E-mail: msthakur@cftri.res.in. Tel: +91 821 2515792.
Fax: +91 821 2517233.

Development of new age and upcoming bio-diagnostic techniques has revolutionized the field of analytical and bioanalytical chemistry. The team efforts of researchers from various fields such as biochemistry, biotechnology and material science has led to the development of several robust and reliable biosensing tools applicable in numerous fields such as health sector, environmental safety, clinical diagnostics and food technology. The usage of nanoparticles such as quantum dots, silver and gold nanoparticles for their efficient tailoring to conjugate with numerous biosensing agents such as enzymes, antibodies, aptamers, cells and tissues which can be used in high throughput and multiplexed analysis of a variety of analytes has greatly improved and replaced the conventional analytical methods. Unique opto-physical properties, surface plasmon resonance and field confinement effects of semiconductor nanoparticles have greatly enhanced the sensitivity and robustness of bio-diagnostics involved in the detection of pathogens/ toxins and other hazardous materials such as pesticides and heavy metals. In the current book chapter, we have tried to deal with the upcoming and novel bioassays for quick and multivariate analytical approach for monitoring pathogens, toxins and other hazardous analytes at ultrasensitive levels. We have also tried to explain the research work carried out by us in our laboratory related with this field.

Introduction

Current biotechnological approach in the field of bioanalytical methods has envisaged the development of simple and reliable biosensing tools with the help of novel strategies such as Förster resonance energy transfer (FRET), fluorescence and chemiluminescence based immunoassays, bioluminescence resonance energy transfer (BRET), surface plasmon resonance (SPR), metal enhanced fluorescence (MEF) and surface Enhanced Raman scattering (SERS) (*1–8*). Commonly used methods for the detection of pathogens, toxins and environmental hazardous materials are mainly based on enzyme linked immunosorbent assay (ELISA), atomic absorption spectroscopy, inductive coupled plasma mass spectrometry (ICP-MS), polymerase chain reaction (PCR), morphological and biochemical characterization and high performance liquid chromatography (HPLC) (*9–11*). Though these methods have classical approach towards analyte detection, are limited mainly by their long preparation time, requirement of skilled personnel and huge economical investment. Moreover, many of these detection methodologies have generally univariate analysis approach which is too cumbersome when a large number of samples need to be examined in a short period of time.

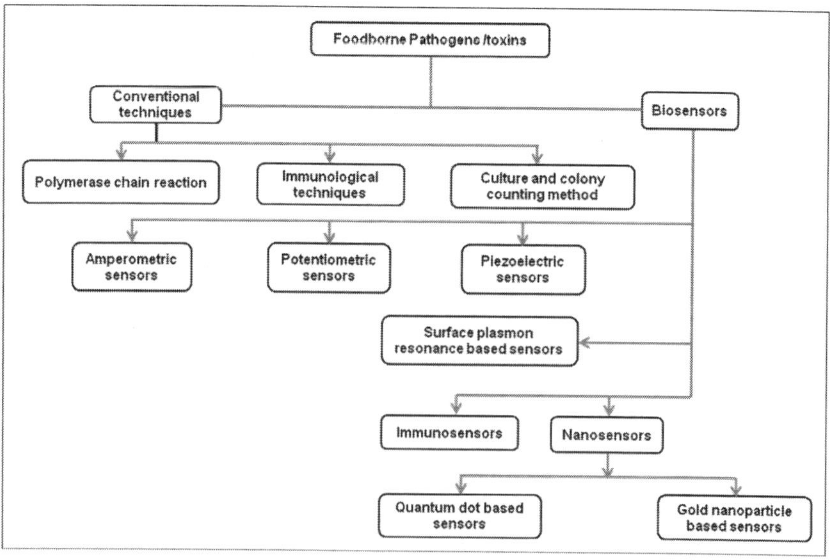

Figure 1. Schematic representation for foodborne pathogen detection.

Biosensors as Alternative Techniques in Food Safety Assessment

Biosensors are the preferred choice over conventional techniques due to their specificity, rapidity, ease for mass fabrication and field applicability as schematically represented in Figure 1 (*12, 13*). Biosensor devices are emerging as one of the relevant tools for food and environmental safety measures (*12*). A schematic representation of a biosensor setup is shown in Figure 2. It has a biorecognition element which mainly consists of biological molecules such as antibodies, whole cells, enzymes and nucleic acids interfacing a transducer (optical, colorimetric, piezoelectric, amperometric, potentiometric and calorimetric). The target analytes are specifically recognized by these biological molecules to generate signals which are greatly enhanced with the aid of amplifiers after their transduction using colorimetric, electrochemical, optical, piezoelectric, potentiometric, calorimetric components of the biosensor system. The enhanced signals are processed and generally digitally get converted into quantitative or readable format.

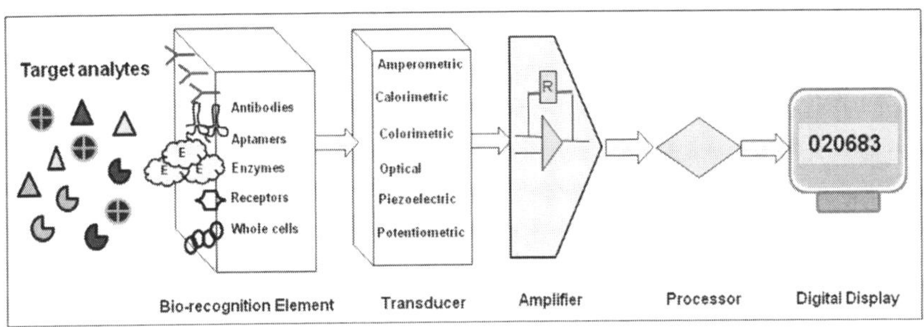

Figure 2. Schematic representation of biosensor setup.

Recent developments in material science and nanotechnology have increased the possibilities of biosensing technologies in food and environmental monitoring immensely. Many compact handheld device designs for field applicability have been investigated for their broad application in the area of pesticide, pathogen/toxin detection (*14, 15*). In the recent past, optical biosensing methods have been gaining considerable interest being ultrasensitive and rapid in nature (*16*). Fluorescence and luminescence based biorecognition elements have been adopted and introduced for the construction of robust and smart biosensor systems meant for analysis of a wide variety of analytes. Nanoparticles such as quantum dots and gold nanoparticles have revolutionized the field of fluorescence resonance energy transfer based assay methods unlike other fluorescent organic dyes owing to their unique opto-physical properties (*17*). Furthermore, independent photon generating properties of luminescent enzyme systems have been utilized to engineer self illuminating quantum dots. Work on fluorescent proteins such as green fluorescent protein (GFP), was extensively carried out by Nobel laureate Osamu Shimomura whose immense clinical and analytical capabilities have recently been recognized. Other fluorophores such as Yellow fluorescent protein (YFP) and lumazine proteins (LumP) have proved to broaden the scope in the field of clinical diagnostics in the near future (*18*). These fluorescent proteins (FPs) being non toxic, proteases resistant (*19*) and stable over a wide range of physical parameters such as temperature and pH gains much interest in modulating FRET/BRET based biosensor system (*20, 21*). Agricultural food products are prone to get infected by a variety of pathogens which include fungus, bacteria and viruses. The utilization of these ultrasensitive labeling agents for the detection of pathogens, toxins, environmental pollutants and pesticides has been well recognized.

Food Contaminants

Public concern over food safety and food quality has gained immense attention over the past decade. According to World Health Organization (WHO) definition, "Foodborne illnesses are diseases, which are either infectious or toxic in nature, caused by agents that enter the body through the ingestion of contaminated food" (*22*). Food poisoning may be of two types namely food infection and food intoxication. Food infection refers to the presence of bacteria or other microbes which infect the body after consumption. Food intoxication however involves the ingestion of toxins including microbially produced exotoxins present within the food even when the microbe that produced the toxin is no longer present or able to cause infection. Presence of a pathogenic microorganism/toxin in food is of main concern especially for food industries that may cause serious public health hazard. Irrespective of overall improvement in food safety issues, many countries report outbreaks of food poisoning due to microbial contamination and their toxins (*23*). According to WHO estimation, in 2005 alone 1.8 million people died from diarrheal diseases caused by contaminated food and drinking water (WHO, 2007a). As per Centers for Disease Control and Prevention (*24*). USA alone reports around 48 million cases of foodborne diseases, resulting in 1,28,000 hospitalizations and ~3,000 deaths each year. The most common foodborne pathogens are *Escherichia coli O157:H7*, some strains of *Staphylococcus aureus, Shigellaspp, Bacillus anthracis, Campylobacter jejuni, Clostridium perfringens, Clostridium botulinum, Salmonella spp., Listeria monocytogenes, Vibrio cholerae, Yersinia enterocolitica, Coxiella brunetii* and few others (*25, 26*). In general, bacterial toxins are of two types. The structural components of the bacteria such as lipopolysaccharide complex released inside the host tissue as a result of bacterial cell lysis are called endotoxins (*26*). Pathogenic bacteria also secrete certain protein based toxin termed as an exotoxin that generally disrupts the normal cellular metabolism. Enterotoxins are chromosomally encoded exotoxins secreted by pathogenic bacteria, which are often water soluble, heat stable and low molecular weight compounds. They generally disrupt the cellular permeability by creating electrolyte imbalance in the cell. Hence, clinical diagnosis, water, food safety and environmental analysis are some of the areas where detection of microbial pathogens is crucial. Need for on-site and quick monitoring of these organisms has always challenged the scientists to come out with novel tools and techniques. Fluorescence and chemiluminescence based immunoreactor columns, FRET, BRET, MEF, SPR and SERS based biosensor systems for the detection of food borne pathogens, toxins pesticides and other environmental contaminants has been a relatively newer area of research which has shown bright future prospect.

In this regard, several research groups have carried out work for the development of efficient biosensor systems for onsite monitoring of food borne toxicants, pesticides and other hazardous chemicals (*1, 27–33*). In recent years, Raman spectroscopy (RS) is becoming a popular tool for the rapid identification of fungi and bacteria and now is widely accepted as a potential identifier of whole cell organisms. However, generation of weak signals through RS decrease the detection sensitivity and can be considered as a time consuming process which limits its application for high throughput analysis (*34*). Fortunately, Raman signals can be enhanced thousand to million fold if the analyte is placed close to a suitable roughened surface such as an aggregated silver colloid substrate as well as gold colloid (*34, 35*). When metallic nanoparticles are used as SERS substrate, random localized plasmons or hot spots are created at the surface giving rise to dramatic signal enhancement. Structural modulation of large local fields of junction plasmons has generated much interest in the fabrication, assembly, and understanding the properties of structures with nanoscale gaps. Silver film over nanosphere (AgFON) was used as a substrate for the detection of Calcium dipicolinate (CaDPA) which is a biomarker for bacillus spores (*33*). The AgFON surface used for the Raman signals enhancement was stable over a period of 40 days. The assay time was required to be less than 15 minutes and as low as 2.6 x 10^3 spores could possibly be detected in real time. A sandwich immunoassay for quick monitoring of *Mycobacterium avium* subsp. *Paratuberculosis* was developed based on the principle of SERS (*32*). An immobilized layer of monoclonal antibodies that target a surface protein on the microorganism was utilized along with extrinsic Raman labels (ERLs) that were designed to selectively bind to captured proteins and produce large SERS signals. By correlating the numbers of *M. avium* subsp. *Paratuberculosis* bacilli present prior to sonication to the amount of total protein in the resulting sonicate, the detection limit determined for total protein was translated to the microorganism concentration which was as low as 500 to 1000 bacterial cells. Moreover, the time required for completion of the assay was less than 24 hours. Because of these unique optical properties, nanoparticles are playing an important role in biosensor applications aiming at improving sensitivity and performance of biosensors (*36–39*). As a result, applications of various nanomaterials such as quantum dots, gold nanoparticles, carbon nanotubes and magnetic nanoparticles in biosensor development are being extensively investigated. Thus, research on nanoparticle based biosensors has opened an interdisciplinary frontier area of nano-biotechnological approach for food, health and environmental monitoring (*40–42*). In this direction, application of nanoparticles for detection of foodborne pathogens and toxins has become a thrust area in food biotechnology. Metal nanoparticles exhibit phenomena like FRET, BRET, LSPR (localized surface plasmon resonance) and SERS that can be used as potent techniques in biosensing. Vinayaka and Thakur (2011) developed bioconjugated fluorescent probe as a model for designing FRET based biosensing tools as shown in Figure 3. Further, CdTe-QD bioconjugates were characterized using absorption spectroscopy, fluorescence spectroscopy, and gel electrophoresis and was suggested that they may be attributed as nanoparticle superstructures for the development of nanobiosensors (7). Vinayaka and Thakur (2010) reviewed and highlighted the usage of quantum dots (QDs) as potential fluorescent probes for monitoring food toxicants and pathogens. They also emphasized on the application of a possible resonance energy transfer phenomenon resulting from nano-biomolecular interactions obtained through the bioconjugation of QDs with biomolecules (30). Recently construction of an immunoreactor based competitive fluoro-immunoassay (Figure 4) for monitoring staphylococcal enterotoxin B (SEB) using bioconjugated quantum dots was carried out (43). Using this method, it was possible to quantify SEB from 1000 ng to 10 ng based on the integrated fluorescence of the SEB–CdTe$_{557}$ bioconjugate eluted from the immunoreactor column with a limit of detection of 8.15 ng. Dong et al. (2004) used synaptobrevin (Syb) or synaptosome-associated protein as a linker to conjugate cyan fluorescent protein (CFP) and yellow fluorescent protein (YFP) for the FRET based detection of Botulinum neurotoxins (BoNTs) (44). BoNT activity characterization conventionally relies on functional assays to monitor the effect of the toxin or using biochemical assays to detect cleaved substrate molecules which are time consuming and laborious task. Therefore, FRET based quick detection system used here for real time monitoring of Botulinum toxin has paved a way to construct robust and reliable biosensor systems using upcoming biosensing tools. Three highly specific, sensitive and reagent free optical signal transduction methods were designed for the detection of polyvalentproteins by directly coupling distance dependent fluorescence self-quenching and/or resonant energy transfer to the protein-receptor binding events (16). Ganglioside (GM1) which is a recognition unit for cholera toxin (CT) was covalently labeled with fluorophores and then incorporated into a biomimetic membrane surface. The presence of CT with five binding sites for GM1 caused dramatic change in the fluorescence signal of the labeled GM1. As reviewed by Ligler et al. (2003) high throughput screening and rapid analysis of a wide variety of toxins viz. staphylococcal enterotoxin B, ricin, cholera toxin, botulinum toxoids, trinitrotoluene and the mycotoxin fumonisin was possible at levels as low as 0.5 ng mL^{-1}. The detection was based on the principle of sandwich and competitive fluoro-immunoassay (45). The optical transducer employed here was charged couple device (CCD) camera to image the individual fluorescence spots. Abhijith and Thakur (2012) have recently used gold nanoparticles (GNPs) for the sensitive detection of aflatoxin B1 based on the principle of metal enhanced fluorescence as shown in Figure 5. In their study, GNPs were conjugated with anti-aflatoxin IgY antibodies which were indigenously generated in poultry. A tenfold enhancement of inherent fluorescence of Aflatoxin B1 was observed in the presence of biologically synthesized gold nanoparticles (BGNPs) which increased the limit of detection for aflatoxin B$_1$ upto 5 picogram (1).

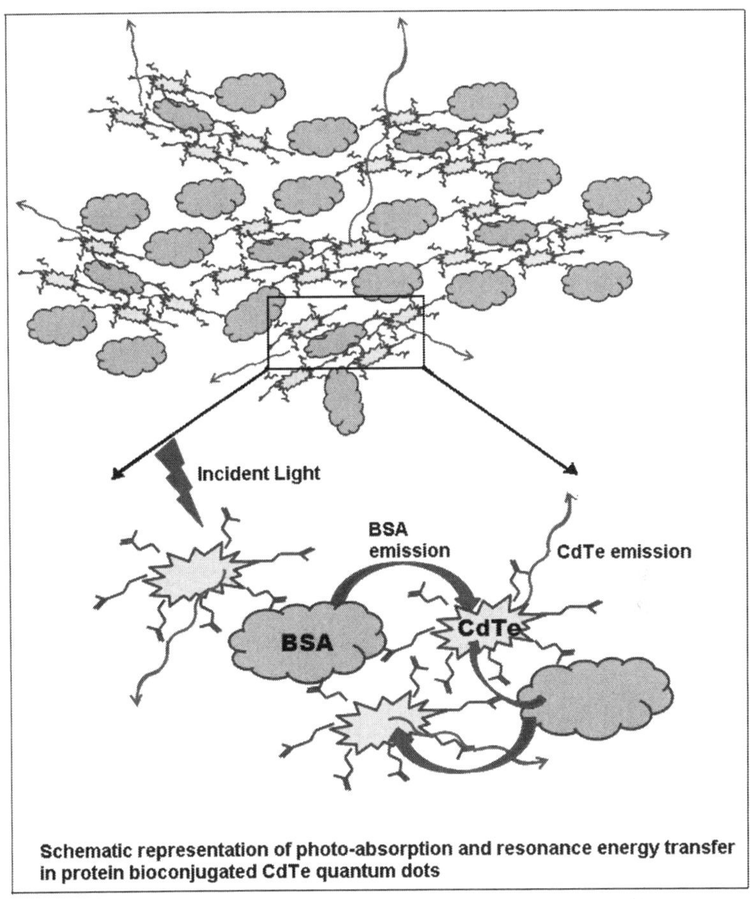

Figure 3. Schematic representation of photo-absorption and resonance energy transfer in protein bioconjugated CdTe quantum dots. Reproduced with permission from Ref. (7). Copyright ACS publication, 2011.

The occurrence of organochlorine pesticides in the environment has been a major concern; due to their high persistence and the possible impacts of their exposure to humans. Dichlorodiphenyltrichloroethane (DDT) is one of most hazardous and widely used organochlorine insecticides. Baker et al. (2012) developed dipstick based immuno-chemiluminescence methods for the detection of DDT with high sensitivity as shown in Figure 6. Anti-DDT antibodies, raised in chicken (IgY), were used as biological sensing elements, by immobilizing onto nitrocellulose membrane strips in a chemiluminescence

(CL)-based dipstick technique. The photons generated during the biochemical interaction were directly proportional to the DDT concentration. Using the proposed dipstick-based immuno-CL method, DDT was detected in the range 0.05–1ng/mL (*46*). Similar studies were carried out by Chouhan et al. (2006) for the detection of a pesticide called methyl parathion at per part trillion (ppt) levels using immuno-chemiluminescence based CCD image analysis (*28*) as shown in Figure 7. Recently, our research group have developed a chemiluminescence based assay format for the detection of formaldehyde in spiked soft drinks, wine and raw fish. The limit of detection was found to be 10pg/mL. The principle of chemiluminescence enhancement (*47*) is schematically represented in the Figure 8. Further, we have studied FRET based phenomenon using QDs and reduced nicotinamide adenine dinucleotide (NADH) as depicted in the Figure 9. Energy routing assay of NADH-QD was applied for detection of formaldehyde as a model analyte in the range of 1000–0.01ng/mL. This promising work seems to bring newer avenues in the detection of a variety of analytes which involves electron transfer processes (*27*).

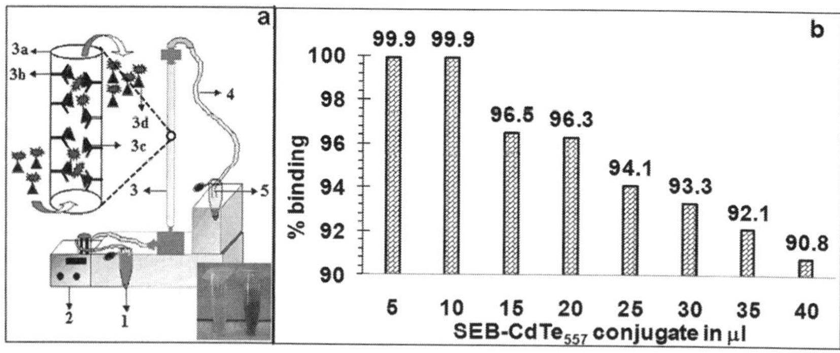

Figure 4. (a) Immunoreactor setup showing competitive binding of SEB and SEB–CdTe557 with immobilized anti-SEB antibody: (1) sample inlet, (2) peristaltic pump, (3) immunoreactor column, (3a) glass capillary column, (3b) immobilized anti-SEB antibody, (3c) SEB, (3d) SEB–CdTe557 bioconjugate, (4) column outlet tubing, (5) sample collection point. TNBS indicated successful immobilization of anti-SEB antibody on Sepharose matrix (inset). (b) Optimization of SEB–CdTe557 conjugate for competitive immunoassay. Various concentrations of conjugate were passed and respective percentage binding was determined. Reproduced with permission from Ref. (43). Copyright RSC Publishers, 2012.

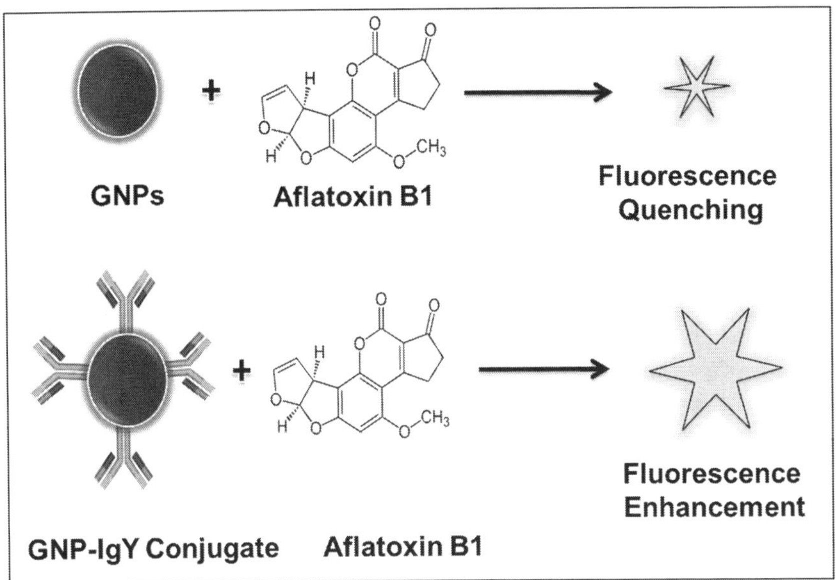

Figure 5. Metal enhanced fluorescence for aflatoxin monitoring with high sensitivity. Reproduced with permission from Ref. (1). Copyright RSC Publishers, 2012.

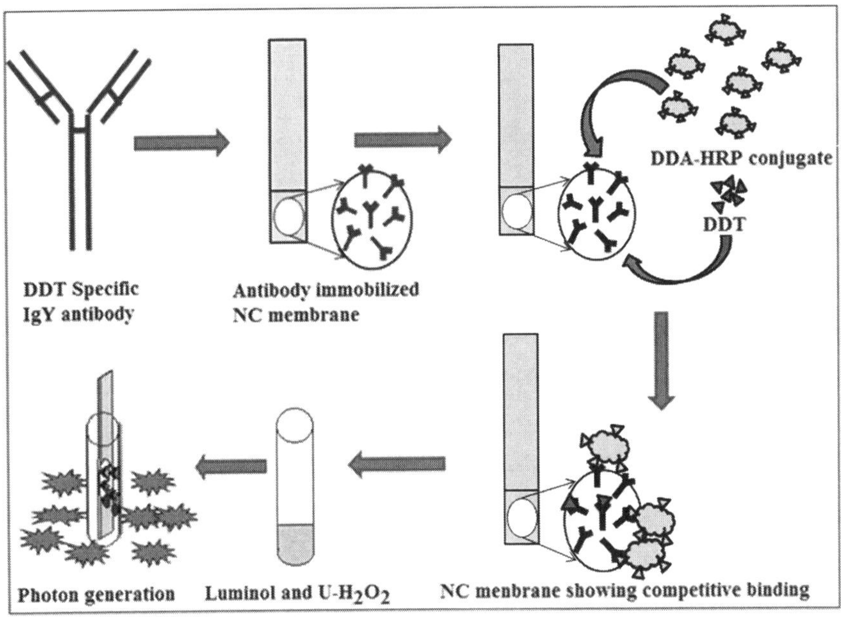

Figure 6. Schematic representation of chemiluminescence based dipstick competitive immunoassay. Reproduced with permission from Ref. (46). Copyright John Wiley & Sons, Ltd., 2012.

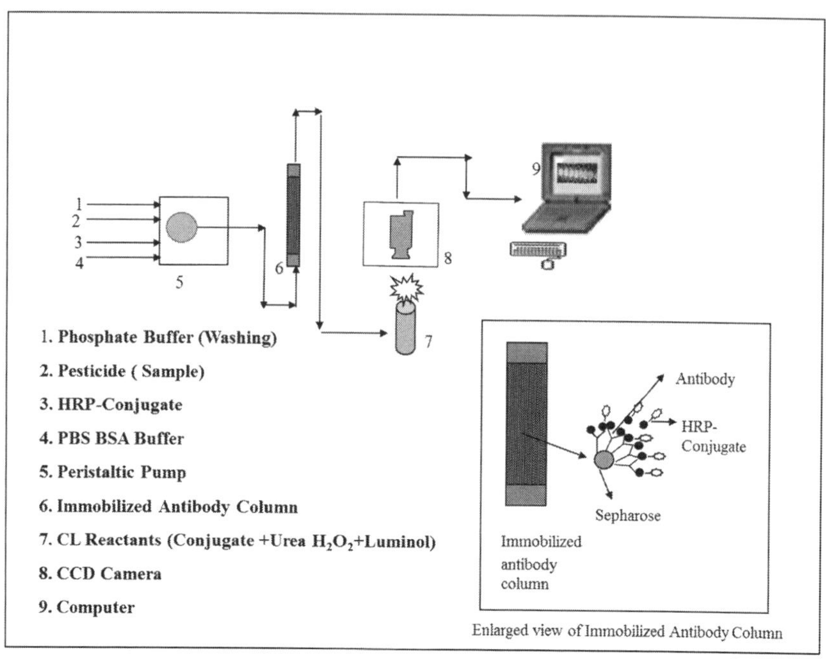

Figure 7. Flowchart of the qualitative determination of methyl parathion using CCD. Reproduced with permission from Ref. (28). Copyright Elsevier B.V., 2006.

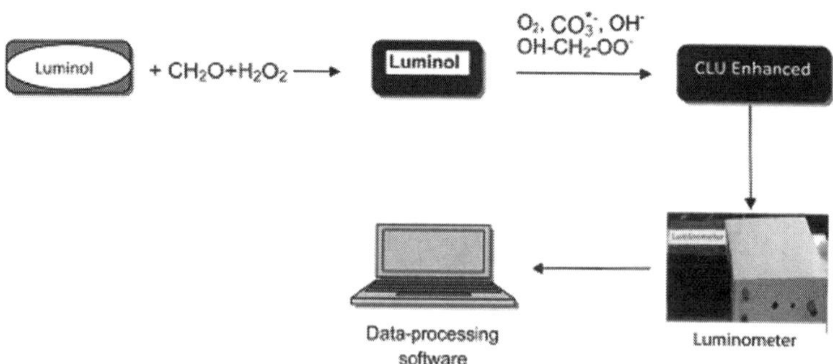

Figure 8. Schematic representation for chemiluminescence based formaldehyde detection. Reproduced with permission from Ref. (47). Copyright RSC Publishers, 2012.

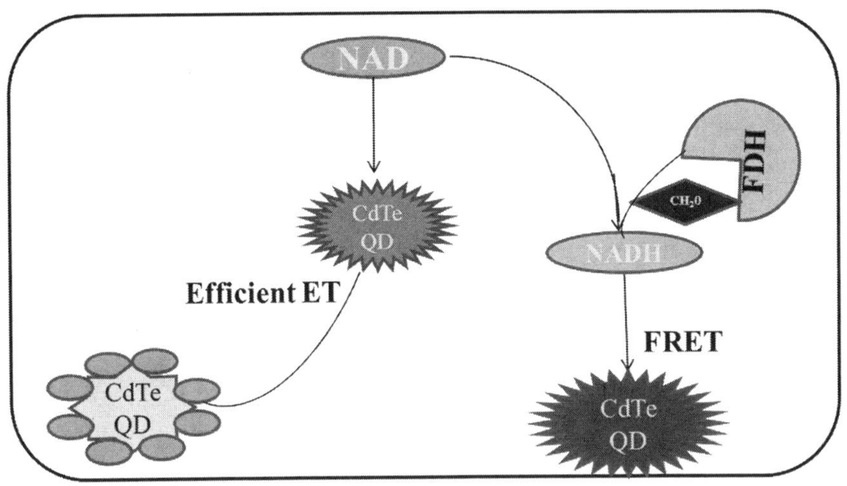

Figure 9. Nicotinamide adenine dinucleotide (NAD) based "turn off" and NADH based "turn on" of QD fluorescence. Reproduced with permission from Ref. (27). Copyright Elsevier B.V., 2012.

Nanoparticles and Their Properties for Biosensing Application

The prospect of controlling matter at molecular and atomic level has inspired many biologists to think about new technological approaches for various applications. Thus, nanotechnology is not only the next step of miniaturization following microtechnology, but also an approach to investigate natural architectures and to mimic them for technological problems (*48*). Though synthesis of nanoparticles is known since centuries, only recently people have started exploring the real possibilities other than cosmetic applications. Nanotechnology is an area of research which deals with design, manufacture and applications of particles at nanometer size typically with dimensions smaller than 100 nm having properties entirely different than the bulk material. Nanotechnology is at the core of advances in biological research through the use of novel materials. Recent advances in these smart nanomaterials such as quantum dots (QDs) and gold nanoparticles (GNPs) have gained a lot of interest in the biological research, because of their unique spectral properties (*49–52*). Engineering the size of particles at nanometer size varies their properties from those of the bulk particles of same material, which is attributed to the variation in bonding state and electronic orbitals of the atoms constructing these particles. Their optical behavior may also differ with respect to the nature of nanoparticle such as insulator, semiconductor or metal. The confinement of electrons in these particles may remarkably change their band structure. For example, in the case of semiconductors, electrons will occupy the valence band completely and conduction bands are left empty. This quantum confinement effect results in discretized density state of electronic orbitals leaving a considerable separation between valence band and conduction band referred to as band gap. However, in bulk particles of the same material, partial fulfillment of conduction band with electrons results in continuous density states (*53*).

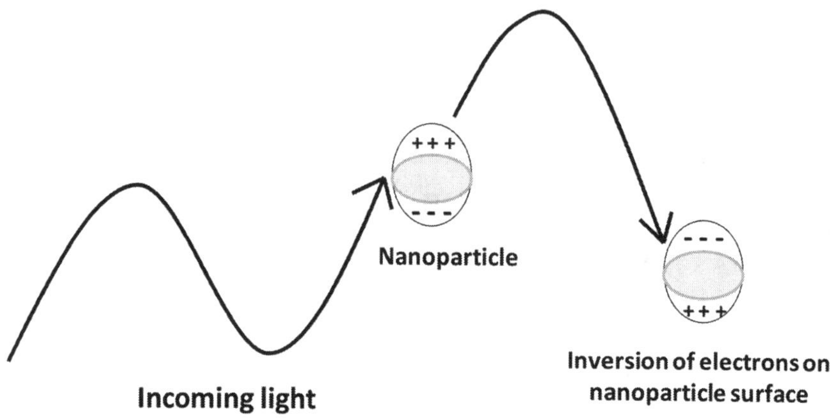

Figure 10. The principle of surface plasmon resonance; where nanoparticles form localized plasmons when exposed to electromagnetic radiation.

Free electrons in metals (which are "d" electrons in noble metals such as silver and gold) travel through the material. Due to the large electron density and delocalization, all the free electrons indeed coherently oscillate under a time dependent electrical field, which gives the well-known bulk plasmon absorption. The free electrons in the large nanoparticles have different oscillation modes toward the time-dependent electrical field of the light. The optical response of the metal nanoparticles is the summation of all the different plasmon frequencies resulting from dipole and higher order oscillation modes. The mean free path in gold and silver is approximately 50 nm. In particles smaller than 50nm, no scattering is expected from the bulk. This means interactions with the surface dominate. When metal nanoparticle size is comparable to the light wavelength, plasmon absorption becomes size-dependent. In particles much smaller than the mean free path, a monotonic absorption transition with size is observed because absorption is mainly due to dipole oscillation mode while higher order oscillations contribute much less to the overall oscillation strength (*54*). When the wavelength of light is much larger than the nanoparticle size it sets up standing resonance conditions as represented in Figure 10. Light in resonance with the surface plasmon oscillation causes the free electrons in the metal to oscillate. As the wave front of the light passes, the electron density in the particle is polarized to one surface and oscillates in resonance with the light's frequency causing a standing oscillation. The resonance condition is determined from absorption and scattering spectroscopy and is found to depend on shape, size and dielectric constants of both the metal and the surrounding material. This is referred to as the surface plasmon resonance, since it is located at the surface. As the shape or size of the nanoparticle changes, the surface geometry changes, causing a shift in the electric field density on the surface. This causes a change in the oscillation frequency of the electrons, generating different cross-sections for the optical properties including absorption and scattering. Changing the dielectric constant of the surrounding material has an effect on the oscillation frequency due to the varying ability of the surface to accommodate electron charge density from the nanoparticles (*55*).

Nanotechnology is playing an important role in the development of smart sensors including biosensors. Nanomaterials provide a promising way to increase the bio-recognition efficiency of a biosensor due to high surface to volume ratio of nanoparticles (*56*). Functional nanoparticles bound to biological molecules (e.g. peptides, proteins, nucleic acids) have been developed for use in biosensors to detect and amplify various (e.g. electronic, optical, and magnetic) signals. Recent studies have shown that biosensors employing nanoparticles provide rapid, sensitive, accurate, and stable measurements (*7, 57, 58*). It is possible to design biosensors for sensitive and label-free detection of foodborne pathogens and their toxins in unprocessed and processed foods. The specific recognition of target organism is done with the help of nano-biorecognition technology, in which the nanoparticles are bio-functionalized by conjugation with a biomolecule (*59, 60*). Metal nanoparticles like gold and silver exhibit very high surface area for bioconjugation. This property paves the way for their use in biosensing. Inorganic nanocrystals (e.g. zinc sulfide, cadmium sulfide, lead sulfide) can be used as electrical tags and also to form quantum dots with fluorescent properties (e.g. cadmium telluride, cadmium sulfide, cadmium selenide) for optical biosensing and imaging. The coupling of inorganic nanoparticles with biomolecules generates hybrid particles possessing unique photophysical properties along with biological activity. Magnetic nanoparticles (e.g. Fe_3O_4, Fe_3S_4) are also powerful and versatile diagnostic tools in biosensors (*61*). Metallic nanoparticles have been used in several forms of bioassays, with detection based on colorimetry, photothermal deflection and scanning electron microscopy (*62*).

Nanoparticles for Food Toxins Detections

It was Theodor Förster who first outlined the quantum-mechanical behavior of non-radiative resonance energy transfer between two molecules called FRET in the year 1946. FRET is a distance dependent nonradiative energy transfer in a dipole-dipole interactions between chromophores over short distances of the order of 10-100 Å. Here quantum of energy is transferred from a donor chromophore to an acceptor chromophore in a close proximity when absorption spectrum of acceptor overlaps with that of emission spectrum of the donor molecule (*63*). During the process, most of the energy emitted in the form of light called fluorescence comes from acceptor molecule. This may also result in decrease of fluorescence from donor molecule called "donor quenching" that in turn depends on efficiency of energy transfer or ability of acceptor molecule to absorb energy non-radiatively (*64*). Therefore, distance between donor-acceptor pair should be within the range from 10 Å to 70 Å that may vary energy transfer efficiency (*64*). Energy transfer efficiency gets affected by donor quantum yield, loss of energy in the form of non-radiative decay, dipolar orientation within the pair and spectral overlap between donor and acceptor. The rate of transfer of this energy is described by the equation (*65*).

$$K_{FRET} = \frac{1}{\zeta_D} \times \frac{9000 \ln(10) K^2 Q_D J}{128 \pi^5 \eta^4 N_A} \times \frac{1}{R^6} \qquad \text{Equation 1}$$

$$= \frac{1}{\zeta_D} \left(\frac{Ro}{R} \right)^6 \qquad \text{Equation 2}$$

$$Ro = \left(\frac{9000 \ln(10) K^2 Q_D J}{128 \pi^5 \eta^4 N_A} \right) \qquad \text{Equation 3}$$

Where R is the distance between the donor and the acceptor. The Förster radius, Ro, is the distance between the donor and acceptor where the FRET efficiency is 50%. It is influenced by the refractive index (η) of the medium; quantum yield of the donor in the presence of the acceptor (Q_D); spectral overlap integral (J), Avogadro's number (N_A); the lifetime of thedonor in the absence of acceptors (ζ_D) and therelative orientation factor (K^2).

Spectral overlap integral (J) is defined by

$$J = \int_0^\infty F_D(\lambda) \, \varepsilon_A(\lambda) \, \lambda^4 \, \delta\lambda \qquad \text{Equation 4}$$

Where $F_D(\lambda)$ is the donor emission profile and $\varepsilon_A(\lambda)$ is the acceptor molar extinction coefficient.

BRET is a commonly observed phenomenon in marine organisms such as jelly fish (*Aequorea victoria*) and sea pansy (*Renilla reniformis*). The bioluminescent proteins emit biophotons by a mechanism called chemically initiated electron exchange luminescence (CIEEL) to excite naturally present fluorophores such as green fluorescent protein (GFP). The resemblance of BRET to FRET in many ways similar except that the former doesn't require an external excitation source. Efficient resonance energy transfer requires that the emission spectrum of the donor must overlap with the excitation spectrum of the acceptor. The BRET efficiency is inversely proportional to the sixth power of the distance between the donor and the acceptor as described in the equation:

$$E = 1 / (1 + R^6 / R_0^6) \qquad \text{Equation 5}$$

Where, R_0 is the distance leading to 50% of energy transfer from the donor to the acceptor.

QDs as Fluorescence Marker

Luminescent QDs are semiconductor fluorescent molecules on an order of 2-10 nm size made up of few hundred atoms. They are made up of II-VI, III-V and IV-VI group elements of the periodic table with a diamcter on the order of the compound's exciton Bohr radius (*66, 67*). QDs show unique optical and electronic properties as a result of strong confinement of excited electrons and corresponding holes called excitons in their structures (*68, 69*). These unique photophysical properties of QDs are unparallel in comparison to conventional fluorophores thereby making QDs all-around attractive tool for biosensing (Table 1). Warner et al. (2009) have developed a fluorescence sandwich immunoassay for the detection of botulinum neurotoxin serotype A (BoNT/A) (*70*). QDs and high affinity antibodies were used as rcporters in the assay. The antibodies will bind to the non-overlapping epitopes present on the toxin and the recombinant fragment. Here the recombinant fragment holotoxin (BoNT/A-Hc-fragment) acts as a structural simulant of the toxin molecule. Immunoassay was carried out in a 96-well microtitre plate using toxin specific primary capture antibody and QD-coupled secondary antibody as the reporter antibody. Toxin was quantified based on the fluorescent signals obtained from QD. The assay was sensitive up to 31 picomolar (pM) concentration for botulinum toxin. In another format of bead-based assay, QD conjugated capture antibody was immobilized on Sepharose beads and immunoassay was carried out in microcentrifuge tubes. A renewable surface flow cell (rotating rod) equipped with a fiber optics system for fluorescence measurement has been made use for the quantification of toxin. The toxin was detected at a sensitive level of 5 pM concentration. QD as a fluorescence probe in this technique has proved better than organic fluorescent dye to attain sensitivity in the assay as reported (*70*). The possibility of having background signals due to the scattering of excitation source from QD immobilized Sepharose beads has to be addressed to avoid any non specific signals.

Mukhopadhyay et al. (2009) has reported a detection technique for *E. coli* based on carbohydrate-functionalised CdS quantum dots (*71*). Here CdS QD was coated with thiolated mannose and was used for the specific recognition of mannose specific lectin Fim H in *E. coli*. Mannose capped QDs have shown luminescent aggregates as few as 10^4 *E. coli* per mL. The absence of aggregation with *E. coli* strains defective in Fim H lectin and also with QDs having capping material other than thiolated mannose justifies the assay specificity. The technique is based on ligand-receptor specific aggregation of bacterial cells and therefore is qualitative in nature.

Goldman et al. (2004) has reported a multiplexed fluoroimmunoassay based on CdSe-ZnS core-shell QDs for the detection of cholera toxin (CT), ricin, shiga like toxin (SLT) and SEB in microtitre plate. Toxin specific antibodies were used as recognition molecules to attain specificity in the assay. Individual toxin assay was sensitive enough to determine CT at 10 µg/L with QD_{510}, ricin at 30 µg/L with QD_{555}, SLT at 300 µg/L with QD_{590} and SEB at 3 µg/L with QD_{610}. The assay sensitivity got decreased in multiplexed analysis where the assay sample was a mixture of all four toxins (*72*).

Table 1. Overview of QD properties (*121*)

Sl. No.	*Properties*	*Quantum dots*	*Organic fluorophores*
1.	Size	2-10 nm	$\leq$ 5 nm
2.	Absorption spectra	Broad spectra, gradually increases towards lower wavelength from first excitonic peak	Variable and narrow, more often mirror image of the emission peak
3.	Emission spectra	Narrow and tunable with size	Broad spectra with red end tailed
4.	Effective Stokes shift	Possible to have > 200 nm	Generally < 100 nm
5.	Tunable emission	Unique property of QDs, can be size-tuned from the UV to IR	Not possible
6.	Quantum yield	Generally high	Variable
7.	Fluorescent lifetime	~ 20-30 ns or greater	Less than 5 nanoseconds
8.	Photostability	Resistant to photobleaching	Poor, susceptible to photodegradation
9.	FRET capabilities	Excellent	Variable
10.	Multiplexing capabilities	Excellent, most suitable	Rare
11.	Bioconjugation	Can be designed	Limited
12.	Multi analyte attachment	Possible	Limited

A FRET based immunoassay was developed for the detection of *Aspergillus amstelodami* using QD. QD was conjugated to anti-*Aspergillus* antibody and incubated with (BHQ-3)-labeled *A. fumigatus* that had a lower affinity for the antibody than *A. amstelodami*. Here *A. fumigatus* and *A. amstelodami* were used as quencher and target analytes respectively (*73*). FRET resulted in diminished fluorescence of QD in presence of (BHQ-3)-labeled *A. fumigates*. As higher affinity target analyte (*A. amstelodami*) displaces *A. fumigates,* fluorescence signal from the QD donor increased. This displacement immunoassay could detect *A. amstelodami* concentrations as low as 10^3 spores/mL.

In all these detection procedures, nonspecific interactions and background fluorescence from QDs along with antigen-antibody cross reactivity have been found to be interfering with the assay. Moreover, interactions between individual QDs leading to spectral overlapping may also interfere during multiplexed analysis.

QDs for the Detection of Toxin-Producing Genes

Recent approaches towards single nucleic acid sequence based detection for pathogens have revolutionized the biological sciences. In general, multiplexed detection strategies employ fluorescence based indicators for various approaches such as fluorescent *in situ* hybridization (FISH) techniques, fiber optic-based biosensors and nucleic acid microarrays. Fluorescence-based DNA detection assays employ coupling of nanoparticles such as luminescent semiconductor quantum dots to biomolecules due to their brighter signal and greater resistance to photobleaching than organic fluorophores. QDs have been effectively used for the single nucleotide detection assays based on QD-oligonucleotide bioconjugate probes. These assays generally employ FRET as the transduction mechanism. Wang et al. (2011) successfully made use of QD as an energy donor in a FRET based detection of toxic shock syndrome toxin-1 (TSST-1) producing *S. aureus* (*74*). An oligonucleotide probe designed to bind specifically with tst gene responsible for the production of TSST-1 was conjugated to QD by covalent linkage. A black hole quencher (BHQ) bound to DNA was used as an energy acceptor. Fluorescence quenching of QD was achieved through FRET after attaching BHQ-DNA to *tst* gene probe by match sequence hybridization. A reversal of fluorescence quenching was observed on addition of target DNA due to the detachment of BHQ-DNA from QD-DNA probe because of the different affinities. The intensity of recovered QD fluorescence was proportional to the target DNA ranged from 0. 2 to 1.2 µM.

In another method, complementary oligonucleotide probe sequence has been used which can form hairpin sequence with its own strand. Here, the two ends of the probe sequence are complementary to one another, wherein, one end of the probe sequence constitutes complementary sequence for target toxin gene sequence (*75–77*). Bioconjugation of two different QDs having different emission maximum on either end of the probe sequence forms the biological detection agent. The oligonucleotide probe sequence that has been bioconjugated to QDs forms a hairpin loop by self-complementary sequences hybridization in the absence of any target toxic gene. Thus, the two QDs come within the requisite energy transfer distance, resulting in spatial dipolar interaction between the two QDs. This will results in resonance energy transfer. Hence, the signals generated as a result of dipolar interactions between the two bioconjugated QDs suggests the concentration of toxin (*75–77*). However, in presence of a target toxin gene, the hairpin loop denatures and hybridizes with toxin gene. This resulted in QDs facing away from each other and loss of spatial dipolar interaction between the two QDs. Therefore, the dipolar interaction between QDs is an indication of the absence of any toxin gene in the samples.

QD-Based FRET as an Efficient Assay Technique for Specific Nucleic Acid Detection

Several FRET-based molecular probes that include complementary oligonucleotide bioconjugated QDs have been developed. These probes have ensured separation free detection of DNA, whose fluorescence signals change as a result of hybridization and subsequent resonance interactions. Indeed, the applications of multiple QDs for resonance interactions during FRET based analysis have overcome the interference of background fluorescence as reported (75–77). Here, the sensor consists of two target-specific oligonucleotide probes those binds to the target gene with close proximity to enable resonance transfer as shown in Figure 11. These probes have been bioconjugated to QDs of different emission spectrum or one of the probe labeled with an organic fluorophore and another to a QD. Here, QD functions as either energy donor or an energy acceptor based on the spectral behavior and the spectral overlap between donor-acceptor pairs. In presence of a target DNA, the sandwiched hybrid of DNA and complementary probes has brought the two chromophores into close proximity. That has resulted into fluorescence emission from the acceptor chromophore by means of resonance energy transfer on illumination of the donor chromophore. Therefore, the fluorescence emission from an acceptor chromophore was an indication of the presence of target analyte, wherein the presence of individual fluorescence was an indication of absence of any target analytes, as the unhybridized probes do not participate in FRET.

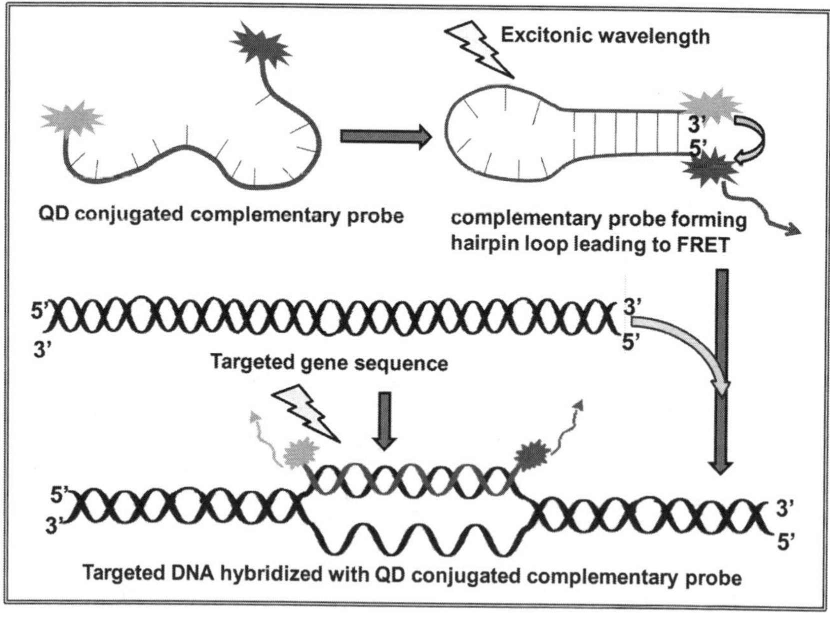

Figure 11. FRET based target gene detection. The technique is based on the principle of fluorescence quenching of donor chromophore in the hairpin configuration of self-complementary target specific oligonucleotide probe.

GNPs as Labels for SPR and RET

GNPs have been used extensively in various biological applications because of many reasons (*78*). Gold being noble metal is inert and is resistant to corrosion. Functionalization of gold surface with biological molecules such as antibodies and nucleic acids results in nanobiomaterials exhibiting suitable optical, electronic and biological properties. Because of the enhanced electromagnetic fields generated by GNPs, their strong plasmon resonance and high extinction coefficient, they have been used tremendously in many biomedical applications (*79–81*) labelling, targeted delivery and various sensing applications (*82, 83*).

Out of many prospective applications possible for GNPs in biosensing research, their ability in either quenching or to enhance the fluorescence of fluorophores such as QDs thereby participating effectively in resonance energy transfer based mechanisms gains much interest. Due to the strong plasmon field and broad absorption spectrum in visible region GNPs can be promising superquenchers for FRET based applications (*84–86*). Here GNPs can act as an efficient energy acceptor from excited fluorophores, which are in close proximity to a gold particle, resulting in a non-radiative relaxation of the fluorophore. In such cases, the presence of an analyte may displace the fluorophore competitively thereby releasing fluorophore from quenched state. The aptitude of GNPs to influence the FRET was well studied to understand the phenomenon and predict the extent of FRET variation. An extensive fluorescence variation can be observed for a fluorophore placed in the vicinity of metal nanoparticles possessing a strong plasmon field. Kang et al. (2011) used Cypate, a near infrared fluorophore, to estimate the level of fluorescence alternation by plasmon field strength. They observed both fluorescence quenching and fluorescence enhancement for Cypate as a result of changing the distance between fluorophore and GNP surface. Fluorescence quenching observed at very close proximity between Cypate and GNP resulted into fluorescence enhancement as opposed to quenching when they were separated up to a particular distance. This feature of GNP to control fluorescence would be of much interest in developing highly sensitive biosensing and imaging techniques (*85*).

Oh et al. (2005) also used GNPs for developing an inhibition assay method based on the modulation in FRET efficiency between QDs and GNPs. They have employed streptavidin-biotin interaction as a model system to assay the avidin concentration in sample. In principle, the assay was based on fluorescence quenching of streptavidin conjugated QDs (SA-QDs) by the biotinylated GNPs (biotin-GNPs). They observed almost 80% quenching in fluorescence intensity of SA-QDs in the presence of biotin-GNPs. Addition of biotin-GNPs that had been pre-saturated with avidin resulted in a full recovery of the fluorescence intensity of SA-QDs. The detection limit of avidin was found to be around 10 nM with dynamic detection range extended up to 2 µM as reported (*87*). In a similar work, $CdTe_{523}$ and GNP bioconjugated to SEB and anti-SEB IgY respectively were used as donor-acceptor pair during FRET to study resonance interactions and fluorescence quenching in an immunocomplex. The mutual affinity between IgY-GNP and SEB-$CdTe_{523}$ was made use to obtain efficient energy transfer between respective chromophores. This phenomenon resulting in fluorescence quenching of $CdTe_{523}$ was used for effective detection of SEB in the range 1000 ng/mL to 20 ng/mL with a regression coefficient of $R^2 = 0.9813$. The limit of detection in this method was 8.53 ng/mL (*43*).

Application of BRET

Since there is no requirement of an excitation light source for the donor, BRET can avoid problems such as light scattering, high background noise, and direct acceptor excitation (*88*). Additionally, since the donor or both the donor and acceptor can be co-expressed in the cell as fusion proteins and the excitation follows a localized event (delivery of the substrate), the target of interest can be specifically excited - a property that comes in very useful for *in vivo* application (*89, 90*).

Bioluminescent enzyme systems such as *Renilla* luciferase (Rluc), bacterial luciferase (Bluc), firefly luciferase (Fluc) and blue bioluminescent aequorein photoprotein systems have been used as efficient BRET donors as shown in Figure 12. Rluc has widely been used in artificial BRET systems as donor while Fluc due to its long wavelength emission has been limited in its application in the excitation of commonly used fluorescent proteins. Green Fluorescent protein (GFP) and yellow fluorescent proteins (YFP) and their variants have been very commonly used as BRET acceptors in conjunction with Rluc (*91*). On the other hand, the firefly luciferase system emits ataround 560 nm, GFP and some of its variants are not suitable as acceptors. Alternative acceptors such as Cy3/Cy5 and the fluorescent protein DSRed can be used with this protein donor. Such acceptors have found use in monitoring antigen–antibody binding (*92*) and protein– protein interactions (*8, 93*). In one study, Rluc was used as donor to assay ligand-binding. The acceptors used were yellow fluorescent protein, the YPet variant and the Renilla green fluorescent protein (RGFP). Ligand induced intramolecular rearrangements of beta-arrestin and recruitment to G protein-coupled receptors was studied with increased sensitivity (*94*).

Rluc has been commonly used as the reporter for *in vivo* imaging, as well. Protein-protein interactions have been studied using Rluc in mice models to study binding of rapamycin induced FK506 binding protein 12 (FKBP12) with FKBP12 rapamycin binding domain (FRB). The acceptors used were two red fluorescent proteins, TagRFP and TurboFP635. The BRET fusion proteins were injected intravenously in mice which were then imaged using an IVIS-200 or IVIS-Spectrum equipped with a charge-coupled device camera. Imaging was performed in sequence luminescence scan mode (*95*). A proprietary BRET assay with Renilla luciferase was developed by Perkin-Elmer; the substrate used was Deep Blue C (emission at 395 nm), the acceptor was an optimized GFP2 acceptor. This configuration had greater spectral resolution between the donor and acceptor pair.

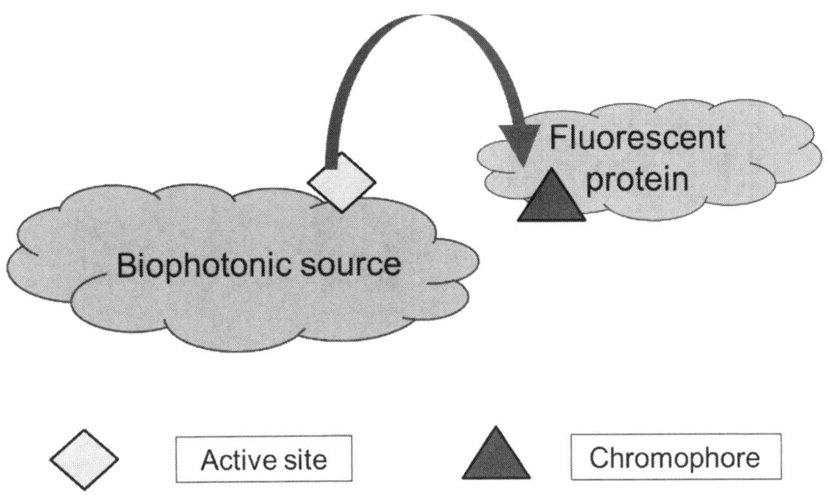

Figure 12. BRET phenomenon in which photonic emission from a biological source is nonradiatively transferred to a fluorescent protein.

Dyes and small molecules can also be used as acceptors. A bioluminescent photoprotein aequorin, (derived from jellyfish) consisting of the luciferase apoaequorin complexed to its coelenterazine substrate is principally used as areporter for Ca^{2+} ions since its bioluminescence is triggered by Ca^{2+} ions (*96, 97*). A modified BRET assay used dyes like Dabcyl or QSY-7-labeled avidin to quench the bioluminescence of biotinylated Aequorin upon exposure to Ca^{2+} ions (*98*). Aequorin has also been used as a BRET donor to monitor the interaction between Streptavidin (fused with Aequorin) and a biotin carboxyl carrier protein (fused with an EGFP acceptor) (*99*). *Gaussia* luciferase was used as a BRET donor, where it was attached to the C-terminus of maltose-binding protein. BODIPY558–aminophenylalanine was incorporated into Tyr210, which is located near the maltose-binding site of maltose-binding protein in response to a four-base codon (CGGG). In addition to the luminescence of luciferase, the double-labeled protein showed the emission of BODIPY558, which was strongly quenched by a neighboring tryptophan in the absence of maltose but was recovered in the presence of maltose (*100*).

In addition to fluorescent proteins and dyes, other potential fluorophores such as fluorescent nanoparticles can serve as potential BRET acceptors as depicted in Figure 13. Semiconductor nanoparticles such as quantum dots (QDs) and gold nanoparticles (GNPs) have widely been used for efficient tailoring of useful biochemical assays involved in monitoring of food borne pathogens, toxins and pesticides. So et al. (2006) investigated the use of fluorescent semiconductor QDs for *in vivo* molecular imaging. The QD_{665} and luciferase conjugated was tested for BRET in both serum and whole blood samples followed by injection into live mice either subcutaneously, intramuscularly or intravenously (*90*).

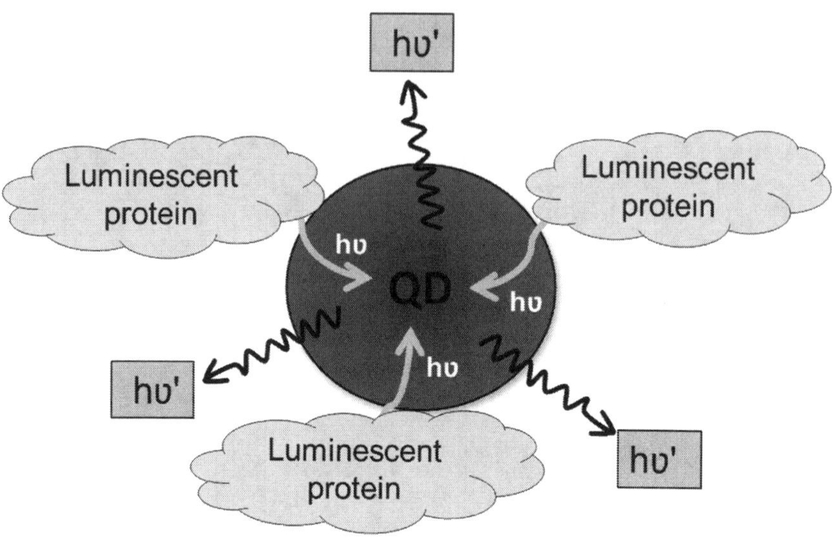

Figure 13. Pictorial representation of self-illuminated quantum dots (QDs) where donor is a luminescent protein. The emission wavelength from QDs is red shifted.

Biophotonics for Food Safety Assessment

The term "Biophotonics" unites biological science with photonics, the latter being the science and technology of generation, manipulation and detection of photons (*101*). Biophotonics has therefore become the established general term for all techniques that deal with the interaction between biological and other organic and inorganic materials having the potential to produce / emit photons. Recently, Moulton et al. (2009) have constructed three stable photon generating plasmids for real time *Salmonella* tracking *in vivo* (*102*). After cloning the firefly luciferase cDNA, scientists have used the luciferase gene as a sensitive reporter of gene expression. Some examples of these applications include using these genes to study the progression and regression of viral and bacterial disease, such as HIV and mycoplasmas, as well as the non-invasive assessment of therapeutic treatment of tumor proliferation and regression in animal models owing to their target specificity and biocompatibility. Finally, luciferases are also successfully used as biosensors for environmental pollutants such as arsenates, mercury, lead, phenols, agrochemicals, and xeno-estrogens that may act as endocrine disruptors to humans and wildlife (*103–105*).

The choice of developing optical based biosensors employs luminescence technology in which the role of bioluminescence is immense. High quantum yield, low background noise, specificity, ease of usage and compact instrumentation possibility has revolutionized bioluminescence based systems. Bioluminescence based biosensor system has been widely applied in various field *viz.* environment monitoring, biomedical science, in-vivo imaging, bio diagnostics and food technology. Yoo et al. (2007) have developed a method to immobilize bioluminescent bacteria using polyvinyl alcohol styrylpyridinium (PVA-SbQ) to detect phenol and hydrogen peroxide which are protein damaging and oxidative respectively (*106*). Presence of naphthalene and salicylate ions at different pH was also investigated using recombinant *Pseudomonas fluorescence* HK44 containing promoterless *lux* gene cassette (*107*). Trajkovska et al. (2005) have investigated the behavior of firefly luciferase enzyme activity in the presence of aromatic pesticides (*108*). Applegate et al. (1998) have investigated the light emission by the *luxCDBAE* bioluminescent bioreporter for Sensing Benzene, Toluene, Ethybenzene and Xylene (BTEX) (*109*). The highly sensitive calcium assays was possible only when the role of calcium to trigger luminescence in aequorin system was discovered by Shimomura half a century ago. The assay of intracellular calcium level has a wide importance in biomedical science since many biochemical reactions involve calcium ions. Creton et al. (1999) have extensively reviewed the measurement of intracellular calcium levels by chemiluminescence method using aequorin system and have emphasized their advantages over the radioactive method which is a biohazard (*110*). The science of in-vivo imaging is not new and the conventional methods employs magnetic resonance imaging (MRI), ultrasound, electroencephalogram (EEG), X-ray computed tomography (CT), X-Rays and Single photon emission computed tomography (SPECT). However these instruments are very expensive and trained personnel are required to handle and processing of the raw images. In recent years, application of fluorescent and luminescent probes for the *in vivo* imaging is gaining importance. High signal to noise ratio, noninvasive nature, cost effectiveness and ease of usage are the few unique characteristics which is the reason for their popularity. In the recent past several studies have been carried out in animal models to study cancer proliferation, cell signaling and gene regulation by the aid of fluorescent and bioluminescent tags. However the light emission using chemiluminescent and bioluminescent probes is in the range of (480-560 nm) which is highly absorbed by the living tissues and hemoglobin. This serious limitation of using luminescent technique in imaging was intensively investigated by several researchers by shifting their light emission to red and far red emission signals via several techniques. Wu et al. (2009) conjugated a far-red fluorescent indocyanine derivative to biotinylated Cypridina luciferase to produce detectable signals in far red regions by bioluminescence resonance energy transfer (*111*). Similarly, protein engineering, substrate modification and different buffer system has also been done to emit the signals in higher wavelength region. Fluorescent proteins such as green fluorescent proteins (GFP) and certain other variants of GFP with higher emission wavelength have been employed as fluorescent tags non-invasively for monitoring gene expression, protein localization and their interactions (*112, 113*). Similarly Lumazine proteins and Yellow Fluorescent Proteins (YFP) have been discovered from genus *Photobacterium* and *Vibrio harveyi* sp. respectively which act as secondary emitter receiving its energy by transfer from the flavin primary emitter (*114*).

Bioluminescence emission from luciferase enzyme systems, being highly specific and sensitive have been widely used in conjugation with a variety of nanoparticles to form self illuminating probes for *in vivo* cell imaging and has revolutionized the field of biophotonics (*90*).

Chemiluminescence is the generation of electromagnetic radiation as light due to a chemical reaction. In this reaction, visible light emission is most common, which can be detected at high sensitivity by optical based systems such as photo multiplier tube (PMT), avalanche photodiode (APD) and Charged couple device (CCD) camera. All these optical transducers enhance the photonic signal to achieve the desired signal output. The basic scheme of PMT and CCD is shown in Figure 14 (a), (b), and Figure 15 respectively.

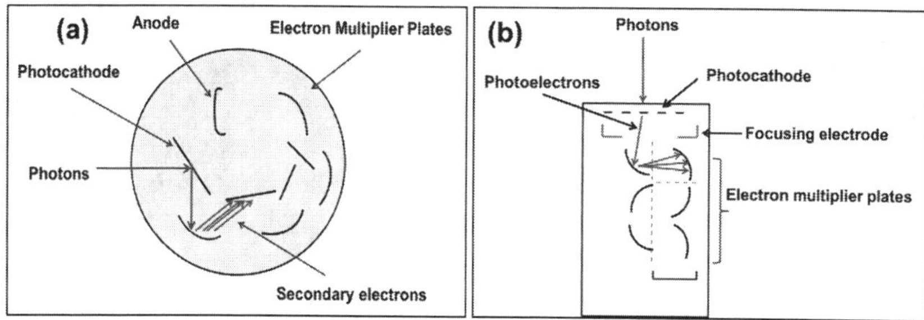

Figure 14. (a): Top view of a side illuminated photomultiplier tube. Photons strike the photocathode, generating a photoelectron. These photoelectrons are accelerated into the electron multiplier plates, or dynodes to produce multiple electrons which are eventually collected by anode and (b) Side on view of an end-on illuminated photomultiplier tube (120).

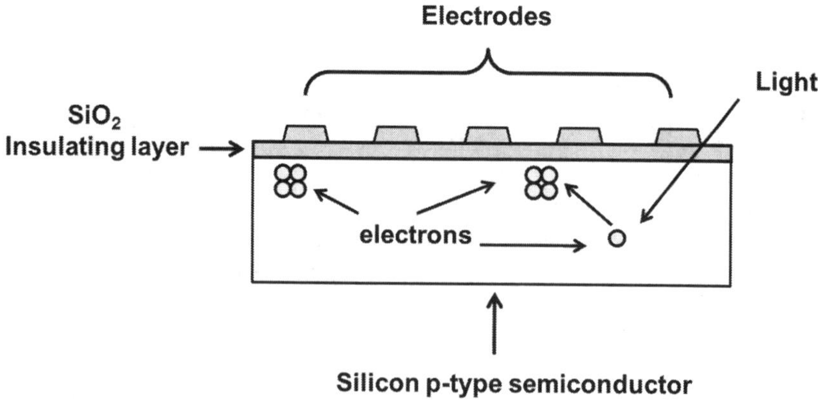

Figure 15. Side view of a charged couple device (CCD) array with a conduction channel (120).

There are numerous examples of chemiluminescent/ bioluminescent reactions and many of them are remarkably powerful to be used for ultrasensitive analyses. During the recent past, the firefly luciferase has been extensively evolved for sensitive applications involving measurement of Adenosine triphosphate (ATP), including monitoring biomass, analyzing biological fluid contamination from microbes, assessing cell viability, and assaying enzymes involving ATP generation. This enzyme system is ultrasensitive and can estimate ATP even at femtogram levels (*115*). The light emission is proportional to the ATP content. ATP being the universal energy currency is present in all viable living organisms. The average ATP content present in microbes is ca.1 femtogram (*116*). Hence, rapid and sensitive viable count of bacterial contaminated food samples can be determined using ATP assay method. Stannard and Wood (1983) developed a microfluidic and dielectrophoretic device for real time detection of intracellular ATP present in food borne bacteria. The device construct uses dielectrophoretic force to capture microorganisms and measures the ATP content by ATP luminescence (*117*). Understanding of host pathogen interaction is an important aspect in order to study the food borne infection. Siragusa et al. (1999) studied the real time monitoring of *Escherichia coli O157:H7* adherence to beef carcasses using recombinant *lux* bioreporters (*118*).

Unfortunately, firefly luciferase enzyme is thermally unstable and henceforth its application is limited. Several works have been carried out to overcome this issue owing to the usefulness of this enzyme. The luciferase enzyme system has widely been used in monitoring surface hygiene and food contaminants. The detection of bacterial contamination in food using firefly luciferase enzyme system requires efficient extraction of ATP from the bacteria. There are several reports on the optimization of ATP extraction buffer which generally consists of a combination of cationic or anionic surfactants, protein precipitating agents, acids or disinfectants. However, the ATP extraction agents tend to inhibit the luciferase enzyme activity during the reaction. Considerable work has been carried out in this area so as to minimize the interference of the ATP extraction agents with the luciferase activity. Hattori et al. (2003) isolated a novel luciferase resistant to benzalkonium chloride which is a cationic surfactant and have been widely used for this purpose (*119*).

Conclusion

Presently in our laboratory at CFTRI, Mysore, India we are involved in using nanotechnological approaches for the development of chemiluminescence/ bioluminescence-based biosensing methods for various applications in food science and technology. In the recent past, we have developed a method based on immuno-chemiluminescence and image analysis using CCD for the detection of methyl parathion (MP) with high sensitivity. Further, semiconductor nanoparticles such as quantum dots (QDs) and gold nanoparticles (GNPs) have been synthesized and utilized for FRET and MEF based novel biosensing methods. Moreover, we are working on enzymes from firefly and marine bacteria applicable in food hygiene and sanitation. The luminescent enzymes and their interaction with nanoparticles are being studied to understand the phenomena of BRET and metal induced enhancement of bioluminescence. The detection of hazardous materials such as heavy metals, aflatoxin, DDT, staphylococcal enterotoxin B, atrazine, formaldehyde, acrylamide, bisphenol A and methyl parathion in food samples respectively have been successfully carried out by us using multiple biosensing methods. The versatility, ease of use, and sensitivity of novel biosensing tools for the detection of harmful chemicals/toxins promisebetter and reliable bio-diagnostics in the near future.

References

1. Abhijith, K. S.; Thakur, M. S. *Anal.Methods* **2012** DOI:10.1039/c2ay25979f.
2. Dacres, H.; Dumancic, M. M.; Horne, I.; Trowell, S. C. *Biosens. Bioelectron.* **2009**, *24*, 1164–1170.
3. Homola, J.; Dosta´lek, J.; Chen, S.; Rasooly, A.; Jiang, S.; Yee, S. S. *Int. J. Food Microbiol.* **2002**, *75*, 61–69.
4. Ince, R.; Narayanaswamy, R. *Anal. Chim. Acta* **2006**, *569*, 1–20.
5. Liang, G-X.; Pan, H-C.; Li, Y.; Jiang, L-P.; Zhang, J-R.; Zhu, J-J. *Biosens. Bioelectron.* **2009**, *24*, 3693–3697.
6. Massey, M.; Algar, W. R.; Krull, U. J. *Anal. Chim. Acta* **2006**, *568*, 181–189.
7. Vinayaka, A. C.; Thakur, M. S. *Bioconjugate Chem.* **2011**, *22*, 968–975.
8. Xia, Z.; Rao, J. *Curr. Opin. Biotechnol.* **2009**, *20*, 37–44.
9. Gooding, J. J. *Anal. Chim. Acta* **2006**, *559*, 137–151.
10. Ko, S.; Grant, S. A. *Biosens. Bioelectron.* **2006**, *21*, 1283–1290.
11. Lazcka, O.; Campo, F. J. D.; Mu˜noz, F. Z. *Biosens. Bioelectron.* **2007**, *22*, 1205–1217.
12. Thakur, M. S.; Karanth, N. G. In *Advances in Biosensors*; Malhotra, B. D., Turner, A. P. F., Eds.; Elsevier Science BV: Amsterdam, AE, 2003; Vol. 5, p 131.
13. Rasooly, A.; Herold, K. E. *J. AOAC. Int.* **2006**, *89*, 873–883.
14. Hall, R. H. *Microbes Infect.* **2002**, *4*, 425–432.
15. Pal, S.; Alocilja, E. C. *Biosens. Bioelectron.* **2009**, *24*, 1437–1444.
16. Song, X.; Swansung, B. I. *Anal. Chem.* **1999**, *71*, 2097–2107.
17. Algar, W. R.; Krull, U. J. *Anal. Chim. Acta* **2007**, *581*, 193–201.
18. Wouters, F. S.; Verveer, P. J.; Bastiaens, P. I. H. *Trends Cell Biol.* **2001**, *11*, 5.
19. Ganesan, S.; Ameer-beg, S. M.; Ng, T. T. C.; Vojnovic, B.; Wouters, F. S. *Proc. Natl. Acad. Sci. U.S.A.* **2006**, *103*, 4089–4094.
20. Piston, D. W.; Kremers, G-J. *Trends Biochem. Sci.* **2007**, *32*, 407–414.
21. Prinz, A.; Diskar, M.; Herberg, F. W. *ChemBioChem* **2006**, *7*, 1007–1012.
22. WHO. *Food safety & food-borne illness*, Fact sheet no. 237 (reviewed March 2007); World Health Organization: Geneva; 2007a.
23. WHO. *The world health report, 2007. Global public health security in the 21st century*; World Health Organization: Geneva; 2007b.
24. CDC Estimates: Findings, 2011; http://www.cdc.gov/foodborneburden/2011-foodborne-estimates.html.

25. Moss, M. O.; Adams, M. R. *Food Microbiology*; Royal Society of Chemistry: Cambridge, U.K., 2008.; Vol 3, p 577.

26. Feng, P. Guide to Foodborne Pathogens. In *Escherichia coli*; Labbe, R. G., Garcia, S., Eds.; John Wiley and Sons, Inc.: New York, 2001; pp 143– 62.

27. Akshath, U. S.; Vinayaka, A. C.; Thakur, M. S. *Biosens. Bioelectron.* **2012**, *38*, 411–415.

28. Chouhan, R. S.; VivekBabu, K.; Kumar, M. A.; Neeta, N. S.; Thakur, M. S.; Amitha Rani, B. E.; Pasha, A.; Karanth, N. G. *Biosens. Bioelectron.* **2006**, *21*, 1264–1270.

29. Lisa, M.; Chouhan, R. S.; Vinayaka, A. C.; Manonmani, H. K.; Thakur, M. S. *Biosens Bioelectron.* **2009**, *25*, 224–227.

30. Vinayaka, A. C.; Thakur, M. S. *Anal. Bioanal. Chem.* **2010**, *397*, 1445–1455.

31. Ranjan, R.; Rastogi, N. K.; Thakur, M. S. *J. Hazard. Mater.* **2012**, *225*, 114–123.

32. Yakes, M. D. P. B. J.; Lipert, R. J.; Bannantine, J. P. *Clin. Vaccine Immunol.* **2008**, *15*, 227–234.

33. Zhang, X.; Young, M. A.; Lyandres, O.; Duyne, R. P. V. *J. Am. Chem. Soc.* **2005**, *127*, 4484–4489.

34. Jarvis, R. M.; Brooker, A.; Goodacre, R. *Faraday Discuss.* **2006**, *132*, 281–292.

35. Talley, C. E.; Jackson, J. B.; Oubre, C.; Grady, N. K.; Hollars, C. W.; Lane, S. M.; Huser, T. R.; Nordlander, P.; Halas, N. J. *Nano Lett.* **2005**, *5*, 1569–1574.

36. Vo-Dinh, T.; Cullum, B. M.; Stokes, D. L. *Sens. Actuators, B* **2001**, *74*, 2–11.

37. Haruyama, T. *Adv. Drug Delivery* **2003**, *55*, 393–401.

38. Jain, K. K. *Expert Rev. Mol. Diagn.* **2003**, *3*, 153–161.

39. Jianrong, C.; Yuqing, M.; Nongyue, H.; Xiaohua, W.; Sijiao, L. *Biotechnol. Adv.* **2004**, *22*, 505–518.

40. Cui, D. *J. Nanosci. Nanotechnol.* **2007**, *7*, 1298–1314.

41. Pan, B.; Cui, D.; Ozkan, C. S.; Ozkan, M.; Xu, P.; Huang, T.; Liu, F.; Chen, H.; Li, Q.; He, R.; Gao, F. *J. Phys. Chem. C* **2008**, *112*, 939–944.

42. Zhang, X.; Guo, Q.; Cui, D. *Sensors* **2009**, *9*, 1033–1053.

43. Vinayaka, A. C.; Thakur, M. S. *Analyst* **2012**, *137*, 4343–4348.

44. Dong, M.; Tepp, W. H.; Johnson, E. A.; Chapman, E. R. *Proc. Natl. Acad. Sci. U.S.A.* **2004**, *101*, 14701–14706.

45. Ligler, F. S.; Taitt, C. R.; Shriver-Lake, L. C.; Sapsford, K. E.; Shubin, Y.; Golden, J. P. *Anal. Bioanal. Chem.* **2003**, *377*, 469–477.

46. Baker, S.; Vinayaka, A. C.; Manonmani, H. K.; Thakur, M. S. *Luminescence* **2012** DOI:10.1002/Bio.1389.

47. Akshath, U. S.; Selvakumar, L. S.; Thakur, M. S. *Anal. Methods* **2012**, *4*, 699–704.

48. Chan, W. C. *Biol. Blood Marrow Transplant* **2006**, *12*, 87–91.

49. Bruchez, M.; Moronne, M.; Gin, P.; Weiss, S.; Alivisatos, A. P. *Science* **1998**, *281*, 2013–2016.

50. Grieve, K.; Mulvaney, P.; Grieser, F. *Curr. Opin. Colloid Interface Sci.* **2000**, *5*, 168–172.

51. Aoyagi, S.; Kudo, M. *Biosens. Bioelectron.* **2005**, *20*, 1680–1684.

52. Hutter, E.; Maysinger, D. *Microsc. Res. Tech.* **2011**, *74*, 592–604.

53. Hosokawa, M.; Nogi, K.; Yokoyama, T. *Nanoparticle Technology Handbook*, 1st ed.; Elsevier: Amsterdam, AE, 2007.

54. Kreibig, U.; Vollmer, M. *Optical Properties of Metal Clusters*; Springer Series in Materials Science 25; Springer: Berlin, Germany, 1995.

55. Mulvaney, P. *Langmuir* **1996**, *12*, 788–800.

56. Khanna, V. K. *Sensor Rev.* **2008**, *28*, 39–45.

57. Chouhan, R. S.; Vinayaka, A. C.; Thakur, M. S. *Anal. Method* **2010**, *2*, 924–928.

58. Huang, L.; Peng, Z.; Guo, Y.; Porter, A. L. *J. Bus. Chem.* **2010**, *7*, 15–30.

59. Yang, H.; Li, H.; Jiang, X. *Microfluid. Nanofluid.* **2008**, *5*, 571–583.

60. Tallury, P.; Malhotra, A.; Byrne, L. M.; Santra, S. *Adv. Drug Delivery Rev.* **2010**, *62*, 424–437.

61. Viswanathan, S.; Radecki, J. *Pol. J. Food Nutr. Sci.* **2008**, *58*, 157–164.

62. Jing, N.; Lipert, R. J.; Dawson, B.; Marc, D. P. *Anal. Chem.* **1999**, *71*, 4903–4908.

63. Förster, T. *Energiewanderung und fluoreszenz. Naturwissenschaften* **1946**, *6*, 166–175.

64. Clegg, R. M. Fluorescence Resonance Energy Transfer. In *Fluorescence Imaging Spectroscopy and Microscopy*; Wang, X. F., Herman, B., Eds.; John Wiley & Sons Inc., New York, 1996; Vol. 137, pp 179–251.

65. Li, I. T.; Pham, E.; Truong, K. *Biotechnol. Lett.* **2006**, *28*, 1971–1982.

66. Chan, W. C. W.; Maxwell, D. J.; Gao, X.; Bailey, R. E.; Han, M.; Nie, S. *Curr. Opin. Biotechnol.* **2002**, *13*, 40–46.

67. Pathak, S.; Davidson, M. C.; Silva, G. A. *Nano Lett.* **2007**, *7*, 1839–1845.

68. Alivisatos, A. P. *Science* **1996**, *271*, 933–937.

69. Murphy, C. J.; Coffer, J. L. *Appl. Spectrosc.* **2002**, *56*, 16–27.

70. Warner, M. G.; Grate, J. W.; Tyler, A.; Ozanich, R. M.; Miller, K. D.; Lou, J.; Marks, J. D.; Bruckner-Lea, C. J. *Biosens. Bioelectron.* **2009**, *25*, 179–184.

71. Mukhopadhyay, B.; Martins, M. B.; Karamanska, R.; Russell, D. A.; Field, R. A. *Tetrahedron Lett.* **2009**, *50*, 886–889.

72. Goldman, E. R.; Clapp, A. R.; Anderson, G. P.; Uyeda, H. T.; Mauro, J. M.; Medintz, I. L.; Mattoussi, H.; Mauro, J. M. *Anal. Chem.* **2004**, *76*, 684–688.

73. Kattke, M. D.; Gao, A. J.; Sapsford, K. E.; Stephenson, L. D.; Kumar, A. *Sensors* **2011**, *11*, 6396–6410.

74. Wang, J.; Tang, W.; Fang, G.; Pan, M.; Wang, S. *J. Chin. Chem. Soc.* **2011**, *58*, 463–469.

75. Tyagi, S.; Landegren, U.; Tazi, M.; Lizardi, P. M.; Kramer, F. R. *Nat. Biotechnol.* **1996**, *14*, 303–308.

76. Liu, X.; Tan, W. *Anal. Chem.* **1999**, *71*, 5054–5059.

77. Steemers, F. J.; Ferguson, J. A.; Walt, D. R. *Nat. Biotechnol.* **2000**, *18*, 91–94.

78. Daniel, M. C.; Astruc, D. *Gold. Chem. Rev.* **2004**, *104*, 293–346.

79. Jain, P. K.; El-Sayed, I. H.; El-Sayed, M. A. *Nanotoday* **2007**, *2*, 18–29.

80. Rayavarapu, R. G.; Petersen, W.; Ungureanu, C.; Post, J. N.; van Leeuwen, T. G.; Manohar, S. *Int. J. Biomed. Imaging* **2007**, 1–10.

81. Yang, P.; Yao, S.; Wei, W.; Cai, J. *Am. J. Anal. Chem.* **2011**, *2*, 484–490.

82. Baptista, P.; Pereira, E.; Eaton, P.; Doria, G.; Miranda, A.; Gomes, I.; Quaresma, P.; Franco, R. *Anal. Bioanal. Chem.* **2008**, *391*, 943–950.

83. Sperling, R. A.; Gil, P. R.; Zhang, F.; Zanella, M.; Parak, W. J. *Chem. Soc. Rev.* **2008**, *37*, 1896–1990.

84. Ghosh, S. K.; Pal, A.; Kundu, S.; Nath, S.; Pal, T. *Chem. Phys. Lett.* **2004**, *395*, 366–372.

85. Kang, K. A.; Wang, J.; Jasinski, J. B.; Achilefu, S. *J. Nanobiotechnol.* **2011**, *9*, 1–13.

86. Rayford, C. E.; Schatz, G.; Shuford, K. *Nanoscape* **2005**, *2*, 27–33.

87. Oh, E.; Hong, M.; Lee, D.; Nam, S.; Yoon, H. C.; Kim, H. *J. Am. Chem. Soc.* **2005**, *127*, 3270–3271.

88. Boute, N.; Jockers, R.; Issad, T. *Trends Pharmacol. Sci.* **2002**, *23*, 351–354.

89. Xu, Y.; Kanauchi, A.; von Arnim, A. G.; Piston, D. W.; Johnson, C. H. *Biophotonics, Part A* **2003**, *360*, 289.

90. So, M-K.; Xu, C.; Loening, A. M.; Gambhir, S. S.; Rao *Nat. Biotechnol.* **2006**, *24*, 339–343.

91. Sapsford, K. E.; Berti, L.; Medintz, I. L. *Angew. Chem., Int. Ed.* **2006a**, *45*, 4562–4588.

92. Yamakawa, Y.; Ueda, H.; Kitayama, A.; Nagamune, T. *J. Biosci. Bioeng.* **2002**, *93*, 537–542.

93. Arai, R.; Nakagawa, H.; Kitayama, A.; Ueda, H.; Nagamune, T. *J. Biosci. Bioeng.* **2002**, *94*, 362–364.

94. Kamal, M.; Marquez, M.; Vauthier, V.; Leloire, A.; Froguel, P.; Jockers, R.; Couturier, C. *Biotechnol. J.* **2009**, *4*, 1337–1344.

95. Andrasi, A-D.; Chan, C. T.; De, A.; Massoud, T. F.; Gambhir, S. S. *Proc. Natl. Acad. Sci. U.S.A.* **2011**, *108*, 12060–12065.

96. Kendall, J. M.; Badminton, M. N. *Trends Biotechnol.* **1998**, *16*, 216–224.

97. Alvarez, J.; Montero, M. *Cell Calcium* **2002**, *32*, 251–260.

98. Adamczyk, M.; Moore, J. A.; Shreder, K. *Org. Lett.* **2001**, *3*, 1797–1800.

99. Gorokhovatsky, A. Y.; Rudenko, N. V.; Marchenkov, V. V.; Skosyrev, V. S.; Arzhanov, M. A.; Burkhardt, N.; Zakharov, M. V.; Semisotnov, G. V.; Vinokurov, L. M.; Alakhov, Y. B. *Anal. Biochem.* **2003**, *313*, 68–75.

100. Yamaguchi, A.; Hohsaka, T. *Bull. Chem. Soc. Jpn.* **2012**, *85*, 576–583.

101. Girkin, J. M. *Global Watch* **2002**, *4*, 14–15.

102. Moulton, K.; Ryan, P.; Lay, D.; Willard, S. *BMC Microbiol.* **2009**, *9*, 152–158.

103. Baumann, B.; Meer, J. R. V. D. *J. Agric. Food Chem.* **2007**, *55*, 2115–2120.

104. Olaniran, A. O.; Motebejane, R. M.; Pillay, B. *J. Environ. Monit.* **2008**, *10*, 889–893.

105. Selifonova, O.; Burlage, R.; Barkay, T. *Appl. Environ. Microbiol.* **1993**, *59*, 3083–3090.

106. Yoo, S. K.; Lee, J. H.; Yun, S-S.; Guc, M. B.; Lee, J. H. *Biosens. Bioelectron.* **2007**, *22*, 1586–1592.

107. Matrubutham, U.; Thonnard, J. E.; Sayler, G. S. *Appl. Microbiol. Biotechnol.* **1997**, *47*, 604–609.

108. Trajkovska, S.; Tosheska, K.; Aaron, J. J.; Spirovski, F.; Zdravkovski, Z. *Luminescence* **2005**, *20*, 192–196.

109. Applegate, B. M.; Kehrmeyer, S. R.; Sayler, G. S. *Appl. Environ. Microbiol.* **1998**, *64*, 2730–2735.

110. Creton, R.; Kreiling, J. A.; Jaffe, L. F. *Microsc. Res. Tech.* **1999**, *46*, 390–397.

111. Wu, C.; Mino, K.; Akimoto, H.; Kawabata, M.; Nakamura, K.; Ozaki, M.; Ohmiya, Y. *Proc. Natl. Acad. Sci. U.S.A.* **2009**, *106*, 15599–15603.

112. Bonsma, S.; Purchase, R.; Jezowski, S.; Gallus, J.; Kçnz, F.; Vçlker, S. *ChemPhysChem* **2005**, *6*, 838 –849.

113. Shimomura, O. *Angew. Chem., Int. Ed.* **2009**, *48*, 5590–5602.

114. Metzler, D. E., *Biochemistry-the chemical reactions of living cells*, 2nd ed.; Elsevier Academic Press: San Diego, CA, 2001; Vols. 1&2, pp 1272–1357

115. Chapelle, E. W.; Levin, G. *Biochem. Med.* **1968**, *2*, 41–52.

116. Sharpe, A. N.; Woodrow, M. N.; Jackson, A. K. *J. Appl. Bacteriol.* **1970**, *33*, 758–767.

117. Stannard, C. J.; Wood, J. M. *J. Appl. Bacteriol.* **1983**, *55*, 429–438.

118. Siragusa, G. R.; Nawotka, K.; Spilman, S. D.; Contag, P. R.; Contag, C. H. *Appl. Environ. Microbiol.* **1999**, *65*, 1738–1745.

119. Hattori, N.; Sakakibara, T.; Kajiyama, N.; Igarashi, T.; Maeda, M.; Murakamia, S. *Anal. Biochem.* **2003**, *319*, 287–295.

120. Voss, K. J. *Methods Enzymol.* **2000**, 305, 53–61.

121. Sapsford, K. E.; Pons, T.; Medintz, I. L.; Mattoussi, H. *Sensors* **2006b**, *6*, 925–953.

Chapter 5

The Use of Silver Nanorod Array-Based Surface-Enhanced Raman Scattering Sensor for Food Safety Applications

Xiaomeng Wu,*,[1] Jing Chen,[1] Bosoon Park,[2] Yao-Wen Huang,[1] and Yiping Zhao[3]

[1]Department of Food Science and Technology, the University of Georgia, Athens, Georgia 30602, U.S.A.
[2]Russell Research Center, USDA Agricultural Research Service, Athens, Georgia 30605, U.S.A.
[3]Department of Physics and Astronomy, the University of Georgia, Athens, Georgia 30602, U.S.A.
*E-mail: xwu@uga.edu.

For the advancement of preventive strategies, it is critical to develop rapid and sensitive detection methods for food safety applications. This chapter reports the recent development on the use of aligned silver nanorod (AgNR) arrays prepared by oblique angle deposition, as surface-enhanced Raman scattering (SERS) substrates to detect food safety related substances such as foodborne pathogens, toxins, pesticides, and adulteration agents. Our investigation demonstrated that the AgNR based SERS can be used to distinguish different types and serotypes of bacteria. It can also be used to discriminate mycotoxins of similar chemical structures, such as aflatoxins, with the help of chemometric analysis of the SERS spectra. Pesticides from real food samples, such as tea, can be identified using this method with the same multivariate statistical analysis. When combining SERS with ultra-thin layer chromatography using the AgNR substrates, our detection system can achieve simultaneous separation and detection of mixture samples, which could be critical for real food detection.

Introduction

Food safety has been a global concern of the food industry and the governments for many years. Nowadays, ensuring food safety has become a complicated task as food safety issues change over time. While many of the food safety challenges of the past have been solved, new issues continue to arise. Although over the past decades the ability to detect foodborne outbreaks has advanced and become more rapid than ever, much remains to be done to address the ever-increasing public awareness on food safety due to the numerous food safety outbreaks and scandals.

Foodborne outbreaks associated with microorganisms are among the greatest concerns in food safety. The Center for Disease Control and Prevention (CDC) estimates that 90% of the illnesses due to known pathogens are caused by seven pathogens: *Salmonella*, norovirus, *Campylobacter*, Toxoplasma, *E. Coli* O157, *Listeria* and *Clostridium perfringens* (*1*). According CDC, approximately 48 million become ill, 128,000 are hospitalized, and 3,000 die in the United States each year from foodborne diseases. Norovirus is the most commonly known cause of human illnesses, responsible for 5.4 million illnesses and 149 deaths each year. In addition, each year more than a million illnesses and 378 deaths are caused by *Salmonella*, 176,000 illnesses and 20 fatalities by *E. coli* toxins, 845,024 illnesses and 76 deaths by *Campylobacter*. *Listeria* is one of the most lethal pathogens, causing 1,591 illnesses and 255 deaths annually. For meat, poultry and egg products, category B pathogens with the greatest impact on public health are the bacterial agents Shiga-toxin producing *E. coli, Salmonella, Campylobacter,* and *L. monocytogenes*. Among them *Salmonella* causes the most infections and incidence cases followed by *Campylobacter*. In 2007, *Salmonella* outbreak linked to peanut butter contamination caused 329 illnesses. In 2008, more than 700 cases of Salmonellosis were reported in the United States, and *Salmonella* tainted eggs caused 1,200 illnesses in 2010. Poultry and poultry produces are the major sources of *Salmonellae* which cause human illness (*2*).

In some cases, the foodborne illnesses are not only related to direct bacterial infection, but also caused by the toxins excreted by the microorganism during metabolism. Exotoxins such as shiga-like toxin produced by *Escherichia coli,* staphylococcal enterotoxins produced by *Staphylococcus aureus*, and botulinus intoxication caused by *Clostridium botulinum* can still cause foodborne illness even when the toxin-producing microbes are inactive. Mycotoxins produced by fungi are important toxins commonly found in contaminated foods and feeds. Major mycotoxins include aflatoxins, citrinin, ergot alkaloids, fumonisins, orchratoxin, patulin, trichothecenes, zeralenone, etc (*3*). High exposure of these mycotoxins has been reported globally, especially in East Africa, China, and parts of South-east Asia. In extreme cases, death may result from the consumption of food that is heavily contaminated with these mycotoxins (*4*).

Apart from the pathogen- or toxin-associated outbreaks, illness associated with harmful chemicals is another food safety concern. The use of chemical fertilizers and pesticides is rising in modern agriculture in order to improve the yield and quality of agriculture products. Pesticides are an integral part of modern agriculture, but their use can lead to residues in agricultural products. Because of their potential adverse human health effects, the Federal Government has set limits on allowable levels of pesticide residues in food and animal feed and monitors these products to enforce those levels. Food safety concerns regarding pesticide residues in several food commodities have arisen. Outbreaks associated with foodborne exposure to pesticides residues have been linked to watermelon (*5*), hydroponic cucumbers (*6*), and other vegetables (*7*).

Illegal substances have also been intentionally added to food and feedstuffs by the food processors driven by short-term economical profits. A notorious example is the 2008 Chinese milk scandal, which involves milk and infant formula adulterated with melamine. In this incident, melamine (2,4,6-triamino-1,3,5-triazine) was added to the dairy products in order to increase the apparent protein content due to its nitrogen-rich (67% per mass unit) property. Excessive intake of melamine causes crystal formation in human body which leads to kidney stones. Therefore, reliable and prompt detection of melamine, as well as other illegal adulteration agents in food such as illegal food preservatives and artificial colors, has become necessary.

Aforementioned foodborne outbreaks associated with pathogens, toxins, harmful chemicals, and adulteration agents constantly challenge the safety of the food supply chain of the United States. Federal monitoring and enforcement action of these is dependent on technical capability to detect the target microorganisms and chemicals. In order to ensure food safety, the industry leaders and governments need rapid, sensitive, and accurate detection methods to respond to the incidents and make amends.

Traditionally, bacterial culture-based methods are the conventional way to detect bacterial pathogens in food. These methods often include incubation and enrichment steps requiring 6-24 hours with an additional 1-3 days for confirmation by biochemical tests (*8, 9*). Though these methods are sensitive, relatively inexpensive, and provide both qualitative and quantitative information on the number and the nature of the microorganisms, their lengthy duration makes them unsuitable to satisfy today's food industry requirements for rapid detection. Several rapid detection methods have been developed over the past years, such as enzyme-linked immunosorbent assay (ELISA) (*10*), which is based on the specific recognition of the antigen by antibodies and is measured with the aid of an enzyme–substrate reaction. Several different ELISA methods including competitive ELISA (*11*), PCR-ELISA (*12*) and Dot-ELISA (*13, 14*) are considered highly sensitive, and provide specific and rapid screening of a large number of samples for the detection of foodborne pathogens, specifically *Salmonella*. However, a disadvantage of the ELISA method is that it requires considerable amounts of antigen to be used for detection. Another effective method for the detection and identification of pathogens is the polymerase chain reaction (PCR), which is based on isolating the bacteria and exponentially amplifying a DNA fragment or sequence of interest via enzymatic replication (*15–17*). Real-time PCR offers the advantage of semi-quantification of the bacterial DNA and no post-PCR handling of the sample, resulting in reducing the risk of false-positive results (*18–21*). Although PCR and its modifications may detect *Salmonella* within a relatively short time, issues, such as primer design and PCR-inhibitory effects of complex food matrices, remain and make

PCR detection of *Salmonella* in food far from being a routine procedure (*19, 22*). Thus, the current rapid methods for microbial pathogen detection focus on immunological or genetic identification of the presence of specific foodborne pathogenic bacteria. Several other immunological and molecular methods, such as immunomagnetic capture, nucleic acid hybridization, PCR, and DNA microarray have been used for detection of pathogenic bacteria (*23*). Although these methods have many advantages, each method also has limitations from a practical point of view. For example, any PCR based methods rely on nucleic acid amplification and consequently cannot discriminate nucleic acid amplified from viable and nonviable bacteria. Therefore, alternative methods to rapidly detect pathogens in food matrices are needed for food processing industries.

The detections of any harmful or illegal chemicals in the food samples, either toxins, pesticides, or adulteration agents, are often performed using chromatography and immunological methods. The various chromatographic methods that can be used to determine mycotoxins are reviewed (*24–26*), including thin layer chromatography (TLC), high performance liquid chromatography (HPLC), gas chromatography (GC), etc. Some other rapid methods for mycotoxin detection are often based on the immunological principles. These methods, such as ELISA, flow through membrane based immunoassay, immunochromatographic assay, fluorometric assay with immunoaffinity, fluorescence polarization method and etc., are also reviewed (*27*). Pesticides residues in food are detected and monitored using similar chromatography methods (*28, 29*), and immunological methods (*30–33*).

Although these methods have their advantages to be sensitive, quantitative, and reliable, they all have specific disadvantages. TLC is traditionally one of the most popular detection methods due to its low cost and easy handling procedure, but the method has been replaced by HPLC in most of the case because of its higher speed, accuracy and reproducibility. However, HPLC method has difficulties in detection of co-elution (two compounds escaping from the tubing at once), which may lead to inaccurate compound categorization. One main disadvantage of GC is it being a destructive method. And one common disadvantage of the chromatographic methods is their need for a high capital investment on the instrument. Immunological detection, such as ELISA, has the advantage of being specific to the target, but it often suffer from lacking definitively confirmation and accurate quantification, and having a limited detection range due to the narrow sensitivity of the antibodies. It also requires multiple steps, varied chemical reagents, and long incubation time, which make this method impractical for "real-time" detection in the field.

Since early 21st century, nanotechnology has been emerging in biological science as well as engineering. Advances in nanotechnology are also significantly impacting on the field of diagnostics in medical science. The fabrication and surface modification of nanomaterials have allowed one to tailor their binding affinities for various biomolecules, thus increasing the specificity, sensitivity and speed for diagnostics. This is an appealing alternative to current molecular diagnostic techniques for food safety applications. Using nanotechnology, it is possible to develop various nanomaterials and devices that have demonstrated potential in improving sensitivity for safety related detections. Subsequently, many different nanotechnologies, such as surface-enhanced Raman spectroscopy (SERS), quantum dots, and antibody conjugated fluorescent nanoparticles, could become next generation diagnostic tools capable of detecting and identifying the target organisms and chemicals rapidly even at trace amounts. Those techniques have found applications in medical diagnostics, national security, defense, and food safety. Thus, nanoscale science and technology for food research has focused its investment on detection and intervention technologies for enhancing food safety. Recently, nanotechnology for food quality, safety, biosecurity, better nutrient delivery system, nanomaterials to enhance packaging performance and improving food processing have emerged as a future direction in food systems.

The development of these nanotechnology based novel sensors rely on the unique optical, electrochemical and catalytic properties of a range of nanomaterials, such as quantum dots (wells or wires), gold nanoparticles, carbon nanotubes, nanocrystals, nanosemiconductors. Quantum dots (QDs), quantum wells, and quantum wires are a class of quantum confined systems, whose semiconductor properties enable them to emit stable fluorescence. QDs are usually used for florescence detection due to their tunable and narrow emission spectra, bright fluorescence, and resistance to photobleaching over a long period of time (*34–36*). Gold nanoparticles are often used as label substance in various detection methods due to the small size, and subsequently unique physical and chemical properties such as localized surface plasmon resonance effect (*37, 38*). Carbon nanotubes are widely used in analytical chemistry for their unique optical properties such as small band-gaps and photoluminescence in the near-infrared region (*39, 40*). There are some excellent reviews on these nanomaterial enabled sensors for the detection of heavy metals (*41*), toxic metal ions (*42*), DNA (*43, 44*), harmful chemicals (*45*), and pathogens (*46*), based on the principles of florescence, electrochemistry, surface plasmon resonance, surface-enhanced Raman spectroscopy, etc (*47–50*).

Specifically, many nanotechnology based sensors have been developed for food safety and food defense applications, examples include bio-functionalized nanosubstrate-based biosensor for *Salmonella* detection (*51*); highly-integrated atomic force microscope nanocantilever-based label-less, fast single molecule recognition system for ricin detection (*52–54*); rapid detection of foodborne pathogenic bacteria using silver biopolymer nanoparticles with surface enhanced Raman scattering; and DNA aptamer-based atomic force microscopy (*55*) for the detection of foodborne pathogens and toxins.

Raman scattering is considered a powerful platform with the potential for rapid detection of chemical and biological substances. Traditional Raman spectroscopy relies on the inelastic scattering interaction of the excitation light and the vibrational modes of the molecular bonds. These molecular vibrational modes possess unique "fingerprint" Raman peaks that can be used to identify the particular molecule(s) being probed. However, the application of Raman spectroscopy for trace analysis has been limited by the extremely low sensitivity Because of the small Raman scattering cross section, typically at least $\sim10^8$ molecules are necessary to generate a measurable normal Raman scattering signal for most relevant analytes (*56*). To further enhance Raman efficiencies to enable trace level detection, surface-enhanced Raman scattering (SERS), the phenomenon of a significantly increased Raman signal from molecules in proximity to appropriate metallic nanostructures, is often employed. The nanostructure-induced enhancement can reach as high as 14 to 15 orders of magnitude compared to bulk Raman, which allows the technique to be sensitive enough to detect small amount of molecules, or even single molecule (*57–61*).

The SERS substrates usually consist of noble metal fine structure with dimensions in the scale of nanometers. Since the morphology of the metallic structure plays a major role in determining the magnitude of signal enhancement and sensitivity of detection, different types of substrates have been fabricated or synthesized, such as a random distribution of roughness features produced by oxidation reduction on a metal electrode (*62*), evaporation of thin metal film on a flat substrate, rough metallic surfaces by chemical etching (*63*), silver films on TiO_2 (*64*), colloidal silver nanoparticles (*59*), silver nanoparticle arrays fabricated by nanosphere lithography (*65*), electro-deposition of silver on silver films at high potential (*66*), aligned monolayer of silver nanowires (*67*). Metal colloidal particles are widely used due to the ease of synthesis and manipulation, but the SERS enhancement highly depends on particle size and shape, as well as the aggregation state. Without uniformity and good reproducibility of the metal substrates, the attainment of reproducible spectra remains a major challenge for SERS. The silver nanorod (AgNR) array substrates fabricated by oblique angle deposition (OAD) method have been proven to have a SERS enhancement factor of more than 10^8, and a batch variance below 15%. In addition, these substrates have been shown to markedly enhance the Raman signal of chemical and biological samples including aflatoxins (*68*), important human viruses such as rotavirus, influenza virus and respiratory syncytial virus (*69, 70*), foodborne pathogens including *E. coli* O157:H7, *Salmonella* Typhimurium, *Staphylococcus aureus* (*71*), pesticides like chlorpyrifos and parathion, intentional adulteration agents like melamine (*72*), inorganic substances like uranyl ion cast films (*73*), and allow for detection and discrimination of microRNA families and family members (*74*).

Fabrication of AgNR Arrays as a SERS Active Substrate

The AgNR array substrates are fabricated by a relatively simple method, the oblique angle deposition (OAD) technique. OAD involves positioning the substrate at a large angle α (> 70°) with respect to the incident metal vapor atoms during a physical vapor deposition process. This process results in the preferential growth of isolated nanocolumnar structures on the substrate tilted towards the direction of the vapor. Directional columnar growth is a result of surface diffusion and atomic shadowing effects that occur at the substrate surface. The experimental setup of OAD is illustrated in Figure 1.

Briefly, glass microscopic slides (Gold Seal® Catalog No.3010, Becton, Dickinson and company, Portsmouth, NH) used for AgNR arrays deposition are cleaned with piranha solution (80% sulfuric acid, 20% hydrogen peroxide v/v) and rinsed with deionized water. The slides are then dried with a stream of nitrogen before being loaded into the e-beam deposition system. A 20-nm titanium film, followed by a 200-nm silver film layer, is first evaporated onto the glass slides at a rate of 0.2 nm/s and 0.3 nm/s, respectively, while the substrate surface is held perpendicular to the incident vapor direction. The substrate normal is then rotated to 86° with respect to the incident vapor direction and AgNRs are grown at this oblique angle at a deposition rate of 0.3 nm/s. The film thickness is monitored *in situ* by a quartz crystal microbalance positioned at normal incidence to the vapor source direction. The general structure of the AgNR substrates is shown in Figure 2. Silver nanorod substrates prepared by the OAD technique with a rod length of ~ 870 nm, a diameter of ~ 100 nm, and a tilting angle of 73° have previously been shown to provide SERS enhancement factors of $>10^8$ (based on Raman reporter molecule trans-1,2-bis(4-pyridyl)ethylene (BPE)) (*75*).

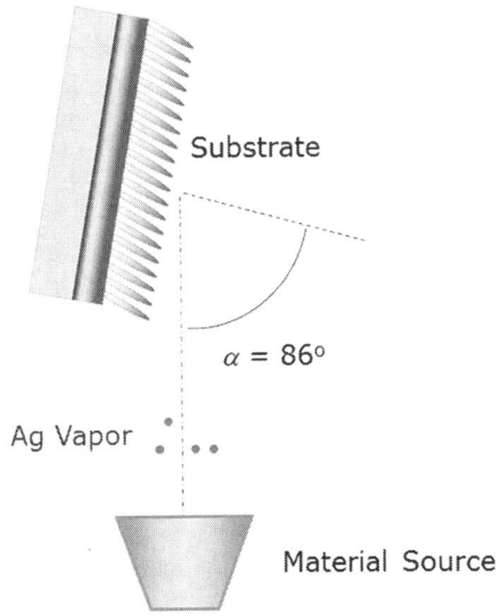

Figure 1. AgNR substrate fabricated by oblique angle deposition: experimental setup.

The deposition conditions will determine the length, tilting angle and nanorod density of the AgNRs, affecting their SERS performance. A systematic study has been performed to explore effects of those parameters on SERS response of the substrates and to optimize AgNR structures for SERS application (*76*). At the same QCM reading length $l = 2300 \pm 50$ nm, but different incident angle α, the measured nanorod length $L = 1190$ nm, 1160 nm, 1290 nm, and 1150 nm, and the measured tilting angle $\beta = 57 \pm 2°$, $61 \pm 3°$, $63 \pm 3°$, and $65 \pm 2°$ for $\alpha = 78°$, 80°, 82°, and 84°, respectively. The diameter of the Ag nanorod D is measured at the very growth end (tip) of the nanorod. With the increase of nanorod length L, the nanorod diameter D also increases, while the nanorod density n decreases. The SERS performance of the substrates is obviously determined by the deposition conditions. The SERS enhancement factor (EF) of the AgNR at these deposition conditions has been studied by Liu et al. (*76, 77*), as shown in Figure 3. The average SERS EF is commonly defined as $EF = \dfrac{I_{SERS}/N_{surf}}{I_{vol}/N_{vol}}$, where N_{vol} and N_{surf} are the number of molecules probed in the liquid sample and on the SERS substrates, respectively, and I_{vol} and I_{SERS} are the corresponding normal Raman and SERS intensities. At a given AgNR length from 200 to 2000 nm, the EF of the AgNR increases as the deposition angle α increases from 78° to 84°. At optimized deposition angle, $\alpha = 84°$, the EF reaches its maximum at 1200 nm AgNR length.

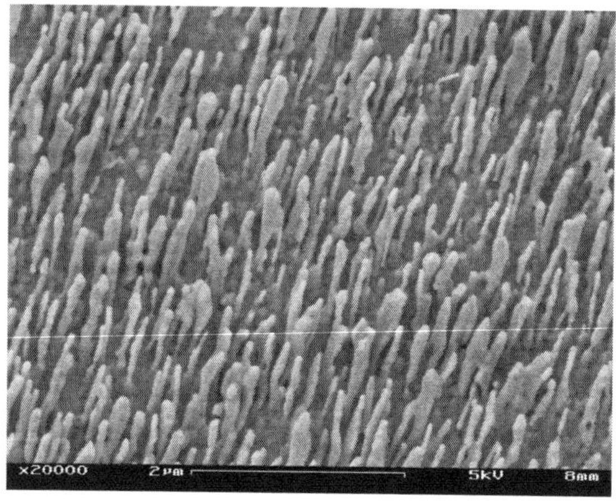

Figure 2. Representative SEM image of the AgNR substrates ($\alpha = 86°$).

AgNR Substrates Based SERS for Foodborne Pathogen Detection

There is an increasing interest of using SERS to detect bacteria. Several forms of nanostructures, such as silver metal deposits (*78, 79*), silver colloid (*80–82*), gold colloid solutions (*83, 84*), and electrochemically roughened metal surfaces (*85*), have been developed over the past decades for this purpose. Although using silver or gold colloids to detect bacteria has achieved a limit of detection (LOD) of 10^3 CFU/ml for *E. coli*, such methods often suffer from a lack of reproducibility because the slight variations of preparation parameters such as cluster size and shape in a colloidal solution could induce significant changes in the SERS enhancement factors by several orders of magnitude. The AgNR substrate fabricated by OAD method has the advantage of high uniformity and reproducibility besides its high SERS enhancement. The ability of AgNR as a SERS substrate to rapidly detect pathogenic bacteria has been studied, and its ability to differentiate between different bacterial species, strains and between viable and nonviable cells based on their characteristic SERS spectra has been demonstrated (*71*).

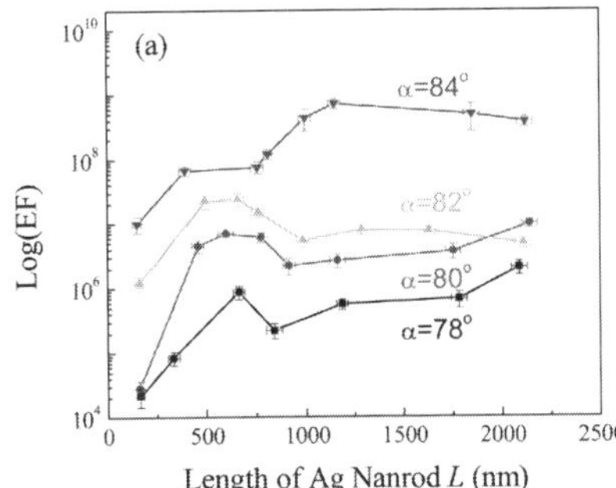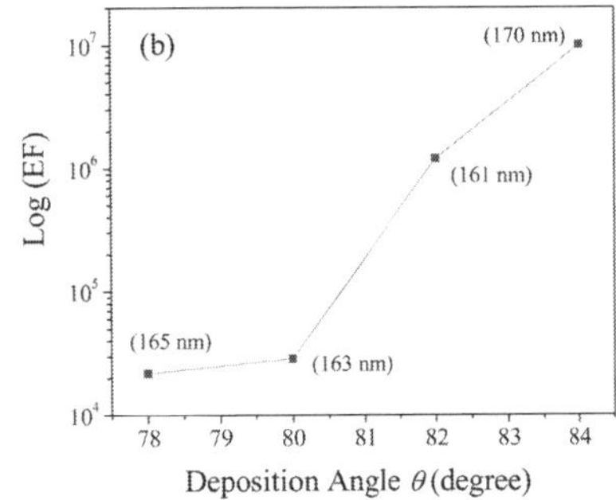

Figure 3. (a) The SERS EF as a function of nanorod length L for samples deposited at α = 78°, 80°, 82°, and 84°, respectively; and (b) the SERS EF as a function of deposition angle α at a fixed nanorod length L = 165 nm. (Reproduced with permission from reference (76). Copyright 2010 SPIE.)

In our study, the pure lab strains of bacteria cultured overnight to the stationary phase (~ 10^9 CFU/ml), are harvested and washed three times with DI water to remove the culture media. Ten µl of the bacterial suspension is pipetted onto the AgNR substrates and let dried under ambient condition, and the SERS spectra are acquired. The spectral differentiation of bacteria from different species can be easily achieved using SERS by examining the presence or absence of unique peaks for a certain species. In Figure 4, highly reproducible SERS spectra of *E. coli* O157:H7, *Salmonella* Typhimurium, *Staphylococcus aureus*, and *Staphylococcus epidermidis* are acquired using a 785 nm excitation laser for 10 s, with relative standard deviations (RSDs) of the peak intensity at $\Delta v = 735$ cm^{-1} and 1328 cm^{-1} being less than 6%, and the significant peaks from these species are identified. Although the specific interpretation of the Raman spectra of bacteria is still controversial and debatable in the aforementioned studies, some general Raman peak assignments are accepted by most researchers. The strong SERS bands at $\Delta v = 735$ and 1330 cm^{-1}, for example, have been attributed to the nucleic acid base adenine in almost all previous SERS studies of nucleic acids and bacteria. The broad band at $\Delta v = 550$ cm^{-1} can be assigned to carbohydrate; the peak at $\Delta v = 1450$ cm^{-1} can be attributed to the CH$_2$ deformation mode of proteins, and the strong band at $\Delta v = 930$ cm^{-1} may be assigned to the C-C stretching modes in proteins. These significant peaks from bacteria are commonly shared by the four species, yet each of the species processes its own unique feature. For example, the protein peak at $\Delta v = 1090$ cm^{-1} appears to be unique for *S. aureus* and *S. epidermidis* but are not present in the spectra of the *E. coli* O157:H7 and *Salmonella* Typhimurium.

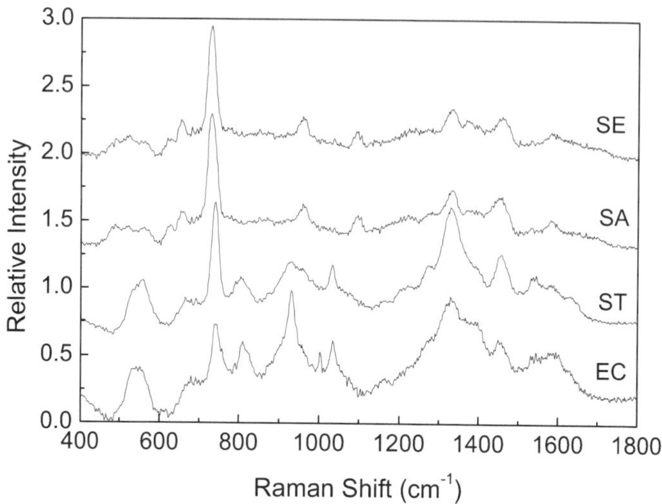

Figure 4. SERS spectra of four bacterial species obtained on AgNR substrates. EC = E. coli O157:H7; ST = Salmonella. Typhimurium; SA = Staphylococcus aureus and SE = Staphylococcus epidermidis. Incident laser power of 24 mW and collection time of 10 s are used to obtain these spectra. Spectra are offset vertically for display clarity. (Reproduced with permission from reference (71). Copyright 2008 The Society for Applied Spectroscopy.)

The AgNR can differentiate bacteria from different species based on the presence of significant peaks; however such method would not be practical to differentiate strains of the same species because their spectral differences would be subtle. Since the spectral data can be viewed as a multi-variant data, usually chemometric analysis can be used to reduce the dimensionality of the dataset, maximize the variance among spectral fingerprints, and classify the spectra from different analytes. Principal component analysis (PCA) is a well-known unsupervised chemometric method for identifying correlations amongst a set of variables and to transform the original set of variables into a new set of uncorrelated variables called principal components (PCs). When these PCs are plotted, the data of similar spectra can be grouped for classification based on the PC scores. Figure 5 shows the PCA results performed using the spectral data of the five different strains of *Salmonella.*, namely *Salmonella* Heidelberg, *Salmonella* Infantis, *Salmonella* Kentucky, *Salmonella* Enteritidis, and *Salmonella* Typhimurium. Five clusters are formed in the Figure 5, and each of the clusters represents a strain of the *Salmonella spp.* with clear separation between them that indicates the feasibility to differentiate bacteria from different strains of the same species.

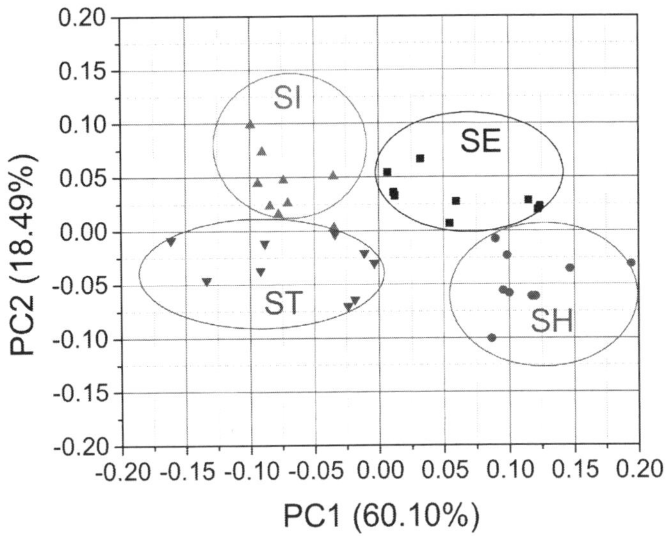

Figure 5. PCA results performed using the spectral data of the four different strains of Salmonella, namely Salmonella Heidelberg (SH), Salmonella Infantis (SI), Salmonella Enteritidis (SE), and Salmonella Typhimurium (ST).

The detection and identification of viability of bacterial cells are crucial for food safety. Although PCR is an effective tool for pathogen detection, one disadvantage of the PCR method is that it could give false negative/false positive identification since it is based on DNA detection, which generates the same signals for both the dead and viable bacterial cells. In contrast, SERS spectra of bacteria is based on the chemical structure, so the structure differences on cell surface between dead and alive cells would display in the spectra. Figure 6 shows spectra of viable and heat prepared cells of *E. coli* O157:H7, and the dead cells show a significantly reduced SERS response at those characteristic bands at Δv = 550 cm^{-1}, 735 cm^{-1}, 1330 cm^{-1} and 1450 cm^{-1} that are presented in the viable cells, hence the differentiation of live and dead cell can be achieved by SERS.

Although the detection and differentiation of different bacterial species or different strains of the same species using AgNR substrate has been successfully demonstrated, two challenges remain for the detection of bacterial pathogens in real food samples (*71*). The first challenge is to improve the LOD of the bacteria, and the second one is to detect the foodborne bacteria collected from its natural environment such as food processing plants instead of a laboratory culture. First of all, even though single cell detection is achieve by using a confocal Raman microscope, the concentration of the bacteria in previous study was usually $10^8 \sim 10^9$ CFU/ml in bulk solution and the result is performed using the pure bacteria strains (*71*). Therefore, a determination and reduction of the LOD is highly demanded. SERS is a surface detection method with limited sample volume and detection area, so linking the surface detection to bulk bacteria concentration is the first step to really understand the detection technique. The relationship between the bulk concentration and the surface detection limit is determined by (1) the volume V of bacterial solution applied; (2) the spreading area A_0 of the sample on the SERS substrate; (3) the size πR^2 of the laser spot, and (4) the bulk concentration C of the bacteria. The amount of bacteria detected on the surface is $N_d = \dfrac{VC\pi R^2}{A_0}$, so, $C = N_d \dfrac{A_0}{V\pi R^2}$. The threshold N_d reflects the true detectability of the SERS method, but C represented the bulk LOD of such detection. Thus, for a fixed surface detection method, the reduction of C can be achieved by decreasing A_0, increasing V, and increasing R. In the equation, R is limited by the Raman instrument used and A_0 could be confined by a narrow well as we reported previously (86).

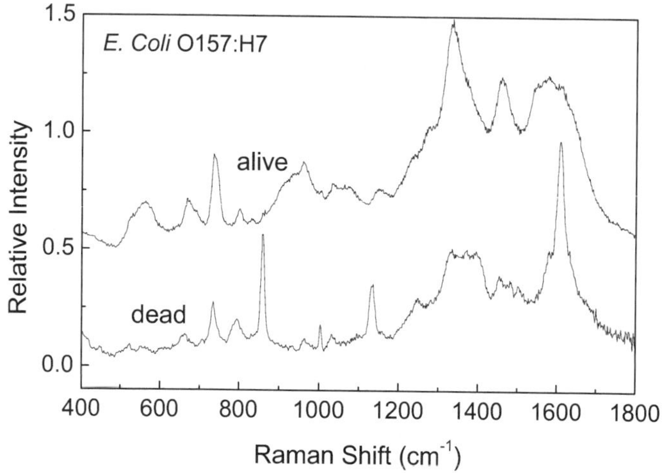

Figure 6. SERS spectra of viable E. coli O157:H7 and non-viable E. coli O157:H7. Incident laser power of 14 mW and collection time of 10 s are used to obtain these spectra. Spectra are offset vertically for display clarity. (Reproduced with permission from reference (71). Copyright 2008 The Society for Applied Spectroscopy.)

Thus, the best strategy to lower C would be to increase V, the volume of the samples put on the substrates, i.e., it can be realized by a pre-concentration method, capture as many bacteria as possible in a fixed surface area with large sample volume. For example, one can use centrifugation or filtration method prior to SERS detection to concentrate the real food sample. Our preliminary results of combining SERS with centrifugation and filtration method for bacteria inoculated mung bean spout have shown that LOD of *E. Coli* can be improved 500 times, about 10^2 CFU/ml. Another strategy to lower the C will be optimizing the substrates, hence increasing the N_d value. It can be realized by functionalization of the substrate surface with bacteria capture agents, such as antibody, aptamer, and antibiotics. Liu *et al.* reported the functionalization of silver nanoparticle with vancomycin could increase the bacteria capture ability of the nanoparticle by 3 orders of magnitude and greatly reduce the distance between bacteria and substrate surface, resulted in dramatically increased SERS intensity (*87*).

As the second challenge is to identify foodborne bacteria in naturally occurring food commodities, attention needs to be applied to sampling methods that are needed for field detection to separate pathogens from food matrix and concentrate pathogens at the same time. The separation method before SERS measurement is crucial to facilitate correctly obtaining bacterial Raman spectra without the interference from other components in food matrix. The main interference in food samples comes from the macromolecules, such as proteins and polysaccharides, which contain similar chemical bonds as the target bacteria. There is no easy method to eliminate the interference entirely from the mixture, but with proper separation methods, it is possible to reduce the interference and aid more prominent bacterial SERS signals. Since bacteria have larger size compared to the macromolecules in food samples, simple physical separation method based on the size of target, such as filtration is used. Wolffs reported a two-step filtration method prior to PCR in detection of *Salmonella enterica* and *Listeria monocytogenes* in food sample, such as chicken rinse and yogurt with a 79.1% recovery rate and 29 min filtration time (*88*). The combination of functionalized AgNR substrate to capture bacteria with two-step filtration method can be used to separate and concentrate pathogens from food samples, as well as further improve the LOD.

Detection and Differentiation of Foodborne Toxins

Aflatoxins (AFs), produced by *Aspergillus flavus, A. parastiticus and A. nomius*, are major mycotoxins of concern due to their highly hepatotoxic, carcinogenic, mutagenic, and teratogenic properties. The four major AFs, AF B1 ($C_{17}H_{12}O_2$), B2 ($C_{17}H_{14}O_6$), G1 ($C_{17}H_{12}O_7$), and G2 ($C_{17}H_{14}O_7$) are naturally occurring and frequently present in human food and animal feed (e.g., cereals, cotton, and groundnuts), and have a regulatory limit for the total AFs of no more than 20 ppb in human food, and no more than 300 ppb in animal feed in the United States. To demonstrate the rapid and sensitive detection of the AFs, SERS spectra of 10 μl AFB$_1$, AFB$_2$, AFG$_1$, and AFG$_2$ generated on AgNR substrates are acquired with a 10-s exposure time at 4.8 mW laser power (*68*).

The SERS spectra of AFB$_1$, AFB$_2$, AFG$_1$, and AFG$_2$ with significant peaks indicated are present in Figure 7A, and assignments of these peaks are essential to differentiate the four AFs from each other. The most significant difference between the AFBs and AFGs is the peak for β(C-H$_2$) ring mode at $\Delta v = 1248$ cm^{-1} for AFBs with a weak intensity, but very strong at $\Delta v = 1234$ cm^{-1} for the AFGs. The AFBs have an extra peak at $\Delta v = 1086$ cm^{-1} of AFB$_1$ and 1085 cm^{-1} of AFB$_2$ for the v(C-C-C) ring deformation mode, and the AFGs also have an extra peak at $\Delta v = 1115$ cm^{-1} for the C-O-C(H$_2$) stretching mode, which represents the extra O- atom in their structures. The structure differences between the AF$_1$s and the AF$_2$s are represented by the peak resulted from the C-C double bond in the AF$_1$s at $\Delta v =$

1620 cm⁻¹, which is absent in AF₂s. These peaks and their assignments are critical in the differentiation among these four AFs; however, it is not always necessary to identify every Raman peaks for differentiation purposes, because the analytes can simply be classified by PCA based on their overall spectroscopic fingerprints. In Figure 7B, we demonstrate the classification of the four AFs through the PCA performed using the spectral data of the four types of AFs.

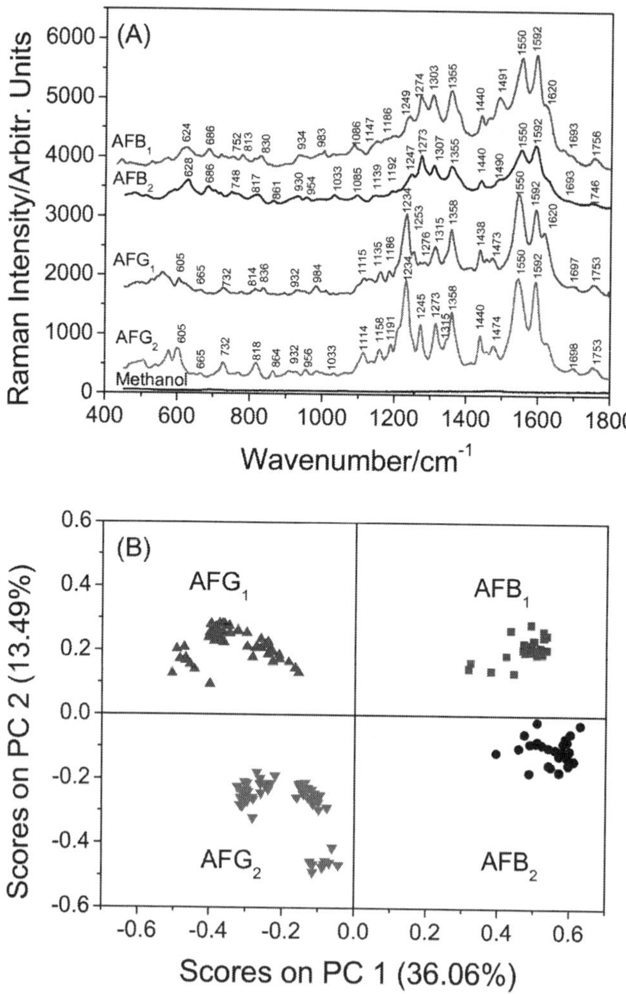

Figure 7. (A) Experimentally obtained SERS spectra of AFs and methanol. The Raman shifts of significant peaks in each AF spectrum are indicated, and the spectra are vertically offset for clear display; (B) PCA score plot of AFs at concentrations above their LODs. (Reproduced with permission from reference (68). Copyright 2012 The Royal Society of Chemistry.)

We have performed a concentration dependent SERS measurement on these four types of AFs to determine the sensitivity SERS by AgNR substrates. The SERS intensity of the peak at $\Delta v = 1592$ cm^{-1} is calculated and used to represent the SERS response; I_{1592} showed a power law relationship against the concentration of the AFs with great goodness of the fitting ($R^2 > 0.95$ for all AFs). As mentioned before, SERS can be viewed as a surface detection method, and the way to determine the LOD is different from conventional bulk detection techniques, such as HPLC or GC. The LOD in bulk solution does not reflect the true detectability of SERS as discussed in the previous section. The LODs of this detection method are determined to be 5×10^{-5} M for AFB$_1$, 1×10^{-4} M for AFB$_2$, and 5×10^{-6} M for both AFG$_1$ and AFG$_2$ in bulk solution (10 μl samples), respectively. In this case, the true LODs inside the laser spot are 6.17×10^{-16} moles(1.93×10^{-4} ng) of AFB$_1$, 1.23×10^{-15} moles(3.88×10^{-4} ng) for AFB$_2$, 6.17×10^{-17} moles(2.03×10^{-5} ng) for AFG$_1$, and 6.17×10^{-17} moles(2.04×10^{-5} ng) for AFG$_2$, respectively.

With the high SERS enhancement of the AgNRs and low background noise of our detection system, high-quality SERS spectra of the AFs have been obtained. Our results demonstrate that AgNR based SERS has the potential to be used as an on-site mycotoxin detection method in the food industry. Yet, the challenges, such as lowering the LOD in bulk solution, pre-concentration and isolation of aflatoxins from food samples remain if the method is used in the field.

Pesticide Detection

Chlorpyrifos is an organophosphate pesticide commonly used in the production of tea, which can cause cholinesterase inhibition, respiratory paralysis and even death in humans. Due to its widespread use, persistency and toxicity, chlorpyrifos has been included in priority list of pesticides within the European Union (EU) (*89*). The residue of a physically similar pesticide, parathion, is also commonly found in tea. Albeit highly toxic, parathion is still illegally used in tea plantations because of its superior insecticidal effect. The analysis of chlorpyrifos and parathion residues using conventional chromatographic methods has been troublesome due to the similarities between these two pesticides. However, due to the superior molecular identification capability, SERS may be used to differentiate these two pesticides. The SERS spectra of chlorpyrifos and parathion are obtained using the AgNR array substrates and are successfully differentiated from each other in the tea samples.

As shown in Figure 8A, the significant peaks of chlorpyrifos are observed at $\Delta v = 691$ (Cl-ring stretches), 851 (P-(O-R)$_2$ stretch), 1343 (C-N stretch), and 1575 cm^{-1} (C=C phenyl stretch) with high reproducibility (RSD of the above mentioned peak intensities $\leq 8\%$). The significant peaks of parathion are at $\Delta v = 647$ cm^{-1} (P=S stretch), 1160 cm^{-1} (C-H wag), and 1328 cm^{-1} (NO$_2$ symmetric stretch). Coupled with a simple Soxhlet extraction and gel permeation chromatography cleanup step, the LOD of chlorpyrifos in tea samples is calculated to be 0.155 ppm, well below the legal maximum residues limit in tea of 2 ppm. It has been demonstrated that using the AgNR-based SERS, one can rapidly screen chlorpyrifos residues in tea. Moreover, the SERS spectra of chlorpyrifos and parathion are mutually distinct, making it possible for specific identification of the two pesticides. Figure 8B shows that the tea samples spiked with parathion and chlorpyrifos are distinctly clustered on the PCA score plot using their SERS spectra, which further demonstrates the feasibility of using SERS for the rapid screening and differentiation of pesticides.

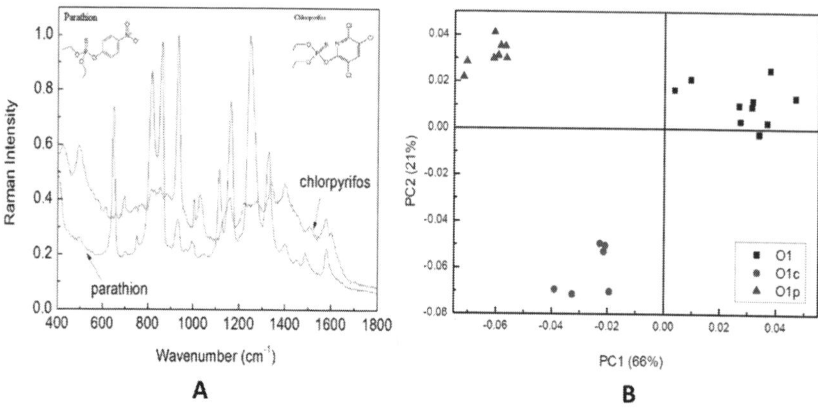

Figure 8. (A) Normalized SERS spectra of chlorpyrifos and parathion at the concentration of 5 ppm. (B) PCA classification of organic tea samples (O1), organic tea samples spiked with chlorpyrifos (O1c) and parathion (O1p). Incident laser power of 40 mW and collection time of 20 s are used to obtain these spectra.

Detection of Intentional Food Adulterants

AgNR-based SERS is also capable of conducting rapid and sensitive detection of melamine (72). In Figure 9, high quality SERS spectra of melamine are acquired using AgNR substrates, and the significant peaks from melamine agreed well with other studies. The following characteristic peaks from melamine are observed at Δv = 498 cm⁻¹, 605 cm⁻¹, 681 cm⁻¹, 704 cm⁻¹, 983 cm⁻¹, 1070 cm⁻¹, 1236 cm⁻¹, and 1396 cm⁻¹, and they reflect true Raman shift bands for melamine in the 500 – 1800 cm⁻¹ range. By studying the correlation between SERS peak intensity and melamine concentration, the LOD for melamine in aqueous solution is determined to be 0.1 mg/l, or 2 pg of melamine under the excitation laser spot, which is one order of magnitude lower than the current food standard. Although such low LOD is achieved with pure melamine solution, many challenges remain in the SERS based detections. For instance, the high protein content in milk poses a major interference for SERS detection as protein macromolecules tend to saturate the sensor at high concentrations. It has been found that the milk contents are likely to quench the SERS signal when the dilution factor is below 1000-fold. Hence the efficient extraction of melamine from milk protein, or the separation of any target compound from the food matrix, becomes critical.

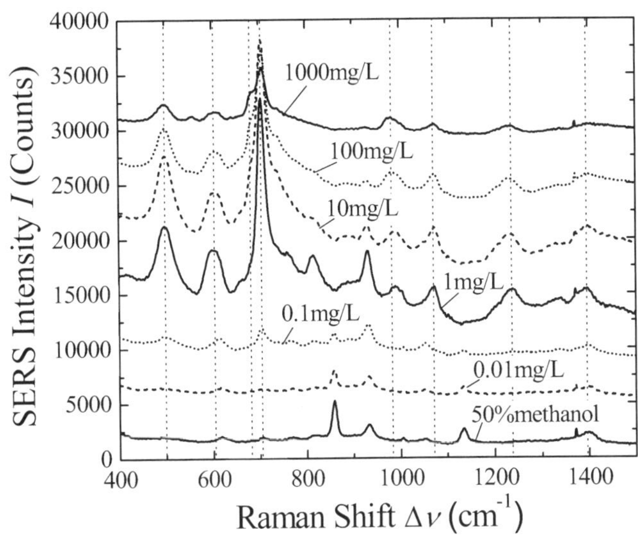

Figure 9. The average SERS spectra of 10 µL melamine solutions with different concentrations: C = 0.01, 0.1, 1, 10, 100, and 1000 mg/L, and the 50% methanol solution. (Reproduced with permission from reference (72). Copyright 2010 The Society for Applied Spectroscopy.)

SERS Coupled with Thin-Layer Chromatography for Detection of Contaminants in Food

Food matrix is a complex system consisting of water, carbohydrates, sugars, proteins, peptides, amino acids, lipids, fatty acids, vitamins, minerals, and other trace constituents. As described earlier, one big challenge of using SERS for food adulteration/toxin/bacteria detection in real-world scenarios comes from the strong interference of various food matrix components with the target signal. The interference becomes inevitable and potent in the label-free intrinsic SERS schemes, in which the target signal is likely to be rather weak due to the small amount of contaminants present. To address this challenge, an on-chip ultra-thin layer chromatography (UTLC) and SERS detection platform has been developed based on the AgNR substrates, which has the ability of simultaneous separation and detection of compounds in mixtures (*90*).

On this platform, a small amount of mixtures is first spotted onto the AgNR substrate, which is then placed into a chamber containing mobile phase solvents. As the mobile phase propagates along the substrate due to capillary action provided by the porous nanostructure, the components in the sample mixture also migrate along the development direction. Since different molecules have different affinity with the AgNRs as well as with the mobile phase solvents, the migration velocity of individual components vary, causing the components to be retained at different locations on the substrate. Thus, utilizing the thin-layer chromatography principles, mixtures can readily be separated on-chip. The substrate is then scanned at different locations along the mobile phase development direction by a Raman probe to obtain a profile of spatially-resolved SERS spectra, from which spectra of individual constituents are resolved (Figure 10).

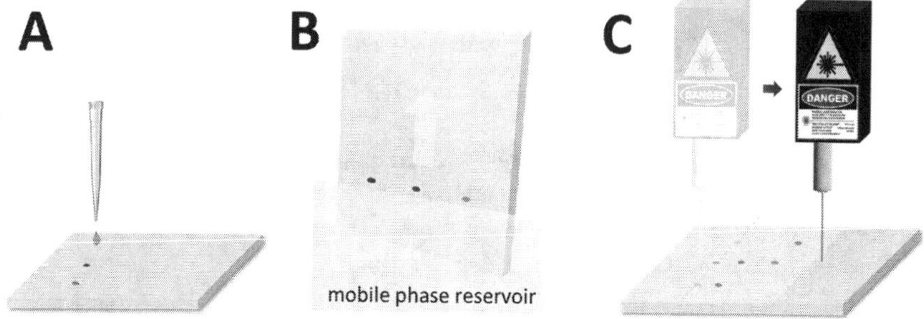

Figure 10. Schematic representation of the UTLC-SERS approach. (A) Apply samples onto the AgNR substrate. (B) Develop the AgNR substrate in a mobile phase. (C) Scan the substrate with a Raman probe along the mobile phase development direction.

Figure 11 shows the spatially-resolved SERS spectra of two melamine-Rhodamine 6G (R6G) mixtures (concentration ratio = 1:1 and 100:1) after UTLC development with methanol as the mobile phase, and the corresponding SERS peak intensity change along the development direction. In the equimolar mixture, melamine is retained near the sample origin while the R6G molecules are carried over to the solvent front (Figure 11A). When melamine concentration is 100× higher than that of R6G, no R6G signal is detected from the massive melamine background. However, after the UTLC separation, R6G spectra are able to be obtained at the solvent front, completely unaffected by the presence of melamine. This implies that using such an on-chip SERS platform, the interference from the food matrix background could potentially be reduced or even eliminated when trace amount of adulterants are to be targeted at, which is highly desirable in practice as it contributes to less complicated sample pre-treatment and improved assay efficiency.

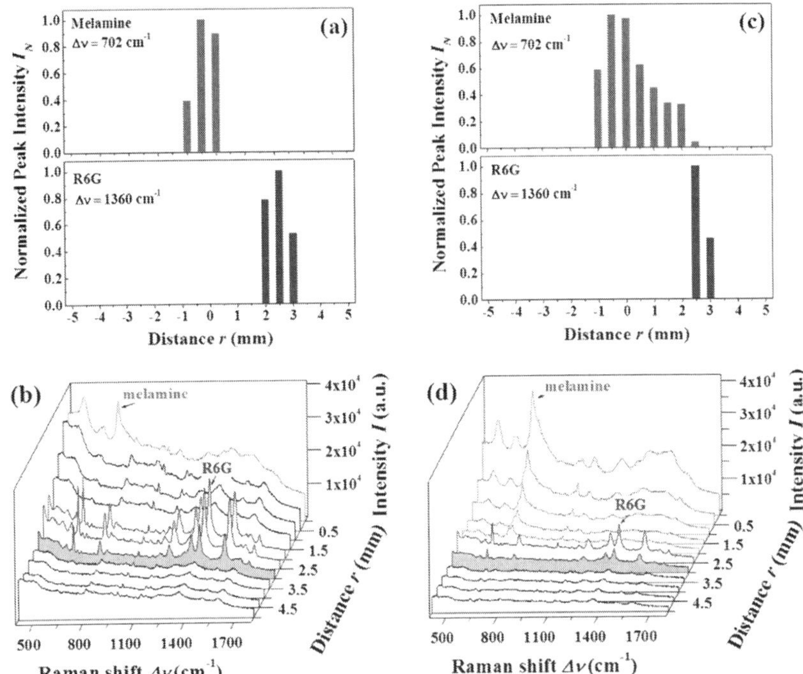

Figure 11. Separation of melamine and R6G (R6G): (a) and (b) equimolar mixture of melamine and R6G (1 : 1); (c) and (d) 100 : 1 mixture of melamine and R6G. In the chromatograms (a and c), 0 indicates the sample origin, whereas the cyan and blue bars indicate the normalized peak intensity of melamine (702 cm⁻¹) and R6G (1360 cm⁻¹), respectively. In the corresponding SERS spectra, the cyan and blue curves represent the melamine and R6G spectra at the locations with the highest signature peak intensity. The slices filled with solid grey indicate the solvent migration front. (Reproduced with permission from reference (90). Copyright 2012 The Royal Society of Chemistry.)

Conclusions

Surface enhanced Raman spectroscopy offers considerable potential in the area of molecular identification and trace element analysis, which has significant applications in both biological and chemical analyses. We have demonstrated that the AgNR based SERS are able to detect and differentiate a wide range of chemical and biological analytes that are closely related to food safety. Good signal-to-noise and reproducible SERS spectra of bacteria are obtained, and the bacterial SERS spectra show clear distinction between different species, and different strains. Its ability to distinguish between nonviable and viable cells is also presented. The four types of aflatoxins are identified and differentiated by SERS at low LODs, so are the two different pesticides, chlorpyrifos and parathion, with similar chemical structure. Food contaminants such as melamine can be detected, and the combination of SERS and TLC illustrates the potential of removing interference and detection of the target at the same time. All these results demonstrate that the application of SERS in real food sample is possible.

These works clearly demonstrate the great potential of the AgNR substrate as a real-time multiplexing SERS-based sensor for chemical and biological applications. The speed, specificity and ease of implementation of SERS techniques represent a valuable alternative to current food safety diagnostic tools and provide the possibility of portable sensor for on-site food inspection.

From a problem-solving standpoint, SERS is not fully developed at present. Much future research work needs to be done on the applicability of this technique to real food samples. Priority should be given to the development of optimized SERS-active substrates that are of consistent quality and suitable for long-time storage for each application. The optimization and functionalization of the substrates are also the keys to lower the LOD in bulk samples. The food samples are often mixed with contaminants and various backgrounds, which will confound the target SERS signals. Therefore additional sampling techniques, such as microfluidics, dielectrophoresis, are also needed to be developed and incorporated into the SERS detection procedure to allow substantial isolation of the target analytes in real sample matrices, and to be integrated into lab-on-a-chip devices. For AgNR based SERS detection methods to be valid for the real-world applications, it should have the capability to process high volume of samples. The aforementioned devices, combining SERS with the sampling methods, would provide such a high throughput. In addition, we have also demonstrated the fabrication of a robust, uniform, and patterned multi-well SERS chip, which has 40 wells on the substrates (*86*). With advanced microfabrication technology, one can even pattern the SERS substrate to a 96 well plate. Using this type of substrate, a device capable of high throughput biosensing and multiplexing is feasible. In operational settings, detection and identification of analytes can be done by comparing a measured SERS spectrum of a target against a SERS signature library. Therefore, a comprehensive, reproducible, and robust SERS spectral library is essential to a successful SERS-based detection method. The signatures of the spectra in the library must be reproducible and exhibit minimal variability induced by sensor uncertainty, biological growth conditions as well as sample preparation conditions.

Acknowledgments

This work was partly supported by USDA CSREES Grant #2009-35603-05001. The authors would like to thank the contributions from our collaborators, Prof. R. A. Dluhy, Prof. R. A. Tripp, Dr. J. D. Driskell, and co-workers, Dr. Y. J. Liu, Dr. V. H. Chu, Dr. S. Hennigan, Dr. J. L. Abell, Mr. X.B. Du, and Ms. H. L. Huang.

References

1. Scallan, E.; Hoekstra, R. M.; Angulo, F. J.; Tauxe, R. V.; Widdowson, M.-A.; Roy, S. L.; Jones, J. L.; Griffin, P. M. *Emerging Infect. Dis.* **2011**, *17*, 7.
2. Zhang, G.; Ma, L.; Patel, N.; Swaminathan, B.; Wedel, S.; Doyle, M. P. *J. Food Prot.* **2007**, *70*, 997.
3. Bennett, J. W.; Klich, M. *Clin. Microbiol. Rev.* **2003**, *16*, 497.
4. Wild, C. P.; Gong, Y. Y. *Carcinogenesis* **2010**, *31*, 71.
5. Goldman, L. R.; Smith, D. F.; Neutra, R. R.; Saunders, L. D.; Pond, E. M.; Stratton, J.; Waller, K.; Jackson, R. J.; Kizer, K. W. *Arch. Environ. Health* **1990**, *45*, 229.
6. Goes, E. A.; Savage, E. P.; Gibbons, G.; Aaronson, M.; Ford, S. A.; Wheeler, H. W. *Am. J. Epidemiol.* **1980**, *111*, 254.
7. Wu, M.-L.; Deng, J.-F.; Tsai, W.-J.; Ger, J.; Wong, S.-S.; Li, H.-P. *Clin. Toxicol.* **2001**, *39*, 333.
8. Gracias, K. S.; McKillip, J. L. *Can. J. Microbiol.* **2004**, *50*, 883.
9. Swaminathan, B.; Feng, P. *Ann. Rev. Microbiol.* **1994**, *48*, 401.
10. Tian, H.; Miyamoto, T.; Okabe, T.; Kuramitsu, Y.; Honjoh, K.-i.; Hatano, S. *J. Food Prot.* **1996**, *59*, 1158.
11. Tan, S.; Gyles, C. L.; Wilkie, B. N. *Vet. Microbiol.* **1997**, *56*, 79.
12. Scheu, P.; Gasch, A.; Zschaler, R.; Berghof, K.; Wilborn, F. *Food Biotechnol.* **1998**, *12*, 1.
13. Tapchaisri, P.; Wangroongsarb, P.; Panbangred, W.; Kalambaheti, T.; Chongsa-nguan, M.; Srimanote, P.; Kurazono, H.; Hayashi, H.; Chaicumpa, W. *Asian Pac. J. Allergy Immunol.* **1999**, *17*, 41.
14. Rahman, H. *Indian J. Med. Res.* **1999**, *110*, 47.
15. Correa, A. A.; Toso, J.; Albarnaz, J. D.; Simoes, C. M. O.; Barardi, C. R. M. *J. Food Qual.* **2006**, *29*, 458.
16. Myint, M. S.; Johnson, Y. J.; Tablante, N. L.; Heckert, R. A. *Food Microbiol.* **2006**, *23*, 599.
17. Trafny, E. A.; Kozłowska, K.; Szpakowska, M. *Lett. Appl. Microbiol.* **2006**, *43*, 673.
18. Wang, X.; Jothikumar, N.; Griffiths, M. W. *J. Food Prot.* **2004**, *67*, 189.
19. Hein, I.; Flekna, G.; Krassnig, M.; Wagner, M. *J. Microbiol. Methods* **2006**, *66*, 538.
20. Uyttendaele, M.; Vanwildemeersch, K.; Debevere, J. *Lett. Appl. Microbiol.* **2003**, *37*, 386.
21. Notzon, A.; Helmuth, R.; Bauer, J. *J. Food Prot.* **2006**, *69*, 2896.
22. Gunaydin, E.; Eyigor, A.; Carli, K. T. *Lett. Appl. Microbiol.* **2007**, *44*, 24.
23. Velusamy, V.; Arshak, K.; Korostynska, O.; Oliwa, K.; Adley, C. *Biotechnol. Adv.* **2010**, *28*, 232.
24. Turner, N. W.; Subrahmanyam, S.; Piletsky, S. A. *Anal. Chim. Acta* **2009**, *632*, 168.
25. Krska, R.; Schubert-Ullrich, P.; Molinelli, A.; Sulyok, M.; Macdonald, S.; Crews, C. *Food Addit. Contam.* **2008**, *25*, 152.
26. Shephard, G. S. *Chem. Soc. Rev.* **2008**, *37*, 2468.
27. Zheng, M.; Richard, J.; Binder, J. *Mycopathologia* **2006**, *161*, 261.

28. Omeroglu, P.; Boyacioglu, D.; Ambrus, Á.; Karaali, A.; Saner, S. *Food Anal. Methods* **2012**, *5*, 1469.

29. Torres, C. M.; Picó, Y.; Mañes, J. *J. Chromatogr., A* **1996**, *754*, 301.

30. Watanabe, E. *JARQ* **2011**, *45*, 359.

31. Kaufman, B. M.; Clower, M. *J. AOAC Int.* **1995**, *78*, 1079.

32. Kaufman, B. M.; Clower, M. *J. - Assoc. Off. Anal. Chem.* **1991**, *74*, 239.

33. Morozova, V.; Levashova, A.; Eremin, S. *J. Anal. Chem.* **2005**, *60*, 202.

34. West, J. L.; Halas, N. J. *Annu. Rev. Biomed. Eng.* **2003**, *5*, 285.

35. Alivisatos, P. *Nat. Biotechnol.* **2004**, *22*, 47.

36. Medintz, I. L.; Uyeda, H. T.; Goldman, E. R.; Mattoussi, H. *Nat. Mater.* **2005**, *4*, 435.

37. Wang, J. *Analyst* **2005**, *130*, 421.

38. Pingarrón, J. M.; Yáñez-Sedeño, P.; González-Cortés, A. *Electrochim. Acta* **2008**, *53*, 5848.

39. Scida, K.; Stege, P. W.; Haby, G.; Messina, G. A.; García, C. D. *Anal. Chim. Acta* **2011**, *691*, 6.

40. Kim, S. N.; Rusling, J. F.; Papadimitrakopoulos, F. *Adv. Mater.* **2007**, *19*, 3214.

41. Aragay, G.; Pons, J.; Merkoçi, A. *Chem. Rev.* **2011**, *111*, 3433.

42. Lee, S. J.; Lee, J.-E.; Seo, J.; Jeong, I. Y.; Lee, S. S.; Jung, J. H. *Adv. Funct. Mater.* **2007**, *17*, 3441.

43. Liu, J.; Liu, J.; Yang, L.; Chen, X.; Zhang, M.; Meng, F.; Luo, T.; Li, M. *Sensors* **2009**, *9*, 7343.

44. Bonanni, A.; del Valle, M. *Anal. Chim. Acta* **2010**, *678*, 7.

45. Zhang, L.; Fang, M. *Nano Today* **2010**, *5*, 128.

46. Vikesland, P. J.; Wigginton, K. R. *Environ. Sci. Technol.* **2010**, *44*, 3656.

47. Jain, P.; Huang, X.; El-Sayed, I.; El-Sayed, M. *Plasmonics* **2007**, *2*, 107.

48. Doering, W. E.; Piotti, M. E.; Natan, M. J.; Freeman, R. G. *Adv. Mater.* **2007**, *19*, 3100.

49. Song, S.; Qin, Y.; He, Y.; Huang, Q.; Fan, C.; Chen, H.-Y. *Chem. Soc. Rev.* **2010**, *39*, 4234.

50. Siangproh, W.; Dungchai, W.; Rattanarat, P.; Chailapakul, O. *Anal. Chim. Acta* **2011**, *690*, 10.

51. Junxue, F.; Bosoon, P.; Greg, S.; Les, J.; Ralph, T.; Yiping, Z.; Yong-Jin, C. *Nanotechnology* **2008**, *19*, 155502.

52. Chen, G.; Zhou, J.; Park, B.; Xu, B. *Appl. Phys. Lett.* **2009**, *95*, 043103.

53. Chen, G.; Ning, X.; Park, B.; Boons, G.-J.; Xu, B. *Langmuir* **2009**, *25*, 2860.

54. Wang, B.; Guo, C.; Chen, G.; Park, B.; Xu, B. *Chem. Commun.* **2012**, *48*, 1644.

55. Wang, B.; Guo, C.; Zhang, M.; Park, B.; Xu, B. *J. Phys. Chem. B* **2012**, *116*, 5316.

56. Kneipp, K.; Kneipp, H.; Itzkan, I.; Dasari, R. R.; Feld, M. S. *Chem. Rev.* **1999**, *99*, 2957.

57. Campion, A.; Kambhampati, P. *Chem. Soc. Rev.* **1998**, *27*, 241.

58. Kneipp, K.; Wang, Y.; Kneipp, H.; Perelman, L. T.; Itzkan, I.; Dasari, R. R.; Feld, M. S. *Phys. Rev. Lett.* **1997**, *78*, 1667.

59. Nie, S. M.; Emery, S. R. *Science* **1997**, *275*, 1102.

60. Xu, H.; Aizpurua, J.; Käll, M.; Apell, P. *Phys. Rev. E* **2000**, *62*, 4318.

61. Xu, H.; Bjerneld, E. J.; Käll, M.; Börjesson, L. *Phys. Rev. Lett.* **1999**, *83*, 4357.

62. Stacy, A. A.; Van Duyne, R. P. *Chem. Phys. Lett.* **1983**, *102*, 365.

63. Carron, K. T.; Xue, G.; Lewis, M. L. *Langmuir* **1991**, *7*, 2.

64. Sudnik, L. M.; Norrod, K. L.; Rowlen, K. L. *Appl. Spectrosc.* **1996**, *50*, 422.

65. Jensen, T. R.; Malinsky, M. D.; Haynes, C. L.; Van Duyne, R. P. *J. Phys. Chem. B* **2000**, *104*, 10549.

66. Sauer, G.; Nickel, U.; Schneider, S. *J. Raman Spectrosc.* **2000**, *31*, 359.

67. Tao, A.; Kim, F.; Hess, C.; Goldberger, J.; He, R.; Sun, Y.; Xia, Y.; Yang, P. *Nano Lett.* **2003**, *3*, 1229.

68. Wu, X.; Gao, S.; Wang, J.-S.; Wang, H.; Huang, Y.-W.; Zhao, Y. *Analyst* **2012**, *137*, 4226.

69. Neyrolles, O.; Hennigan, S. L.; Driskell, J. D.; Dluhy, R. A.; Zhao, Y.; Tripp, R. A.; Waites, K. B.; Krause, D. C. *PLoS One* **2010**, *5*, e13633.

70. Shanmukh, S.; Jones, L.; Zhao, Y. P.; Driskell, J.; Tripp, R.; Dluhy, R. *Anal. Bioanal. Chem.* **2008**, *390*, 1551.

71. Chu, H.; Huang, Y.; Zhao, Y. *Appl. Spectrosc.* **2008**, *62*, 922.

72. Du, X. B.; Chu, H. Y.; Huang, Y. W.; Zhao, Y. P. *Appl. Spectrosc.* **2010**, *64*, 781.

73. Leverette, C. L.; Villa-Aleman, E.; Jokela, S.; Zhang, Z.; Liu, Y.; Zhao, Y.; Smith, S. A. *Vib. Spectrosc.* **2009**, *50*, 143.

74. Driskell, J. D.; Seto, A. G.; Jones, L. P.; Jokela, S.; Dluhy, R. A.; Zhao, Y. P.; Tripp, R. A. *Biosens. Bioelectron.* **2008**, *24*, 917.

75. Chaney, S. B.; Shanmukh, S.; Dluhy, R. A.; Zhao, Y. P. *Appl. Phys. Lett.* **2005**, *87*, 031908.

76. Liu, Y.; Zhao, Y. In *Proc. SPIE 7757, Plasmonics: Metallic Nanostructures and Their Optical Properties VIII*; 2010; p 77570Q.

77. Zhao, Y. In *Proc. SPIE 8401, Independent Component Analyses, Compressive Sampling, Wavelets, Neural Net, Biosystems, and Nanoengineering X*; 2012; p 84010O.

78. Zeiri, L.; Bronk, B. V.; Shabtai, Y.; Czege, J.; Efrima, S. *Colloids Surf., A* **2002**, *208*, 357.

79. Efrima, S.; Bronk, B. V. *J. Phys. Chem. B* **1998**, *102*, 5947.

80. Jarvis, R. M.; Brooker, A.; Goodacre, R. *Anal. Chem.* **2004**, *76*, 5198.

81. Sengupta, A.; Laucks, M. L.; Davis, E. J. *Appl. Spectrosc.* **2005**, *59*, 1016.

82. Jarvis, R. M.; Goodacre, R. *Anal. Chem.* **2004**, *76*, 40.

83. Montoya, J. R.; Armstrong, R. L.; Smith, G. B. *Proc. SPIE 5085, Chemical and Biological Sensing IV* **2003**, 144.

84. Premasiri, W. R.; Moir, D. T.; Klempner, M. S.; Krieger, N.; Jones, G.; Ziegler, L. D. *J. Phys. Chem. B* **2004**, *109*, 312.

85. Guzelian, A. A.; Sylvia, J. M.; Janni, J. A.; Clauson, S. L.; Spencer, K. M. *Proc. SPIE 4577, Vibrational Spectroscopy-based Sensor Systems* **2002**, 182.

86. Abell, J. L.; Driskell, J. D.; Dluhy, R. A.; Tripp, R. A.; Zhao, Y. P. *Biosens. Bioelectron.* **2009**, *24*, 3663.

87. Liu, T. Y.; Tsai, K. T.; Wang, H. H.; Chen, Y.; Chen, Y. H.; Chao, Y. C.; Chang, H. H.; Lin, C. H.; Wang, J. K.; Wang, Y. L. *Nat. Commun.* **2011**, *2*, 538.

88. Wolffs, P. F. G.; Glencross, K.; Thibaudeau, R.; Griffiths, M. W. *Appl. Environ. Microbiol.* **2006**, *72*, 3896.

89. Lacorte, S.; Lartiges, S. B.; Garrigues, P.; Barcelo, D. *Environ. Sci. Technol.* **1995**, *29*, 431.

90. Chen, J.; Abell, J.; Huang, Y.-w.; Zhao, Y. *Lab Chip* **2012**, *12*, 3096.

Chapter 6

Electrical Capture and Detection of Microbes Using Dielectrophoresis at Nanoelectrode Arrays

Foram Ranjeet Madiyar,[1] Lateef Uddin Syed,[1] Prabhu Arumugam,[2] and Jun Li[*,1]

[1]Department of Chemistry, Kansas State University, Manhattan, Kansas 66506-0401
[2]Advanced Diamond Technologies Inc., 48 East Belmont Drive, Romeoville, Illinois 60446
*E-mail: junli@ksu.edu.

A nanostructured dielectrophoresis (DEP) device based on vertically aligned carbon nanofibers versus a macroscopic indium tin oxide counter electrode has been demonstrated for capture of bacterial cells and virus particles. Computer simulation by finite element modeling illustrates the strong DEP force by the highly focused electric field at the nanoelectrode tips. Experiments on DEP capture of two types of microbes, i.e. *E. coli* bacterial cells (~1-2 micron in size) and bacteriophage virus particles (~80-200 nm in size), further validated the prediction. The DEP capture was found reversibly controlled by the AC voltage from 100 Hz to 1 MHz. The comparable size of the nanoelectrode produced stronger interaction with virus particles, generatingstriking lightning patterns. This technique can be potentially utilized as a fast sample preparation module in a microfluidic chip to capture, separate, and concentrate microbes in analyzing small volume of dilute samples as a part of a portable detection system for field-deployable applications.

Introduction

There is a strong need of portable systems for rapid detection of pathogenic microbes (*1, 2*). Microfluidic devices have been recognized for their great potential in such applications. Since it is desired to detect specific microbes at very low concentrations, on-chip capture, sorting, and concentration is critical (*3, 4*). Dielectrophoresis (DEP) is very attractive for this purpose since all polarizable particles can be easily manipulated using programmable electric fields (*5, 6*). Various microscale DEP devices have been demonstrated in effective manipulation of single cells with the size varying from tens of microns (mammalian cells) to about 1 micron (bacterial cells) (*7–9*). Since the DEP force is proportional to the volume of the target particles, it decreases rapidly when the particle size is reduced to only ~100 nanometers for viral particles. Previous studies have demonstrated that it was feasible to trap and accumulate viral particles using micro-DEP devices (*10–12*), but more sophisticated control requires nanostructured DEP electrodes (*13*). Recently, we have developed a simple nanoscale DEP device based on a nanoelectrode array made of vertically aligned carbon nanofibers versus a macroscopic indium tin oxide counter electrode, which can capture either single or large ensembles of bacterial cells and viral particles from high-velocity fluidic flows (*14–16*). The electric field was found to be highly focused at the nanoelectrode tip, interacting differently with bacterial cells and viral particles due to their sizes. It is particularly effective in capturing viral particles whose sizes are comparable to the spatial distribution of the high electric field generated at the nanoelectrode tips.

Principles, Device Design, and Fabrication

The principles of DEP were previously described by Pohl (*6*). When a polarizable microparticle (either charged or neutral) is placed in a non-uniform electric field, a net electrical force will be generated on the particle, causing it to move along the field. Generally, the electric field is produced by applying an alternating current voltage to a pair of electrodes. The time averaged DEP force (F_{DEP}) exerted on a spherical particle is given by (*6*):

$$\langle F_{DEP} \rangle = 2\pi r^3 \varepsilon_m \, \mathrm{Re}[K(\omega)]\nabla E^2 , \tag{1}$$

where r is the radius of the particle, ε_m is the permittivity of the suspending medium, ∇E^2 is the gradient of the square of the applied electric field strength, and $\mathrm{Re}[K(\omega)]$ is the real component of the complex Clausius-Mossotti (CM) factor given by (*6*):

$$K(\omega) = \frac{\varepsilon_p^* - \varepsilon_m^*}{\varepsilon_p^* + 2\varepsilon_m^*}, \text{ where } \varepsilon^* = \varepsilon - j\frac{\sigma}{\omega} \tag{2}$$

With ε^* representing the complex permittivity and the indices p and m referring to the particle and medium, respectively. Parameter σ is the complex conductivity, ω is the angular frequency ($\omega = 2\pi f$) of the applied electric field, and $j = \sqrt{-1}$. In this study, the proper medium is chosen to give $\mathrm{Re}[K(\omega)] > 0$ so that the particles experience a DEP force directing toward higher electric field strength, i.e. positive DEP (pDEP) (6).

According to *eq* 1, the DEP force (F_{DEP}) is proportional to the volume or cube of the radius (r^3) of the particle. On the other hand, the hydrodynamic force to carry the particles with flow (i.e. Stokes drag force F_{Drag}) is directly proportional to the radius of the particle by

$$F_{Drag} = 6\eta\pi r k \upsilon \tag{3}$$

where η is the dynamic viscosity of solution, k is a small factor accounts for the wall effects, and υ is the linear flow rate (flow velocity) (8). Sedimentation force and Brownian force are negligible. The advantage of nanostructured DEP devices is that the magnitude of ∇E^2 can be enhanced by orders of magnitude so that even small viral particles can be captured.

Figure 1 schematically illustrates the design with a nanoelectrode array as the 'points' electrode and a macroscopic indium tin oxide slide as the 'lid' electrode in a "points-and-lid" configuration (7) for DEP experiments. The nanoelectrode array consists of vertically aligned carbon nanofibers embedded in SiO_2 matrix with only the tip exposed. The average diameter of the vertically aligned carbon nanofibers is ~100-120 nm (17, 18). Either a randomly distributed array with an average spacing of ~1-2 microns or a regular patterned array can be used (17, 18). The nanoelectrode array is covered with a 2-μm SU-8 photoresist layer, a negative photoresist widely used in microelectronics industry, with only 200 x 200 μm² active area exposed. This area is aligned at the center of a 1 mm diameter circular chamber connected with 500 μm wide microfluidic channels etched in a 18-μm SU-8 photoresist layer on the top indium tin oxide slide. These two pieces are then permanently bonded together through the SU-8 photoresist layers.

Finite Element Model (FEM) Simulation

According to *eq* 1, the DEP force on a particle depends on ∇E^2 defined by the electrode arrangement and the Clausius-Mossotti factor $\mathrm{Re}[K(\omega)]$ defined by the dielectric properties of the particles and the surrounding medium. This leads to several possible electrode array designs. The FEM simulation is useful to guide the design and performance evaluation. A general-purpose Computational Fluid Dynamics code CFD-ACE+ (ESI Group Inc.) was used to numerically solve all the governing equations subjected to appropriate boundary and initial conditions. Microbial particles are treated as spheres. The time-averaged DEP force on each particle in an alternating current AC electric field is calculated according to *eq* 1 and the particle position is traced from initial configuration by solving the momentum equation

$$m_p\, d\mathbf{u}/dt = \langle \mathbf{F}_{DEP} \rangle + \mathbf{F}_{drag}, \tag{4}$$

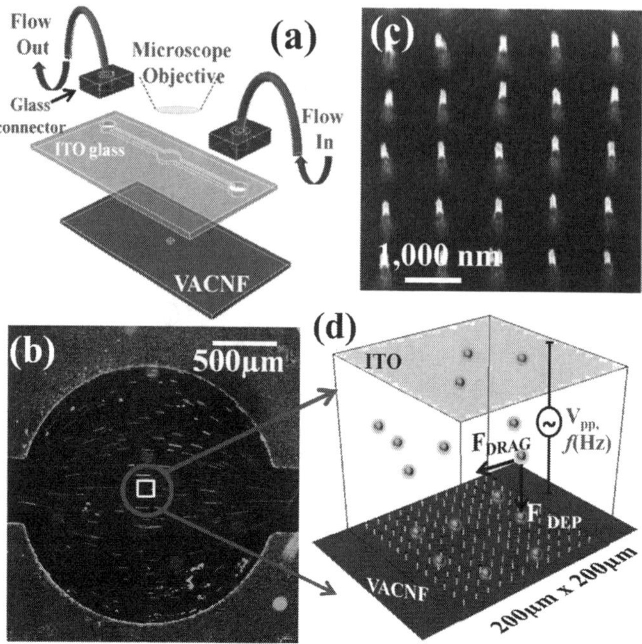

Figure 1. Schematic of the DEP device. (a) The components of the device, including a lid electrode (indium tin oxide (ITO) coated glass) with a 18-μm SU-8 photoresist layer containing a microfluidic channel, a nanoelectrode array chip covered with 2-μm SU-8 photoresist except exposing a 200 x 200 μm² area, glass fluidic connectors, and microbore tubes. (b) A low-magnification optical microscope image showing the flow profile of particle distribution of fluorescent labeled bacteriophage solution passing through the bonded device. (c) SEM image of a nanoelectrode array made of e-beam patterned regular vertically aligned carbon nanofibers. (d) Schematic diagram of microbial particles in the active nano-DEP area, which are subjected to the hydrodynamic drag force (F_{Drag}) along the flow direction and the dielectrophoretic force (F_{DEP}) mostly perpendicular to the nanoelectrode array surface. (Adapted from reference (16). Copyright 2013 Wiley-Blackwell.)

With the velocity $\mathbf{u} = ds/dt$ and the initial conditions $\mathbf{u}(t=0) = \mathbf{u}_0$ and the position $s(t=0) = s_0$. Here m_p is the mass of each particle. The particles are assumed small compared to the scale of field non-uniformity and the particle density is low so that the inter-particle effects can be neglected. Inertial force and history forces (Basset force) are neglected and particle-boundary interaction is neglected.

Figure 2 shows the two-dimensional (2D) finite element method simulation in our previous study (*14*) which demonstrated effective pDEP trapping of 1-μm diameter spherical particles at a nanoelectrode array at various voltages (of 1 MHz sine wave) and flow velocities. The physical properties of the particles in the solution are set to those of *E. coli* cells. The microfluidic channel is 20 μm in height and 102 μm in length, with a linear array of 12 NEs of 200 nm in width and 2 μm in spacing at the bottom. The 200 nm width is used because the smallest grid size that the FEM package can handle is 100 nm. The continuum theory may not be valid below 100 nm. The 2-μm separation was chosen to eliminate the electric field overlapping among nearby nanoelectrodes.

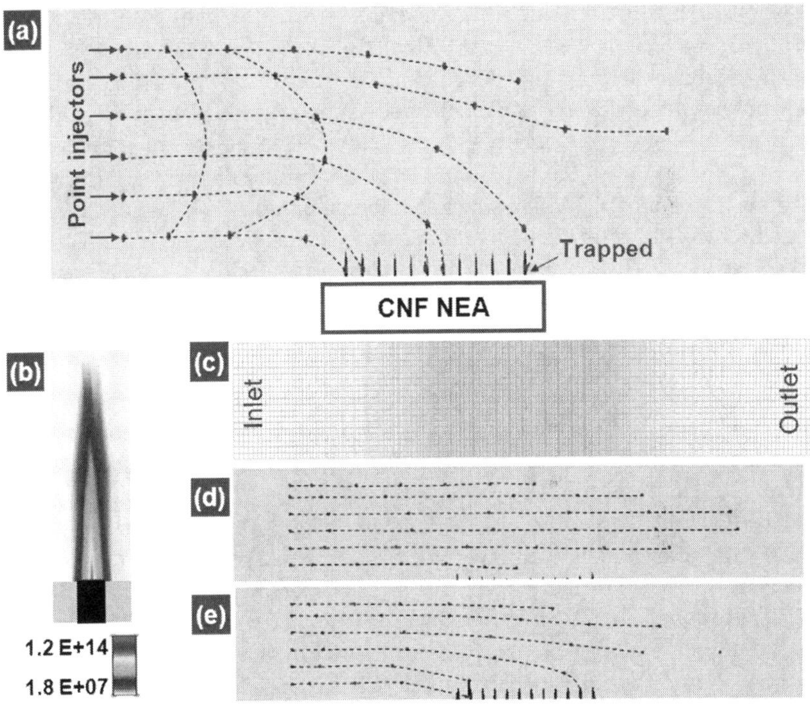

Figure 2. 2D Finite element method modeling of positive DEP trapping at the carbon nanofiber nanoelectrode array. (a) Trapping of 1-μm diameter spherical particles under the influence of Stokes drag and dielectrophoretic forces at 26 V_{pp} 1 MHz altenrating current bias in a flow with the velocity of 10 mm/s. The dashed lines show the particle trajectories. (b) The distribution of the square of electric field strength (E^2) around the nanoelectrode surface in a linear color scale. (c) Structured grid of the channel geometry consisting of varied cell size. (d) Particle trajectories at 2 V_{pp} and (e) 9 V_{pp}. The flow velocity is 2 mm/s. Note: The particles are introduced from point injectors at different heights (3, 6, 9, 12, 15 and 17 μm). (Reproduced with permission from reference (14). Copyright 2007 American Chemical Society.)

As shown in Figure 2a, the trajectories indicate that all 1-μm particles injected within 12 μm from the bottom are trapped, while others (including those near the ceiling) are deflected downward at 26 V_{pp} and 10 mm/s flow velocity. A larger array may trap all particles. The flow velocity (10 mm/s) is much higher than those used in a micro- "points-and-lid" device (~0.1-0.5 mm/s) (*19*) or micro-inter-digitated device (0.04-2 mm/s) (*8*) at the same voltage, reflecting the higher trapping efficiency with nanoelectrods. The $E(r)^2$ map around a nanoelectrode tip (Figure 2b) shows a maximum of ~1.2×10^{14} V^2m^{-2}, 200 times higher than that of the micro- "points-and-lid" device (*19*). Figure 2c shows the structured grid of the microchannel geometry that was divided into 7790 cells with the smallest grid size being 100 nm located at the nanoelectrode tip. Figures 3d and 3e shows the particle trajectories inside the microchannel at lower V_{pp} (2 V and 9 V) and flow velocity (2 mm/s). At 2 V_{pp}, only particles injected at the height ≤ 3 μm are trapped and those above are not influenced by the DEP. At $V_{pp} = 9$ V, particles injected at height ≤ 9 μm are trapped and those above are deflected downward, indicating that they can be captured if longer nanoelectrode array is used. A sufficiently high V_{pp} is critical in generating a large ∇E^2 for effective pDEP capture of the particles.

As expected from the FEM analyses, the capture efficiency (i) decreases as the flow rate is increased, since the hydrodynamic force F_{Drag} experienced by the particles increases with the flow rate; (ii) increases as the applied electric field (or V_{pp}) is increased; (iii) decreases as the initial particle position is higher, since ∇E^2 drops rapidly from the electrode surface; and (iv) increases as the particle size is increased, since F_{DEP} is proportional to r^3 (by *eq* 1) while F_{Drag} is proportional to r (by *eq* 3).

Figure 3 further shows the quantitative electric field gradients ∇E^2 and the magnitudes of F_{Drag} and F_{DEP} at different heights above the nanoelectrode tip, which are derived from the 2D simulation. The lateral x-axis is along the microfluidic channel and the vertical y-axis is normal to the nanoelectrode array surface. Clearly, the vertical electric field gradient (∇E_y^2) is proportional to V_{pp}^2 and drops rapidly within ~5 μm from the nanoelectrode surface and then decreases much slower beyond this range. The vertical component dominates the field gradient for the height over 5 μm, while the lateral component (∇E_x^2) dominates below ~3 μm, as indicated by the overlapping of ∇E^2 and ∇E_y^2 points at the height >3 μm (Figure 3b). These analyses indicate that particles are first pulled down toward the nanoelectrodes by the vertical pDEP force which is orthogonal with F_{Drag}. Once it is within ~3 μm from the nanoelectrode, the particle encounters a lateral pDEP force which is orders of magnitude higher than the drag force (see Figure 3c). As a result, they are firmly trapped at the nanoelectrode site.

Estimation of Clausius-Mossotti (CM) Factor

Though bioparticles have complicated internal structure and inhomogeneous composition, which may affect the DEP via the frequency-dependent Clausius-Mossotti factor as described in *eq* 1, a smeared-out multishell model (*20, 21*) using the average conductivity and permittivity for the whole particle was found to be sufficient in providing quick guide for experimental designs. The bacterial cells are thus modeled as a sphere covered by two shells accounting for inner cytoplasmic membrane and outer cell wall. An effective complex permittivity $\varepsilon_{eff}{}^*$ for *E. coli* cells is adapted from previous reports (*22–24*). Most DEP experiments were done using DI water ($\varepsilon_m = 80$) as the suspending medium. But adding non-polar agents like mannitol may lower the medium permittivity by 70% and enhance the Clausius-Mossotti factor (*4*). The model for viral particles is simpler due to absence of a cell wall and cell membrane. The parameters for $\varepsilon_{eff}{}^*$ are adapted from previous reports on vaccina virus (*25*).

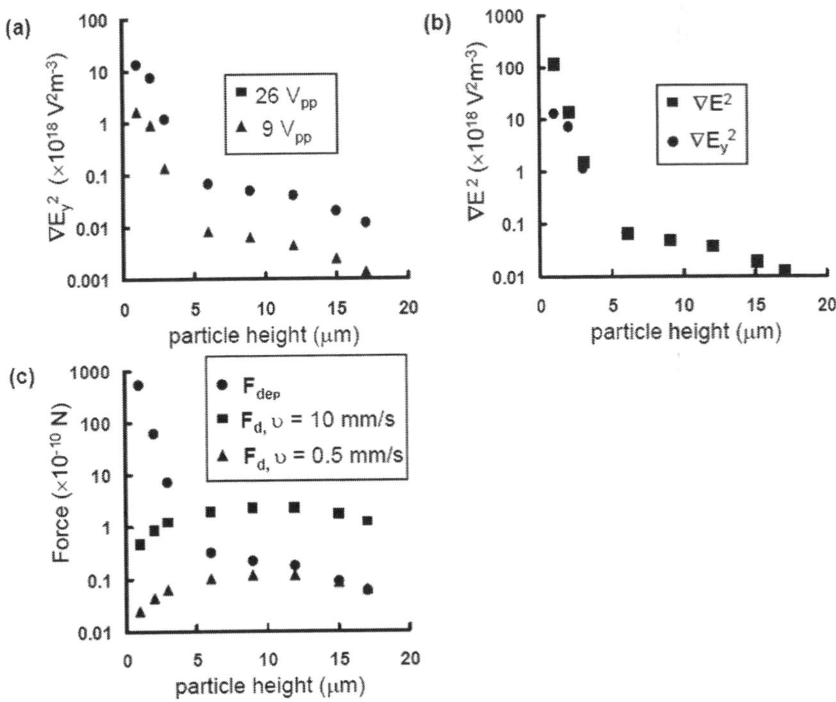

Figure 3. (a) The magnitude of ∇E_y^2 for particles at different positions with V_{pp} = 26 V and 9 V, respectively. (b) The comparison of ∇E_y^2 and total ∇E^2 at V_{pp} = 26 V vs. particle position. (c) The magnitude of DEP and drag forces at V_{pp} = 26 V with the flow velocity υ = 0.5 and 10 mm/s, respectively. (Reproduced with permission from reference (14). Copyright 2007 American Chemical Society.)

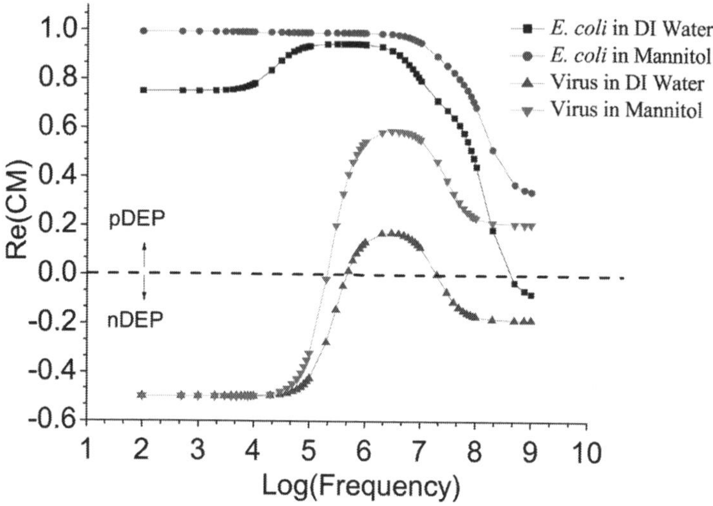

Figure 4. The calculated Clausius-Mossotti factors vs. the logarithm of the frequency of alternating current voltage for 1-μm bacterial and 100-nm viral spherical particles in pure DI water and 280 mM aqueous mannitol solution.

Figure 4 shows a plot of the value of the real part of Clausius-Mossotti factor, Re [CM], vs. the logarithm of alternating current frequency for bacterial and viral particles suspended in pure DI water and 280 mM aqueous mannitol solution. The conductivities of *E. coli* suspended solutions were measured to be 6.74 μS/cm in DI water and 0.2 μS/cm in mannitol solution. In the case of viruses, these values were slightly higher at 23.5 and 11.6 μS/cm, respectively. As seen in Figure 4, the Re [CM] for *E. coli* cell shows similar trend in both media, with positive values nearly over the full frequency range. The curve shifts slightly positive in mannitol solutions. This indicates that *E. coli* cells will experience pDEP force across the broad frequency range. In contrast, virus particles experience nDEP at low and high frequencies but pDEP in the moderate frequency range. Mannitol is necessary to ensure a large positive Re [CM] for pDEP capture. These estimations are very useful in designing DEP experiments to separate bioparticles from a heterogeneous mixture.

DEP Capture of Bacteria

DEP capture of bacterial cells was demonstrated with nontoxic *E. coli* strain DHα5 (18265-017, Fisher Scientific). Normally, 20 μl of grown culture was incubated in 2.0 ml of fresh media to reach a cell concentration of ~1x10^9 cells/ml. The cells were centrifuged at 5000 rpm. The collected cells were resuspended and washed three times with 1X phosphate buffer saline (PBS) to remove leftover media. Labeling *E. coli* cells was done in two steps. First, ~3x10⁹ cells/ml were incubated with FITC conjugated rabbit anti-*E. coli* Ab (AbD Serotech, NC) at 330 µg/ml for 1 hr. at room temperature (RT). The cells were then washed twice with the PBS buffer. Second, *E. coli* cells were incubated with Alexa 555 conjugated goat anti-rabbit second Ab (Invitrogen, CA) at 130 µg/ml for 1 hr. at RT. The labeled *E. coli* cells were then washed three times with PBS buffer and then with DI water. The cells were finally resuspended in DI water to a concentration of ~1x10⁹ cells/ml for DEP experiments.

The DEP device was placed under an upright fluorescence optical microscope (Axioskop II, Carl Zeiss, NY, USA) using 50X objective lens. The nanoelectrode array employed in this study had an exposed carbon nanofiber (CNF)density of ~2x10⁷ CNFs/cm², with an average spacing of ~2.0 µm. A filter set with excitation wavelength of 540-552 nm and emission wavelength of 567-647 nm (filter set 20HE, Carl Zeiss, NY,USA) was used in connection with an Axio Cam MRm digital camera to record fluorescence videos at an exposure time of 0.4 s using multi-dimensional acquisition mode in the Axio-vision 4.7.1 release software (Carl Zeiss MicroImaging, Inc). One of the major concerns while performing bacterial capture experiments is non-specific adsorption of bacteria. To overcome this issue, the microfluidic channel was injected with 1.0 ml Bovine serum albumin (BSA) solution (2% w/v) at a flow rate of 0.2 µl/min before performing DEP experiments. This step helped to passivate the surface of SU-8 photoresist and SiO₂ in the fluidic channel and substantially reduced the non-specific adsorption of *E. coli* cells. The channel was then rinsed with 2 ml DI water at a flow rate of 5.0 µl/min.

DEP experiments were carried out by injecting labeled *E. coli* suspension into the passivated channel at a specified flow velocity. When no DEP force was applied on the bacteria, they flow with the media due to the hydrodynamic drag force. Figure 5a shows *E. coli* cells flowing through the channel at a linear flow velocity of 1.6 mm/sec. Each *E. coli* cell appears as a stretched line corresponding to the travel distance over the finite exposure time. This was used to calculate the linear flow velocity at the nanoelectrode array surface. When an alternating current AC voltage of 10 V_{pp} at 100 kHz frequency is applied between the nanoelectrode array and the macro-ITO electrode, pDEP force is generated, which pulled *E. coli* cells toward the carbon nanofiber nanoelectrode array at the bottom of the microfluidic channel and trapped them at isolated carbon nanofiber tips. Once trapped, the stretched *E. coli* moving lines changed to bright fixed spots as shown in Figures 5b. Varying the parameters of alternating current voltage has concluded that the optimum DEP capture conditions for *E. coli* cells are 100 kHz, though any frequency in the range of 50 kHz to 1 MHz can generate DEP effects. Higher V_{pp} gives stronger DEP force and normally reliable results can be obtained with ≥ 9 V_{pp}.

Quantitatively, the number of captured *E. coli* cells was measured by counting the fixed bright spots using the auto measure module of the Carl Zeiss software. Figure 5c shows the capture kinetics at three different flow velocities. No *E. coli* cell was trapped when the alternating current voltage was not applied. But as soon as the 10 V_{pp} alternating current voltage was applied, the count of bright spots instantaneously jumped up, indicating DEP capture of *E. coli* cells at the nanoelectrode array. The count remained almost the same during the time the voltage was applied, but the value clearly depended on the flow velocity, dropping from ~300 at 0.11 mm/sec to ~70 at 1.6 mm/sec. The flow velocity at 1.6 mm/sec matched the highest flow velocity (0.04 - 2.0mm/sec) used in interdigitated micro-DEP device (*8, 19*). The number of saturated bright spots is further plot vs. the flow velocity in Figure 5d. A monotonically decreasing curve was obtained as the flow velocity was increased. This indicates that the pDEP force was sufficient to attract many cells (mostly those close to the nanoelectrode array surface) toward the nanoelectrode array. Once they were at the carbon nanofiber tip, the lateral DEP force became larger than the hydrodynamic drag force along the flow direction and thus the *E. coli* cell was captured at the fixed tips of the nanoelectrode array.

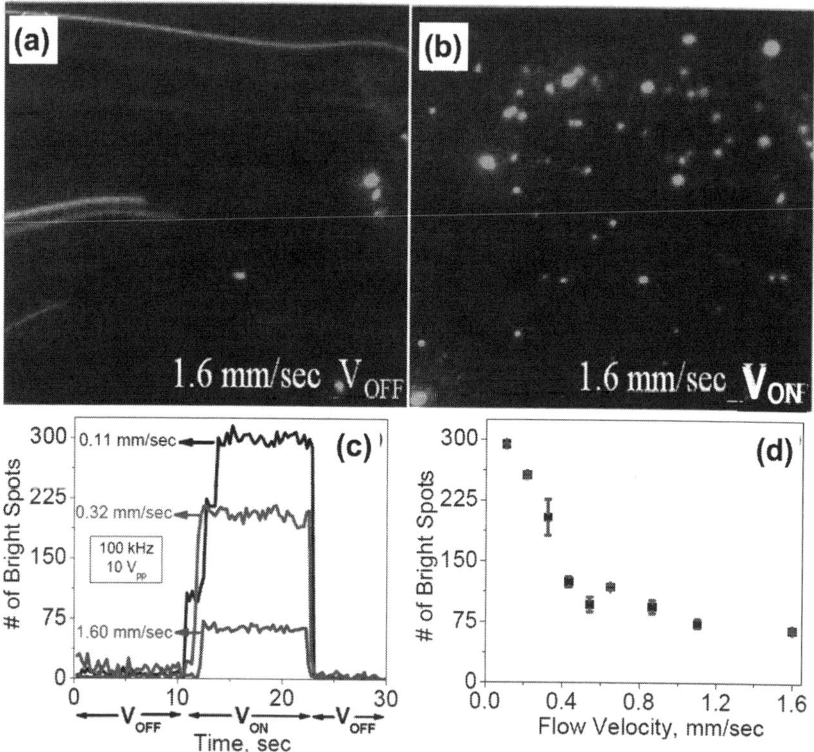

Figure 5. Snapshot images of the carbon nanofiber nanoelectrode array as E. coli cells flowing through at 1.6 mm/sec, with the 100 kHz 10 V_{pp} AC bias turned (a) off and (b) on, respectively. The size of the frame is about 100 x 100 μm^2. (c) Number of captured E. coli cells (correlated to the number of bright spots in (b)) with respect to time at three flow velocities of 0.11, 0.3, and 1.6 mm/sec respectively. (d) The correlation of the number of bright spots (representing captured E. coli cells) and the flow velocity. (Reproduced with permission from reference (15). Copyright 2011 Wiley-Blackwell.)

DEP Capture of Virus Particles

Bacteriophage T4r (Carolina Biological Supply Company, Burlington, NC) and and T1 (ATCC, Manassas, VA) were chosen to demonstrate the capability to capture small virus particles using the nano-DEP device. The bacteriophage was co-cultured with *E. coli* as described in a previous report (*16*). The final solution was filtered with a 0.2 μm filter (Fisher, PA) to remove live bacteria or bacterial debris. Double layer agar method was used to determine the titer of the phages (*16*). Washing and labeling were carried out by centrifugation using Amicon® Ultra 0.5 centrifugal filter devices (Millipore, Billerica, MA). Labeling of phages was carried out using a 500X working solution of SYBR® Green I Nucleic Acid Gel Stain (Lonza, Rockland, ME) in Tris EDTA (TE) buffer. The labeled and washed phages were dispensed in double DI water. From the discussion on Clausius-Mossotti factors, addition of 280 mM mannitol was necessary to enhance the efficiency of pDEP capture of virus particles (*10, 26*). The final concentration of the phages for the normal DEP experiments was ~5x10^9 pfu·ml^{-1} except in some concentration-dependent experiments.

The DEP experimental setup is similar to that for bacteria capture in the previous section. The flow of the labeled *Bacteriophage* T4r was first examined at low magnification (with a 10X objective lens) as shown in Figure 1b. The stretched lines represent the movement of individual bacteriophage particles carried by the hydrodynamic flow of the media during the exposure time. The figure gives a good indication of the distribution of the particles as they entered from the narrow straight channel (500 μm in width) into the larger circular microchamber (2 mm in diameter) and only a fraction of the bacteriophage particles passed above the active nanoelectrode array area. Fluorescence videos over the 200 μm x 200 μm active nanoelectrode array NEA area was recorded with 50X objective lens as the labeled virus flowing through. The capture efficiency for bacteriophage was much higher than that of bacteria causing them to overlap after capture. Hence it was difficult to distinguish individual bacteriophages in many experiments. To overcome this, the integrated fluorescence intensity over the 200 μm x 200 μm active nanoelectrode array area was used in place of counts of isolated bright spots (except in some later experiments) to quantify the capture efficiency during the kinetic DEP process (*16*).

As shown in Figure 6, the integrated fluorescence intensity rose to a saturated level in less than 10 secs as a 10 V$_{pp}$ alternating current voltage was applied on the DEP device while flowing 5x10^9 pfu/ml *Bacteriophage* T4r solution through the channel at the flow velocity υ varying from 0.085 to 3.06 mm/sec. Interestingly, a plot of the captured amount vs. the flow velocity showed a maximum at 0.73 mm/sec (see Figure 6b). This is in drastic contrast with the monotonic dependence on flow velocity in bacterial capture (see Figure 5d) (*15*).

At υ < 0.73 mm/sec, isolated bright spots were seen (Figure 6c), similar to bacteria capture. However, At υ ≥ 0.73 mm/sec, the snapshot images showed fractal-like lightening patterns (Figures 6d&6e). These patterns are called Lichtenberg figures and are commonly generated under conditions where a high electric field is produced at a sharp electrode surrounded by charged or polarizable materials as is the case here. The generation of such patterns requires a relatively high concentration of polarizable particles and so it was seen only when the particle flux was sufficiently high. Even though similar "pearl-chain-like" pattern was observed by Suehiro et al. in DEP trapping of *E. coli* cells between interdigitated microelectrodes (*27*), our previous DEP studies showed that only isolated *E. coli* cells were captured at the nanoelectrode array (*14, 15*).

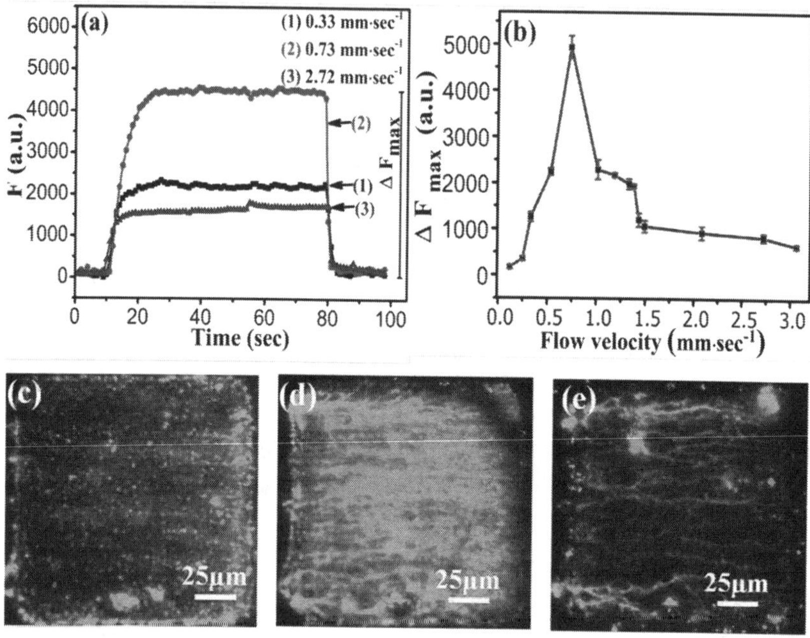

Figure 6. The effects of flow velocity on DEP capture at 10 kHz and 10 V$_{pp}$ with 5x10^9 pfu/ml Bacteriophage T4r in 280 mM aqueous mannitol solution. (a) The kinetic curves of the integrated fluorescence intensity over the 200 µm x 200 µm NEA area as the alternating current voltage is turned on (at ~10 secs) and off (at ~ 80 secs). (b) The quantity of DEP capture, represented by the maximum increase of the integrated fluorescence intensity (ΔF $_{max}$), versus the flow velocity, which is peaked at 0.73 mm/sec. (c)- (e) are the representative snapshots from the videos at ~80 sec (just before the AC voltage was turned off) at flow velocity of 0.33, 0.73 and 2.72 mm/sec, respectively. (Reproduced with permission from reference (16). Copyright 2013 Wiley-Blackwell.)

In contrast to DEP capture of *E. coli* cells (*14, 15*), DEP capture of *Bacteriophage* T4r requires lower frequency (from ~100 Hz to ~100 kHz with the maximum performance at 10 kHz) (*16*). Considering that mannitol had to be added to adjust the permittivity and conductivity of the media (i.e. water) (*10, 26*), it is clear that the small virus particles (*Bacteriophage* T4r, 80-200 nm in size) have very different CM factor comparing to much larger bacterial cells (*E. coli*, ~1-2 µm in size). It seems that pDEP capture of virus particles was much more efficient than that of bacteria.

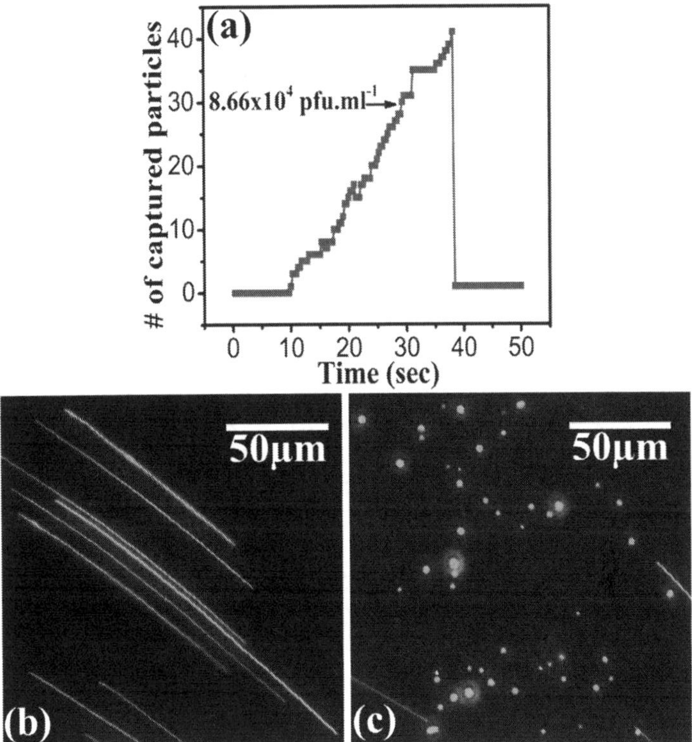

Figure 7. DEP capture of virus particles at low concentrations. (a) The kinetic curve of DEP capture with 8.9x10⁴ pfu.ml¹ of Bacteriophage T1 in 280 mM aqueous mannitol solution flowing through the 200 µm x 200 µm active nanoelectrode array area at 0.87 mm/sec. (b) and (c) are representative snapshots from the video with the alternating current voltage off and on, respectively. The flowing virus particles in solution are seen as streaks while the captured ones are focused bright spots. The alternating current AC voltage was fixed at 10 kHz and 10 V_{pp} in this set of experiments. (Reproduced with permission from reference (16). Copyright 2013 Wiley-Blackwell.)

Most interestingly, the DEP kinetics dramatically changed when a very dilute solution of *Bacteriophage* T1 (8.7x10⁴ pfu/ml) was passed through the nano-DEP device. At such a low concentration, the DEP capture was found to be fully limited by mass transport. The capture of individual virus particles onto the nanoelectrode array was observed as random single events and the number of captured virus particles linearly increased with time over the ~30 secs timeframe (as shown in Figure 7a). Similar results were obtained at various flow velocities from 0.59 to 0.94 mm/sec. The representative snapshot in Figure 7b shows the streaks over the active nanoelectrode array area when the alternating current voltage was off, indicating the distance that the phage particles moved at 0.87 mm/sec flow velocity in the 0.2 sec exposure time. The longest streaks represent the particles which were within the focus depth (~0.6 µm (15)) in the whole exposure time and thus can be used to calculate the flow velocity near the nanoelectrode arraysurface. Similar to earlier sections, the captured virus particles show focused isolated spots in the snapshot image (Figure 7c). These captured virus particles can be precisely counted. From the experiment video at 0.87 mm/sec flow velocity, it was observed that 40 out of 67 particles were captured giving a capture efficiency ~60%. This is very encouraging and can be further enhanced by fabricating elongated active nanoelectrode array area across the full width of a straight microfluidic channel so that all virus particles are forced to pass through the zone with strong electric field. With proper design, the nanoelectrode array based DEP device may capture virus particles at concentrations potentially approaching 1-10 cfu/ml. By coupling with highly sensitive detection methods (such as surface enhanced Raman spectroscopy), it is very promising to develop an ultrasensitive portable microfluidic system for rapid viral pathogen detection.

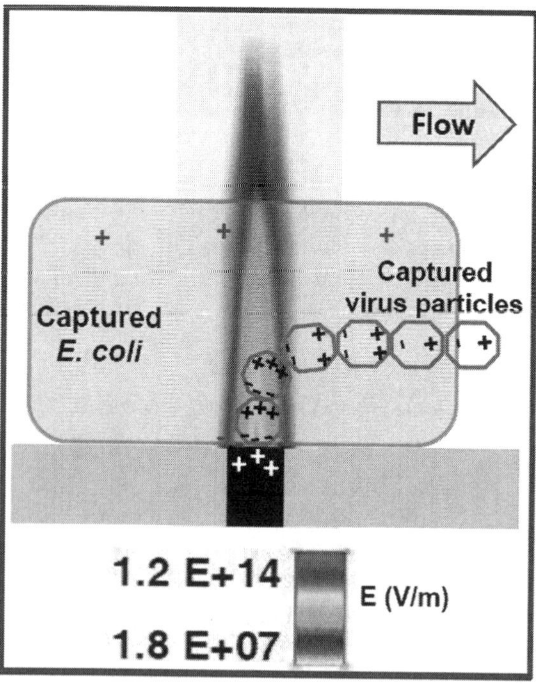

Figure 8. The schematic of the relative polarization effect on a virus particle (~150 nm) and an E. coli cell (~1-2 µm) based on the finite element method simulated electric field profile at the tip of a nanoelectrode (200 nm in dia.) which is embedded in SiO₂ matrix.

Discussion

From the modeling, simulation, and experimental studies, it is clear that there are differences between the DEP effects on bacteria and virus particles. The size of the particle relative to the spatial distribution of the electrical field strength at the nanoelectrode tip could be the main reason, as schematically illustrated in Figure 8. It is noted that, in finite element method simulation, the particles were assumed small compared to the scale of field non-uniformity and the inter-particle interaction was neglected. But, in fact, the electric field drops rapidly at the nanoelectrode (see Figures 2b and 8), the particles cannot be treated as points and the electric field variation across the particle has to be taken into account. The strong polarization to smaller-sized virus particles is expected to greatly increase the inter-particle interaction. As a result, the captured virus particles may behave as extended nanoelectrode tips to capture other particles, forming chain-like lighting patterns shown in Figures 6d and 6e. In contrast, a relatively large bacterial cell screened the electric field at the nanoelectrode tip and limited capturing only one cell per nanoelectrode site. These factors may also account for the discrepancies in the optimum AC frequencies for pDEP between calculated Clausius-Mossotti factors and experimental results. Refined finite element method simulation and Clausius-Mossotti model is needed to develop precise understanding of the DEP processes at nanoscale.

Conclusion

A nano-DEP device based on a "points-and-lid" configuration of a Nanoelectrode array against a macroscopic Indium tin oxide counter electrode in a microfluidic channel has been designed, fabricated, and tested for pDEP capture of bacterial cells and viral particles. Reversible capture of both types of microbial particles was observed at high flow velocities. The experimental results are in general agreement with finite element method simulation and simple calculation of Clausius-Mossotti factors. In the case that high densities of viral particles were captured by pDEP, the highly polarized particles formed chained structures showing interesting lighting patterns. These studies revealed interesting interplay between the highly focused electric field at the Nanoelectrode with bioparticles of comparable sizes. It is promising to develop such nano-DEP devices as an on-chip sample preparation module in a portable microfluidic system for rapid detection of microbes.

References

1. Fung, D. Y. C. *Compr. Rev. Food Sci. Food Saf.* **2002**, *1*, 3–22.
2. Lazcka, O.; Del Campo, F. J.; Munoz, F. X. *Biosens. Bioelectron.* **2007**, *22*, 1205–1217.
3. Cabrera, C. R.; Yager, P. *Electrophoresis* **2001**, *22*, 355–362.
4. Hughes, M. P. *Electrophoresis* **2002**, *23*, 2569–2582.
5. Pohl, H. A. *J. Electrochem. Soc.* **1960**, *107*, 386–390.

6. Pohl, H. A. Dielectrophoresis: The Behavior of Neutral. *Matter in Non-uniform Electric Fields*; Cambridge University Press: New York, 1978.

7. Voldman, J. *Annu. Rev. Biomed. Eng.* **2006**, *8*, 425–454.

8. Li, H.; Zheng, Y.; Akin, D.; Bashir, R. *J. Microelectromech. Syst.* **2005**, *14*, 103–112.

9. Markx, G. H.; Dyda, P. A.; Pethig, R. *J. Biotechnol.* **1996**, *51*, 175–180.

10. Hughes, M. P.; Morgan, H.; Rixon, F. J. *Eur. Biophys. J.* **2001**, *30*, 268–72.

11. Morgan, H.; Green, N. G. *J. Electrost.* **1997**, *42*, 279–293.

12. Ermolina, I.; Milner, J.; Morgan, H. *Electrophoresis* **2006**, *27*, 3939–48.

13. Akin, D.; Li, H.; Bashir, R. *Nano Lett.* **2003**, *4*, 257–259.

14. Arumugam, P. U.; Chen, H.; Cassell, A. M.; Li, J. *J. Phys. Chem. A* **2007**, *111*, 12772–12777.

15. Syed, L. U.; Liu, J.; Price, A. K.; Li, Y.-f.; Culbertson, C. T.; Li, J. *Electrophoresis* **2011**, *32*, 2358–2365.

16. Madiyar, F.; Lateef, U. S.; Culbertson, C. T.; Li, J. *Electrophoresis* accepted for publication.

17. Li, J.; Ng, H. T.; Cassell, A.; Fan, W.; Chen, H.; Ye, Q.; Koehne, J.; Han, J.; Meyyappan, M. *Nano Lett.* **2003**, *3*, 597–602.

18. Arumugam, P. U.; Chen, H.; Siddiqui, S.; Weinrich, J. A. P.; Jejelowo, A.; Li, J.; Meyyappan, M. *Biosens. Bioelectron.* **2009**, *24*, 2818–2824.

19. Gray, D. S.; Tan, J. L.; Voldman, J.; Chen, C. S. *Biosens. Bioelectron.* **2004**, *19*, 1765–1774.

20. Huang, Y.; Joo, S.; Duhon, M.; Heller, M.; Wallace, B.; Xu, X. *Anal. Chem.* **2002**, *74*, 3362–3371.

21. Huang, Y.; Holzel, R.; Pethig, R.; Wang, X. B. *Phys. Med. Biol.* **1992**, *37*, 1499–1517.

22. Suehiro, J.; Hamada, R.; Noutomi, D.; Shutou, M.; Hara, M. *J. Electrost.* **2003**, *57*, 157–168.

23. Asami, K.; Hanai, T.; Koizumi, N. *Biophys. J.* **1980**, *31*, 215–228.

24. Sanchis, A.; Brown, A. P.; Sancho, M.; Martinez, G.; Sebastian, J. L.; Munoz, S.; Miranda, J. M. *Bioelectromagnetics* **2007**, *28*, 393–401.

25. Park, K.; Akin, D.; Bashir, R. *Biomed. Microdevices* **2007**, *9*, 877–883.

26. Hughes, M. P.; Morgan, H.; Rixon, F. J.; Burt, J. P. H.; Pethig, R. *Biochim. Biophys. Acta* **1998**, *1425*, 119–126.

27. Suehiro, J.; Yatsunami, R.; Hamada, R.; Hara, M. *J. Phys. D: Appl. Phys.* **1999**, *32*, 2814–2820.

Chapter 7

Food Toxin Detection with Atomic Force Microscope

Guojun Chen,[1] Bosoon Park,[*,2] and Bingqian Xu[3]

[1]Senior Development Scientist, Bruker Nano Surfaces Division,
112 Robin Hill Road, Santa Barbara, California 93117
[2]United State Department of Agriculture, Agricultural Research Service,
Russell Research Center, Athens, Georgia 30602
[3]The University of Georgia, Nanoscience and Engineering Center,
Athens, Georgia 30602
*E-mail: bosoon.park@ars.usda.gov.

Food safety is a very important issue for daily life. External introduced toxins or internal spoilage correlated pathogens and their metabolites are all potential sources of food toxins. To prevent unsafe food from circulating in the market, many food toxin detection techniques have been developed to detect various toxins for quality control. Although several routine techniques address the most important food safety issues, many challenges still exist in terms of sensitivity, cost and speed, especially for lethal proteinaceous toxins. Atomic force microscope (AFM) is being widely used for nanotechnology imaging or as a force measurement tool for biophysics and molecular biology. Its super-sensitivity in probing specific binding interactions between molecules is also applicable for food toxin detection. This chapter provides a brief summary about the development of AFM and different imaging operation modes from applications and characteristics. AFM detection applications for the cantilever sensing and recognition image techniques are discussed in detail for two categories of surfaces: silicon and metal surface. Furthermore, the recent progress in various aspects of AFM based sensor techniques are discussed with successful applications in food toxin detection provided as examples. The super-sensitivity and easy user accessibility make these AFM techniques good candidates for applications in the field. Challenges in the further application of AFM techniques are similar to those of the general field of biosensing, and include chemical scheme strategies, sensing modulus and substrate development. Through the mutual advances in several scientific disciplines, including materials, chemistry and molecular biology, AFM-based food toxin detection techniques will see more broad applications.

Introduction

As a daily-consumed product, food supports the normal life and progress of our society. However, in certain instances unsafe food can result in very serious consequences, such as contamination by externally introduced toxins or spoilage pathogens and their toxic metabolites. For pathogenic contamination in food, the well-established microbiological and biochemical testing procedures for detection have been widely used by the food industry effectively and several test kits for this purpose are commercially available (*1*). The common techniques adopted include direct plating, culture enrichment and PCR (polymerase chain reaction), especially real-time PCR (*2*). Although there are obvious drawbacks associated with each method, existing methods solve most problems related to pathogens with limited specificity and sensitivity. Currently, the challenging issues lie in certain proteinaceous toxins in food. These proteinaceous toxins are lethal at very low levels that go beyond the detection capabilities of current state-of-the-art techniques. In the following section, some general difficulties and challenges associated with these techniques will be exemplified.

Challenges in Food Toxin Detection and Its Current Status

Complex Physical Form and Chemical Compositions of Food Matrices

The variety of different food matrices and compositions in the food industry make sampling extremely difficult. Typically, there are only trace amounts of target analytes that could be pathogens or toxins. For example, the sampling procedure for solids or liquids are thoroughly different (*3*). Moreover, the nontoxic food matrix interferes with detection and deactivated pathogens could result in false positive results, complicating the accurate determination of toxins in food.

Labor Intensive and Cost

Most of culture based detection methods require days to weeks to enrich the suspected target analytes. These techniques are very costly and require certain analytes and specific media for successful toxin detection (*4*).

Due to these challenges associated with traditional methods, many new detection methods are under development, such as electroanalytical fluorescence, and luminescence based methods, as well as, nanotechnology based techniques including microfluidics and atomic force microscopy (AFM). Among them, AFM-based methods are an attractive choice for toxin detection, because of its high sensitivity with a single molecule level as well as cost-effectiveness.

AFM History

Four years after the invention of the scanning tunneling microscope (STM) at IBM Zurich by G. Binning and H. Rohrer in 1982, the AFM was invented by Binning et al. (*5*). Without the limitation of conductive surfaces associated with the STM, AFM opened the door to visualize the nanomaterials in liquid. Shortly after its invention, the unique capability of AFM received the attentions from biological fields. However, the applications are very limited due to the technical bottlenecks. Only two years later the micro-fabrication techniques made the batch manufacturing of AFM probes possible. This was a big step forward; because it solved the fundamental problem of the transducer in AFM system (*6*). Subsequently, the optical beam deflection (OBD) sensing was introduced to read out the deflection of the AFM probe more accurately and easily (*7*). Another big obstacle for AFM was to image biomolecules, since the constant force applied induces unwanted displacement of objects weakly attached to the surface. In 1993, the tapping mode was invented and addressed the difficulty mentioned and enabled immediate applications in many bio-materials, such as DNA, proteins, lipid bilayers and polymers (*8*). Up to this moment, AFM was a fully equipped tool for material science. However, the scanning speed limitation remains a challenging issue for image resolution and dynamic process observation for most commercial AFM systems. Continuing developments and brilliant designs from several research groups, such as Quate, Hansma and Ando, greatly contributed to the success of fast scanning AFM (FSAFM) (*9–11*). FSAFM enables the direct observation of many once unreachable biological process and generated several amazing results. This technology greatly decreases the frame time from several minutes to milliseconds, which changes AFM from slow slide show to video speed imaging. Meanwhile, several derivative tapping modes were generated, such as magnetically actuated AFM (MAC) (*12*), peak force tapping (PFT) (*13*) and others. The MAC tapping mode improved the driving efficiency of probe actuation, but it is still confined by this step. Moreover, MAC also inherits the same flaws in difficult force control and data interpretation of phase and amplitude imaging from acoustic actuated tapping. It is worthwhile to mention that PFT advanced the traditional tapping mode significantly since it doesn't need the troublesome actuation process of the AFM probe in liquid and eventually solved the ambiguity problems in data interpretations of the generated phase and amplitude images from tapping mode by completely using the repulsive force as feedback signal.

Basic Principle of AFM and Its Application

Basic Principle

The principle of AFM imaging process features a sharp tip of a nanometer in diameter that performs a laser-scan over the sample by using optical beam deflection (OBD) as a feedback signal, as shown in Figure 1. AFM enables the non-destructive imaging of soft biomaterials in their close-to-natural environment with high resolution (~nm), offering a way to observe single molecule without the need for fixation and staining. The high resolution image and time-lapse image can provide direct evidences to support and reveal the interaction and structure-functionality relationship of biomolecules (*14*). In addition, the collected imaging signal can be split to distinct components to delineate special characteristics of samples, such as topography image for size and morphology (*15*), phase image for compositional mapping (*16*) and recognition image for uncovering the specific binding site (*17*). The imaging capability clearly makes AFM emerge as a powerful nanoscale tool for biological applications including food toxin detection. Therefore, it is worthwhile to retrospect its evolution and development.

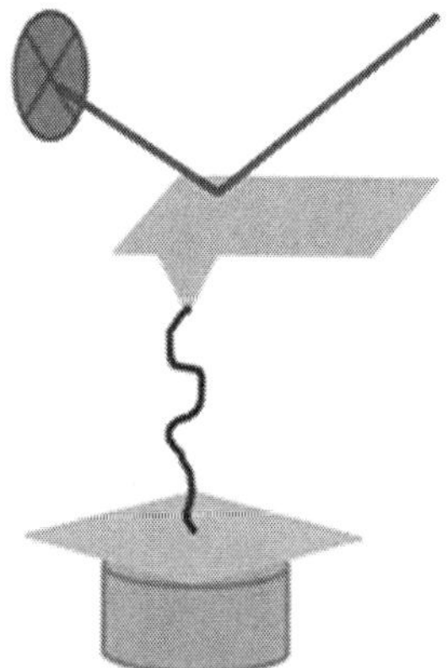

Figure 1. The schematic configuration of atomic force microscope.

Evolution of AFM Imaging

High-Resolution Topography Image

As a general rule, the stiffer and flatter the sample, the higher spatial resolution can be achieved (*18*). For isolated protein on a surface in aqueous solution, the best resolution is a few nanometers to tens of nanometers. In comparison, the real high resolution image at sub-nanometer resolution can also be acquired on 2D-crystallized (*19–21*) or densely packed protein (*22, 23*), and native membranes.

In one pioneering study (*21*), streptavidin two-dimensional (2D) crystal formed on biotin-containing planar lipid layer can be imaged by AFM with sub-nanometer resolution, and its high-affinity binding capability and good flatness of the lipid layer make it suitable as a platform for biomolecule attachment in AFM studies (*20, 24*). This platform provides an optimal attachment strategy with a controlled orientation of the immobilized molecule and less perturbation to the biological function of the sample.

The high-resolution topography images of native photosynthetic membranes from several purple bacteria species (*15, 25*) are also milestone studies in AFM high-resolution imaging. They provided detailed views and rationales of membrane protein assembly adaptation in response to environmental factors, such as light intensity. It was found that photosynthetic complexes in native membrane of *Rsp. Photometricum* could rearrange in response to different light intensities. For example, the assembly of light-harvesting 2 and core complexes showed up in high-light-adapted membrane and light-harvesting 2-only antenna domains in low-light-adapted membranes (*14, 15*).

Phase Image

The topography image offers the evidences of conformation information for biomaterials. However, it is impossible to differentiate the heterogeneous compositions of biomaterials. Amplitude-modulation atomic force microscopy (AM-AFM), one kind of tapping mode, is widely applied for composition mapping.

In brief, the mechanically or magnetically driven tip oscillation is used to probe sample variations through the interaction between tip and sample surface. With a fixed excitation frequency, the amplitude and phase lag of oscillation reflect tip-surface conservative (elastic) and dissipative interaction (inelastic), respectively. While coupling between amplitude and phase lag of the tip can be separated through operating the instrument with constant amplitude, they can further be applied for construction of topography and phase images. The correlation between phase-shift measurement and energy dissipation values can correspond the phase-lag to properties of the materials, such as stiffness, elasticity, and viscosity or surface-adhesion energy qualitatively (*26, 27*) or quantitatively (*28–30*). Therefore, the 'phase imaging' empowers AFM to map out the compositional variations of the specimen's surface through the phase lag of the probe in response to the force between tip and sample surface (*31*).

Numerous applications in the biological field demonstrated the potential of phase imaging in nano-scale characterization. In one of these applications, the phase image was applied to observe the temperature induced phase transition of a lipid bilayer *in-situ*. The solid supported dipalmitoylphosphatidylcholine (DPPC) bilayer shows a broad L_β-L_α transition (*32*). In another application, phase images were applied to uncover the detailed structures of bacterial surfaces, including flagella encapsulated by capsular extracellular polymeric substance (EPS) inside (*33*).

Chemical Force Microscope (CFM)

Although the phase image is able to distinguish compositions with different mechanical properties of sample surfaces, it is a challenge to differentiate the compositions, which have similar mechanical properties. For variant chemical components, such as self-assembled monolayers of mixtures (SAM), a chemical force microscope was developed which extends AFM imaging with specific chemical sensitivity through hydrophobic effects or hydrogen bonding to identify different chemical components. In order to take advantage of the specific chemical interaction for imaging, the AFM tip needs to be functionalized with SAM either by physically absorption (*34, 35*) or covalent chemical bonding. It has already been demonstrated that a chemically modified tip can sensitively map out the spatial distribution of specific functional groups through frictional image (*36–39*).

In these studies, the SAM in the center square region terminates with carboxylic acid-COOH groups and the SAM surrounding the square terminates with methyl groups. Topographical images sometimes fail to reveal this pattern, because the regions of two SAM have similar flatness. The tips containing different chemical groups exhibit the chemical information about the surface. The friction image collected by carboxylic acid functionalized tips displays high friction on the carboxylic acid regions and low friction on the methyl regions. In the opposite way, images recorded with methyl derived tips show a reversal in friction contrast (*36*). In addition, chemical force microscopy can obtain chemically-sensitive images in tapping mode through the relationship between phase lag and adhesion (*40*).

Recognition Image (REC)

To some degree, chemical force microscope offers AFM the capability to map the distribution of specific functional groups on the sample surface. However, improvements in resolution, sensitivity and specificity are still under development due to its low spatial resolution between 100 and 200 nm. (*37*). Extra efforts are needed to improve chemical sensitivity of samples by altering the solvent characteristics, such as composition (*41*) or pH (*42*). In addition, the intrinsic imaging mechanism based on hydrophobic effects or hydrogen bonding hinder the CFM for better specificity (*36*). Hence, most studies by CFM focus on simple modeling of the system in the proof-of-concept stage.

Recognition imaging techniques can be taken as an advanced version of CFM, which features the specific interaction between probes and target molecules as the imaging mechanism (*43*). It offers much higher resolution (usually several nanometers) and better specificity (*44*). Since this technique is based on the magnetic driven AM-AFM, the gentle intermittent contact between tip and sample makes it suitable for imaging soft biological samples, such as DNA complexes (*17, 45*) and proteins (*17, 43, 46*). In the REC technique, the probe molecule is attached on the tip surface through a polymer cross-linker. The introduction of extra long polymer cross-linker in tip surface is helpful to improve the flexibility and specificity of interaction. When the modified tip scans over the sample surface, the damp of oscillation wave in both the top and bottom part can be decoupled through special designed electronics, as shown in Figure 2. The variation in the top and bottom part of the oscillations generates topography images and recognition images accordingly. This technique has effectiveness for nanometer-scale epitope mapping of biomolecules and localizing receptor sites during biological processes.

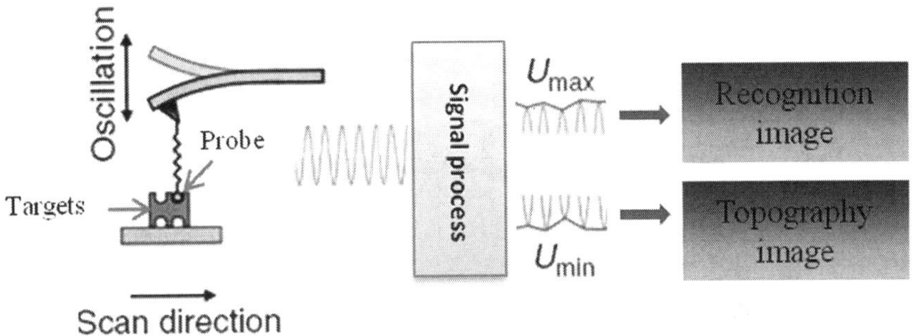

Figure 2. The cantilever oscillation signal is split into minima (U_{min}) and maxima (U_{max}) for generating topography and recognition image respectively.

After H.E. Gaub proved the concept that the specific interaction force can be applied for mapping the distribution of binding partners on samples in the streptavidin pattern study by biotinylated tip (*47*), the seminal work with formal nomenclature of 'recognition image' was performed by P. Hinterdorfer (*48*). In this work, lysozyme molecules fixed on the surface were recognized by the antibody on a scanning tip with a few nanometers lateral resolution. Later, chromatin immobilized on substrate was imaged by its antibody on the tip surface, and a series of control experiments by Bovine serum albumin (BSA) and Adenosine-5′-triphosphate (ATP) addition confirmed that the recognition images are specific (*17*). In the follow-up work, an *in-vitro* selected DNA aptamer specific for histone H4 protein acetylated at lysine 16 was used to image synthetic nucleosomal arrays with precisely controlled acetylation by the AFM recognition technique. The effect of acetylation on chromatin structure was then investigated in nanometer scale through the precise location of these post-translational modifications by recognition images (*17, 45*).

The recognition imaging technique has been proved to be a sensitive method to explore the specific interaction between biomolecules. Due to its high sensitivity, recognition imaging can be applied for a good two-tier detection approach to confirm initial presumptive screening as well as preventing false positive results. To successfully accomplish the toxin detection by AFM based recognition image techniques; there are several important aspects in terms of sample preparation.

Sample Preparation

Substrate

For AFM imaging at single molecule level, the flatness of substrate is critical. Ultraflat substrates are necessary to ensure the success of experiments. There are several choices available, including highly oriented pyrolytic graphite (HOPG), mica, single crystal p (111) silicon wafer, and ultraflat gold. HOPG and mica are two natural materials, which are cleavable to generate 2D-crystalline surfaces with sub-nanometer flatness and perfect cleanness. The other two are artificial materials with comparable flatness. Although HOPG provides perfect physical characteristics, it is relatively hard to be chemically modified for molecule attachment since highly toxic or corrosive agents, such as HF, H_2SO_4 and HNO_3, are needed. Therefore, its application as a substrate for AFM- based single biomolecule study is rarely reported (*49*). Comparatively, mica and silicon wafer can be easily functionalized for immobilization of biomolecules through silanization. However, the drawback of silanization is the difficulty to control surface properties. For example, usually derivatization through silanization deteriorates surface flatness dramatically, from initial flatness with several angstroms to nanometers or even worse. The problem can be easily solved when ultraflat gold is used as a supporter. The gold surface offers not only great flexibility to be functionalized in various ways without compromise in flatness, but also it is chemically stable (*50*). It is an excellent candidate for AFM based single molecule imaging and force studies except for its relatively high cost. On the other hand, the AFM probe is usually made of silicon or silicon nitride. It shares the similar chemical functionalization strategies as mica and silicon wafer. Sometimes, it can also be coated with a thin layer of gold. Although gold-coated AFM probes also have great functionalization versatility, tip radius is possibly enlarged as a consequence of coating, resulting in the loss of image resolution.

Chemical Functionalization

After the appropriate substrate with expected characteristics is chosen, the molecule of interest and/or its cognate receptor should be attached to one or two solid surfaces, either substrate or probe surface. Ideally, the molecule attachment should be performed through specific sites of the molecule to minimize the nonspecific adsorption, support infinite loads, and keep the mechanical or biological properties of the attached molecule (*51*). These criteria are offered by a variety of immobilization schemes ranging from physical adsorption to chemical covalent bonding. The physical adsorption is limited to certain molecules such as BSA, avidin, and lysozymes, and also triggers complications in surface functionalization as well as force data interpretation. Therefore, it is not a general method applicable to all cases and won't be discussed here. As for the chemical immobilization schemes, it can be exemplified for two categories according to the surface chemical characteristics including gold surfaces and silicon related surfaces.

Gold Surface

Gold is the most useful substrate for single molecule study since it is a chemically inert surface with ultra-flatness and easy-functionalization capability (*49*). The basic mechanism of decoration methods is the affinity between thiol related groups and the gold surface. The covalent bond between sulfur and gold is very strong and the bond breaking force was shown to be greater than 1nN, i.e., an order of magnitude higher than the non-covalent interactions between the biomolecules. Moreover, the sulfur containing molecules can form uniform assembled monolayers on gold without significant changes in the flatness of the surface. All these advantages make this scheme a superb candidate for molecule attachment in AFM based single molecule measurements.

There have already been a large number of applications reported in related fields. First, many molecules have been directly chemisorbed to bare gold surface via their endogenous thiol or disulfide groups. For example, Fab fragments of antibodies (*52*), azurin (*53*) and fibronectin (*54*) all bear a disulfide bond in their structure, which is accessible to gold surface. Besides, the bioactivities of these molecules were also maintained during attachment. The thiol group can also be introduced into biomolecules through manual synthesis. For instance, some manual synthetic peptides with cysteine moiety were chemisorbed onto gold surfaces via their carried thiols (*55*). Samples lacking accessible disulfide or thiols can be derivatized with thiol or disulfide containing tags for chemisorption to gold surfaces (*56, 57*). On the other hand, the molecules can be immobilized stepwise on formed SAM through amide or thiolester bonds as well as NTA chelating. The protein with endogenous surface lysine residues can be anchored on the SAM surface through amide coupling reactions (*58*). For molecules with available thiol groups, the maleimide residue can be applied to immobilize molecules through maleimide-thiol coupling (*59*). The well-known specific interaction of hex-histidine peptides or His_6-tagged proteins with NTA-Ni^{2+} also provided a strong and lasting attachment scheme for biomolecules to the surface (*43, 60*).

Silicon-Related Surface

Several procedures for the modification of silicon related surfaces have been well established. They can be summarized into three main categories, each one has its own pros and cons, and a mechanistic insight is essential for selecting the appropriate one for its specific application.

Methods based on organo-silanes are the most versatile with easy operation but require high quality reagents. Various functional groups including NH_2, CH_3, SH, alkene, and halogen can be introduced (*61–65*). In different research groups, various operation conditions were used for silanization. Some relied on vapor deposition, others operated in solution, either under rigorously dry conditions or in neat water as solvent, which gives the misleading impression that 'anything goes'. The mechanism can be helpful for understanding these differences. The silanol groups on surface are the key to organosilane-based decoration. In quartz, each silicon atom carries one silanol group. Less regular surfaces contain germinal silanols (two OH groups per Si atom) such as glass and silica gel. There is a layer of silicon dioxide covered on silicon nitride and silicon surface. For mica, water plasma is essential to introduce silanol groups on its surface by removing the K^+ ion. Once there is silanol groups available on the surface, the organosilane can form the covalent siloxane bridges with silanol groups through hydrolysis and condensation (*49*). Base catalysis is the initiator for hydrolysis of the methoxy/ethoxy groups in organosilane. It was found that the organosilane with monomethoxy/ethoxy was not active enough to perform the surface modification (*66, 67*) and the di- (*68, 69*) and tri-methoxy/ethoxy (*70, 71*) forms were capable. The dimethoxy/ethoxy form was the best choice to provide a defined monolayer on silicon (*69*). Comparatively, the trimethoxy/ethoxy can form a monolayer or multilayers on surfaces, which is dependent on the relative humidity. For example, at 7% relative humidity the amount of adsorbed water was sufficient to form a stable monolayer of aminopropyl triethoxysilane from the gas phase, whereas above 25% relative humidity resulted in a phase transition yielded of a double layer of aminopropyl triethoxysilane (*66*). Thus, the surface silanol groups, traces of adsorbed water, base catalysis, and specific organosilane (tri- or di-) were requisites for successful and quality surface modification.

The ethanolamine based method was developed in 1994 (*72*), when it was found that 100 nm long tentacles are possibly formed on the AFM tip during modification by organosilane (*73*). This method also makes use of the silanol groups on surface. In comparison, it has two distinct drawbacks. The first one is that its chemical mechanism is still vague. The other one is the discrepancy between expected and observed unbinding length was observed in force spectroscopy. On the other hand, hydrogen-terminated groups can be introduced on the silicon surface through treatment by 2% HF in water, which can be utilized to covalently link with α,ω-oligo(ethylene glycol) alkene solution in mesitylene by refluxing under nitrogen (*74*). Subsequently, eletrooxidation was applied to selectively oxidize the chain termini into carboxyl groups. The carboxyl groups can be activated with EDC/NHS to immobilize the biomolecule with amine groups. This method is not widely applied since it deals with very toxic chemicals and has strict operational conditions.

Therefore, the organosilane-based method is preferred method for molecule immobilization on solid surfaces for the AFM force spectroscopy and recognition imaging applications (*75*).

Early Exploration of AFM Techniques in Food Toxin Detection

To be a highly sensitive biological sensor, the AFM probe should be chemically functionalized with the sensing modulus on its surface to specifically detect its cognate toxin in solution or on the substrate through directly using the deflection signal of probe or image processing. The sensing modulus, chemical scheme for immobilization, and output signals are three vital determinants for the sensor performance.

Sensing Modulus

Briefly, the AFM apparatus can be understood as a force transducer. To functionalize AFM as a biological sensor, the appropriate sensing modulus for various target toxins should be selected first. Currently, the most widely used sensing moduli are antibodies which have strong interaction with toxin as well as high specificity (*76*). However, the antibodies are easy to be deactivated either by the external environment or by the complex components in the real food extraction sample. Because of these challenges, only the mimic samples were investigated in most studies (*76–79*). Moreover, the production of specific antibodies requires long periods of time with high cost, and each toxin requires its own specific antibody. In the cases where there are several target toxins to be detected, time and antibody costs can increase dramatically. To address easy degradation and difficulties in large-scale production of antibody sensing modulus, DNA/RNA aptamers are good candidates for sensing purpose since aptamers are relatively stable and can be chemically synthesized in large scale. Another virtue of aptamers is that they can be general or specific to the whole category of target toxins through simple modification in a DNA/RNA sequence. Several selected aptamers have been successfully applied for the detection of hepatitis C virus helicase (*76*), flu-virus nucleoprotein (*80*) and microbial pathogens (*78*). As an emerging technique, numerous efforts are continuously needed to select better quality aptamers against biotoxins to improve affinities and specificities for reliable and accurate biosensing applications. However, aptamer-based sensors are on the horizon for next generation biosensing methods and their incomparable advantages have the interest of current researchers.

Chemical Scheme for Immobilization

Certain chemical schemes are needed to immobilize selected antibodies or aptamers on the interface as a sensing modulus. As mentioned previously, the ideal chemical scheme should keep the biological activity intact and make firm conjugation to minimize nonspecific interaction. The firm connection and minimization of nonspecific interactions are helpful for decreasing background noise and maintaining the sensitivity of intact biological activity. Currently, in most situations, the matured amide chemistry is adopted for bioconjugation for amino group of lysine residues on the antibody surface. Normally, 1-ethyl-3-(3-dimethylaminopropyl)carbodiimide hydrochloride (EDC) and sulfo-N-hydroxysuccinimide (NHS) are used to activate the carboxylic acid group to improve the efficiency of conjugation (*76*). However, this extra chemistry step brings in experimental ambiguousness since the activation procedure is complex and requires technical expertise. Therefore, the pre-synthesized NHS ester form of carboxylic acid is directly used for bioconjugation as an advanced version (*49*). Although the amide chemistry provides firm covalent bonding, it could adversely affect the biological activity of antibody because the random reaction on lysine residues could block the binding site of antibody. On the other hand, the surface thiol group from cysteine residues can be applied to make conjugation with maleimide group (*59*). However, similar issues could occur for this scheme.

With the advent of 'click chemistry', this problem was mostly resolved. There are no azide group in natural biological system, the man-made amino acid residue with azide sidechain can be purposely and accurately expressed in the specific position of protein (*81*). Through copper free bioorthogonal click chemistry, the sensing modulus can be immobilized on the solid surface with firm connection and controlled reaction location without deteriorating biological activity. There is potential to further improve immobilization conditions as more sophisticated chemical schemes are developed in the bioconjugation research field (*82, 83*).

Output Signal

In the AFM system, all interactions or bindings between sensing moduli and food toxins are reported as the deflection in signal changes of the AFM probe. This modification in deflection signals can be further translated into different forms of output signals such as resonance frequency shifts or special image processing, which correspond to the cantilever sensor and REC detection. The accomplishments and challenges associated with each technique are discussed as follows:

Cantilever Sensor

The physics behind the cantilever sensor are straightforward. Basically, the externally introduced mass of captured food toxins on a soft cantilever can induce the change of cantilever deflection as well as resonance frequency, which can be monitored by an optical laser projecting on the cantilever (*84, 85*). By calibrating with a standard sample, the method can also characterize the amount of toxin quantitatively from food matrix assuming the toxin in the sample tested is much less than the sensing modulus on cantilever surface. The procedure of this method is schematically described in Figure 3. The sensing moduli are chemically attached on the cantilever surface. After the probe is immersed in the test samples, the change of signals is monitored. Initially, this technique was mainly applied for clinical purposes such as detection of botulinum toxins (*79*). Recently, similar technique was introduced for food toxin and pathogen detection including large bacteria (*76*), toxic proteins (*86, 87*) and small chemicals molecules (*88*). The samples containing *E. coli* and *Vibrio* cholera with a low concentration ranging from $1*10^3$ CFU/mL to $1*10^5$ CFU/mL could be detected with an antibody functionalized cantilever (*76, 89*). *Staphylococcal* enterotoxin B, one of enteric proteineous toxins produced by *Staphylococcus* aureus, in buffer solution can be detected at extremely low concentration from picogram per milliliter (pg/mL) level to femtogram per milliliter (fg/mL) using a polyclonal antibody as a sensing modulus (*86, 87*). It was also demonstrated that cantilever sensor can effectively detect small chemical toxins such as mycotoxin in buffer solution at low concentration from 3 ng/mL to 6 ng/mL by antibodies as the sensing modulus (*90*). Ultrasensitivity has been repeatly addressed in these reports. However, only simple mimic samples either in water or buffer solutions were tested in most of these detection reports and antibodies were the only choice as a sensing modulus in these works. As discussed previously, however, unstable antibody sensing modulus limits the application of this technique in real samples such as food matrix. To the best knowledge of the authors, there are few reports to demonstrate successful detection of *Staphylococcal* enterotoxin B in food matrix such as milk and apple juice by using an antibody. The deficiency of antibody has intrigued people to seek better sensing modulus. Recently, it was reported that a small peptide was applied for detection of trimethylamine as a sensing modulus (*88*). Additionally, DNA aptamers attract much attention from the research community and are expected to replace antibodies in the future. Some early explorations have been reported in the field of toxin detection with other techniques (*78*). It is expected that more efforts need to be contributed to new sensing modulus such as DNA aptamers and small peptide ligands to detect toxins in complex food matrices.

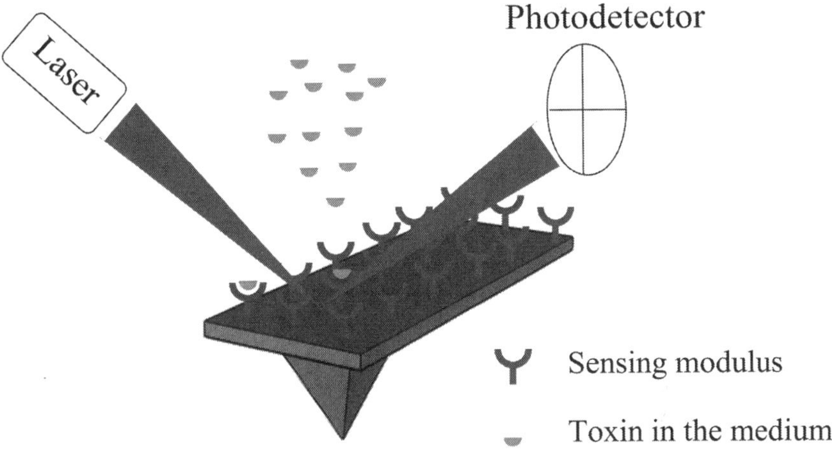

Figure 3. The schematic of cantilever sensor.

REC Image

As discussed earlier, the weak interaction force between the cognate pair on the probe and substrate can be translated into a recognition image. Normally, the black spots on the recognition image indicate the location of the toxin molecules on the surface where relatively stronger interactions with the sensing moduli on probe surface happen. Further, by counting the 'black spots' on the recognition image, the concentration of toxin can be estimated semi-quantitatively. This technique has been widely applied for investigating the interaction force and epitopes between biomolecules such as proteins, DNA and lipid bilayers in biophysics and molecular biology fields (*43*). However, applications in toxin detection have recently started and only a couple of reports of pure toxin samples were investigated (*77, 91, 92*). In that work, the ricin toxin was first immobilized on gold surface through mature amide chemistry, shown in Figure 4. Ricin toxin, which is a byproduct of castor oil production, was discovered in the seeds of the castor bean plant, *Ricinis communis*. It is more lethal than rattlesnake venom by weight and there is no antidote for ricin toxin once introduced above the lethal dosage. Subsequently, the AFM probe is functionalized with anti-ricin antibodies through the similar chemistry scheme as shown in Figure 5 and then scans over the substrate with ricin attached for imaging. By comparing the topographical and recognition images, the features belonging to a ricin molecule can be easily identified. Each bright spot in the topographic image, if it represents a ricin molecule, will result in a dark spot in the recognition image at the same location as seen in Figure 6. Based on the results obtained from this experiment, detection with the AFM REC method can easily achieve femto- or sub femto-level detection limits. In other studies, DNA aptamers, as the sensing modulus on the probe, were also investigated for the ricin toxin detection (*92*). Unsurprisingly, excellent sensitivity, usually fg/mL, can be easily achieved by this technique due to the strong imaging capability of AFM. However, all these experiments are based on the pure samples, not toxin in food matrix. Components in food matrix can contaminate the imaging substrate through unspecific adsorption. In this case, the recognition process during imaging will inevitably be disturbed by these unwanted contaminants, which may produce 'false positive' results. Therefore, sample preparation is a big challenge for this technique. A substrate with antifouling properties is expected to address the situation through preventing the unspecific adsorption on substrate from food matrix, which is possible to achieve by special physical and chemical treatments (*93, 94*).

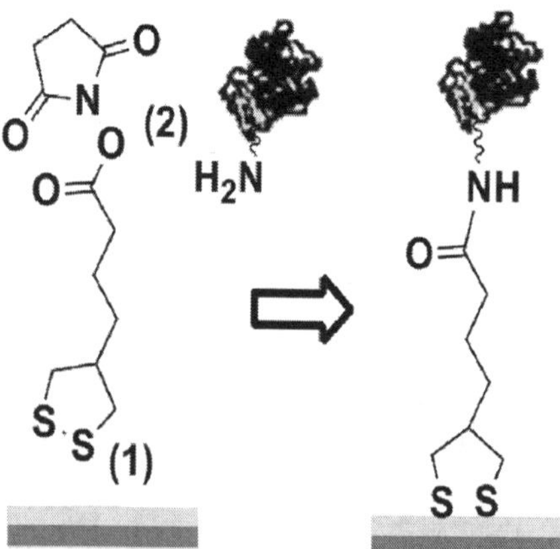

Figure 4. Schematic representation of ricin immobilization on the gold surface. (1) active ester attached to the gold coated mica surface by the thioctic acid moiety; (2) ricin binds to active ester by forming amide bond with amine groups of the protein.

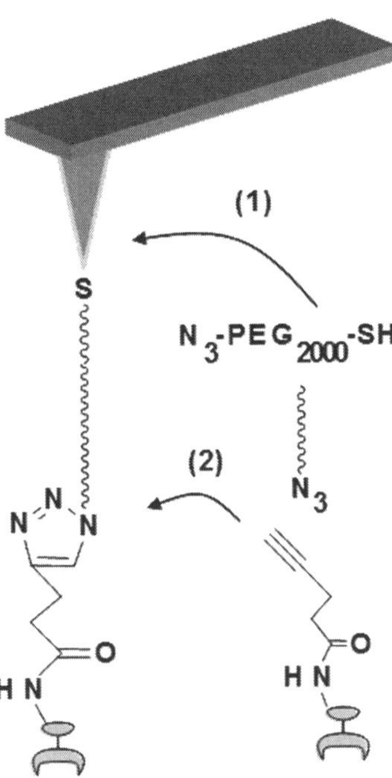

Figure 5. Schematic of AFM probe modification. (1) The thiol moiety of PEG2000 formed a SAM on the gold-coated tip; (2) the anti-ricin Ab with an alkyne group attached to the tip through click chemistry.

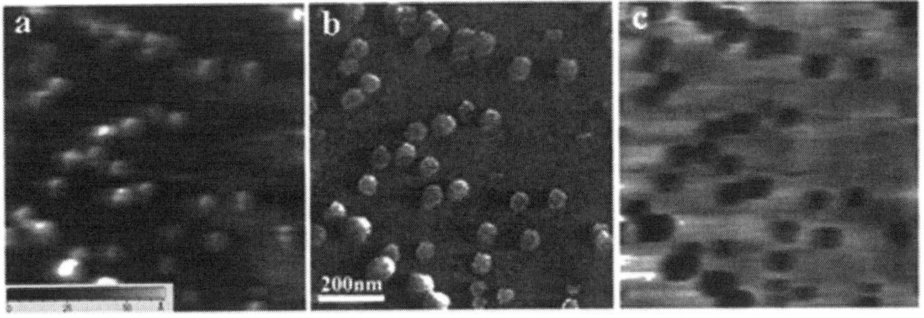

Figure 6. The detection results through AFM imaging. (a) Topographical image for 24 fg/mL; (b) Corresponding amplitude image; (c) Corresponding recognition image.

Summary

Although applications of food toxin detection based on AFM still faces many difficulties, the super sensitivity and easy user accessibility of these techniques are very attractive characteristics to develop new detection methods for food toxins. The results from these initial studies are very encouraging. However, it is necessary for more research to be carried out with real samples in food matrix, since real samples can reveal limitations needed for practical application of this technology in the field. Food toxin detection with AFM can benefit from advances in basic science in the areas of new chemistry schemes, more powerful sensing modulus, better antibodies, and improved functionalized substrate. It is strongly believed that AFM based food toxin detection methodologies will become important complementary techniques for current culture-based detection methods in the food industry and benefit everybody's healthy life.

References

1. Brett, M. M. *J. Appl. Microbiol.* **1998**, *84*, 110S–118S.
2. Watson, D. H.; Lindsay, D. G. *J. Sci. Food Agric.* **1982**, *33*, 59–67.
3. Ge, B.; Meng, J. *J. Lab. Autom.* **2009**, *14*, 235–241.
4. Baeumner, A. *Anal. Bioanal. Chem.* **2008**, *391*, 449–450.
5. Binnig, G.; Quate, C. F.; Gerber, C. *Phys. Rev. Lett.* **1986**, *56*, 930–933.
6. Binnig, G.; Gerber, C.; Stoll, E.; Albrecht, T. R.; Quate, C. F. *EPL (Europhysics Letters)* **1987**, *3*, 1281.
7. Meyer, G.; Amer, N. M. *Appl. Phys. Lett.* **1988**, *53*, 1045–1047.
8. Zhong, Q.; Inniss, D.; Kjoller, K.; Elings, V. B. *Surf. Sci.* **1993**, *290*, L688–L692.
9. Manalis, S. R.; Minnc, S. C.; Quate, C. F. *Appl. Phys. Lett.* **1996**, *68*, 871–873.
10. Walters, D. A.; Cleveland, J. P.; Thomson, N. H.; Hansma, P. K.; Wendman, M. A.; Gurley, G.; Elings, V. *Rev. Sci. Instrum.* **1996**, *67*, 3583–3590.
11. Ando, T.; Kodera, N.; Takai, E.; Maruyama, D.; Saito, K.; Toda, A. *Proc. Natl. Acad. Sci. U.S.A.* **2001**, *98*, 12468–12472.
12. Han, W.; Lindsay, S. M.; Jing, T. *Appl. Phys. Lett.* **1996**, *69*, 4111–4113.
13. Shi, J.; Hu, Y.; Hu, S.; Ma, J.; Su, C. *Method and Apparatus of Using Peak Force Tapping Mode to Measure Physical Properties of a Sample*; Patent WO/2012/078415, 2011.
14. Scheuring, S.; Boudier, T.; Sturgis, J. N. *J. Struct. Biol.* **2007**, *159*, 268–276.
15. Scheuring, S.; Sturgis, J. N. *Science* **2005**, *309*, 484–487.
16. Tamayo, J.; Garcia, R. *Appl. Phys. Lett.* **1997**, *71*, 2394–2396.
17. Stroh, C.; Wang, H.; Bash, R.; Ashcroft, B.; Nelson, J.; Gruber, H.; Lohr, D.; Lindsay, S. M.; Hinterdorfer, P.; Quate, C. F. *Proc. Natl. Acad. Sci. U.S.A.* **2004**, *101*, 12503–12507.
18. Alessandrini, A.; Facci, P. *Meas. Sci. Technol.* **2005**, *16*, R65–R92.
19. Schabert, F.; Henn, C.; Engel, A. *Science* **1995**, *268*, 92–94.

20. Yamamoto, D.; Nagura, N.; Omote, S.; Taniguchi, M.; Ando, T. *Biophys. J.* **2009**, *97*, 2358–2367.

21. Scheuring, S.; MÜLler, D. J.; Ringler, P.; Heymann, J. B.; Engel, A. *J. Microsc.* **1999**, *193*, 28–35.

22. Seelert, H.; Poetsch, A.; Dencher, N. A.; Engel, A.; Stahlberg, H.; Muller, D. J. *Nature* **2000**, *405*, 418–419.

23. Muller, D.; Baumeister, W.; Engel, A. *J. Bacteriol.* **1996**, *178*, 3025–3030.

24. Darst, S. A.; Ahlers, M.; Meller, P. H.; Kubalek, E. W.; Blankenburg, R.; Ribi, H. O.; Ringsdorf, H.; Kornberg, R. D. *Biophys. J.* **1991**, *59*, 387–396.

25. Gonçalves, R. P.; Bernadac, A.; Sturgis, J. N.; Scheuring, S. *J. Struct. Biol.* **2005**, *152*, 221–228.

26. Tamayo, J.; Garcia, R. *Appl. Phys. Lett.* **1997**, *71*, 2394–2396.

27. Magonov, S. N.; Elings, V.; Papkov, V. S. *Polymer* **1997**, *38*, 297–307.

28. Garcia, R.; oacute; mez, C. J.; Martinez, N. F.; Patil, S.; Dietz, C.; Magerle, R. *Phys. Rev. Lett.* **2006**, *97*, 016103.

29. Martínez, N. F.; García, R. *Nanotechnology* **2006**, *17*, S167.

30. Xu, W.; Wood-Adams, P. M.; Robertson, C. G. *Polymer* **2006**, *47*, 4798–4810.

31. Garcia, R.; Magerle, R.; Perez, R. *Nat. Mater.* **2007**, *6*, 405–411.

32. Leonenko, Z. V.; Finot, E.; Ma, H.; Dahms, T. E. S.; Cramb, D. T. *Biophys. J.* **2004**, *86*, 3783–3793.

33. Suo, Z.; Yang, X.; Avci, R.; Kellerman, L.; Pascual, D. W.; Fries, M.; Steele, A. *Langmuir* **2006**, *23*, 1365–1374.

34. Tsao, Y.; Evans, D.; Wennerstrom, H. *Science* **1993**, *262*, 547–550.

35. Biggs, S.; Mulvaney, P. *J. Chem. Phys.* **1994**, *100*, 8501–8505.

36. Noy, A.; Frisbie, C. D.; Rozsnyai, L. F.; Wrighton, M. S.; Lieber, C. M. *J. Am. Chem. Soc.* **1995**, *117*, 7943–7951.

37. Noy, A.; Vezenov, D. V.; Lieber, C. M. *Annu. Rev. Mater. Sci.* **1997**, *27*, 381–421.

38. Frisbie, C. D.; Rozsnyai, L. F.; Noy, A.; Wrighton, M. S.; Lieber, C. M. *Science* **1994**, *265*, 2071–2074.

39. Green, J.-B. D.; McDermott, M. T.; Porter, M. D.; Siperko, L. M. *J. Phys. Chem.* **1995**, *99*, 10960–10965.

40. Noy, A.; Sanders, C. H.; Vezenov, D. V.; Wong, S. S.; Lieber, C. M. *Langmuir* **1998**, *14*, 1508–1511.

41. Sinniah, S. K.; Steel, A. B.; Miller, C. J.; Reutt-Robey, J. E. *J. Am. Chem. Soc.* **1996**, *118*, 8925–8931.

42. Vezenov, D. V.; Noy, A.; Rozsnyai, L. F.; Lieber, C. M. *J. Am. Chem. Soc.* **1997**, *119*, 2006–2015.

43. Kienberger, F.; Ebner, A.; Gruber, H. J.; Hinterdorfer, P. *Acc. Chem. Res.* **2006**, *39*, 29–36.

44. Hinterdorfer, P.; Dufrene, Y. F. *Nat. Methods* **2006**, *3*, 347–355.

45. Lin, L.; Fu, Q.; Williams, B. A. R.; Azzaz, A. M.; Shogren-Knaak, M. A.; Chaput, J. C.; Lindsay, S. *Biophys. J.* **2009**, *97*, 1804–1807.

46. Lin, L.; Wang, H.; Liu, Y.; Yan, H.; Lindsay, S. *Biophys. J.* **2006**, *90*, 4236–4238.

47. Ludwig, M.; Dettmann, W.; Gaub, H. E. *Biophys. J.* **1997**, *72*, 445–448.

48. Raab, A.; Han, W. H.; Badt, D.; Smith-Gill, S. J.; Lindsay, S. M.; Schindler, H.; Hinterdorfer, P. *Nat. Biotechnol.* **1999**, *17*, 902–905.

49. Ebner, A.; Wildling, L.; Zhu, R.; Rankl, C.; Haselgrubler, T.; Hinterdorfer, P.; Gruber, H. J. *Top. Curr. Chem.* **2008**, *285*, 29–76.

50. Love, J. C.; Estroff, L. A.; Kriebel, J. K.; Nuzzo, R. G.; Whitesides, G. M. *Chem. Rev.* **2005**, *105*, 1103–1170.

51. Neuman, K. C.; Nagy, A. *Nat. Methods* **2008**, *5*, 491–505.

52. Harada, Y.; Kuroda, M.; Ishida, A. *Langmuir* **1999**, *16*, 708–715.

53. Bonanni, B.; Kamruzzahan, A. S. M.; Bizzarri, A. R.; Rankl, C.; Gruber, H. J.; Hinterdorfer, P.; Cannistraro, S. *Biophys. J*. **2005**, *89*, 2783–2791.

54. Bustanji, Y.; Arciola, C. R.; Conti, M.; Mandello, E.; Montanaro, L.; Samori, B. *Proc. Natl. Acad. Sci. U.S.A.* **2003**, *100*, 13292–13297.

55. Kienberger, F.; Pastushenko, V. P.; Kada, G.; Gruber, H. J.; Riener, C.; Schindler, H.; Hinterdorfer, P. *Single Mol.* **2000**, *1*, 123–128.

56. Dean, D.; Han, L.; Grodzinsky, A. J.; Ortiz, C. *J. Biomech.* **2006**, *39*, 2555–2565.

57. Gad, M.; Itoh, A.; Ikai, A. *Cell Biol. Int.* **1997**, *21*, 697–706.

58. Wagner, P.; Hegner, M.; Kernen, P.; Zaugg, F.; Semenza, G. *Biophys. J*. **1996**, *70*, 2052–2066.

59. Gruber, H. J.; Kada, G.; Pragl, B.; Riener, C.; Hahn, C. D.; Harms, G. S.; Ahrer, W.; Dax, T. G.; Hohenthanner, K.; Knaus, H.-G. *Bioconjugate Chem.* **2000**, *11*, 161–166.

60. Kienberger, F.; Kada, G.; Gruber, H. J.; Pastushenko, V. P.; Riener, C.; Trieb, M.; Knaus, H.-G.; Schindler, H.; Hinterdorfer, P. *Single Mol.* **2000**, *1*, 59–65.

61. Khatri, O. P.; Adachi, K.; Murase, K.; Okazaki, K.-i.; Torimoto, T.; Tanaka, N.; Kuwabata, S.; Sugimura, H. *Langmuir* **2008**, *24*, 7785–7792.

62. Ratto, T. V.; Rudd, R. E.; Langry, K. C.; Balhorn, R. L.; McElfresh, M. W. *Langmuir* **2006**, *22*, 1749–1757.

63. Ratto, T. V.; Langry, K. C.; Rudd, R. E.; Balhorn, R. L.; Allen, M. J.; McElfresh, M. W. *Biophys. J*. **2004**, *86*, 2430–2437.

64. Ebner, A.; Kienberger, F.; Kada, G.; Stroh, C. M.; Geretschläger, M.; Kamruzzahan, A. S. M.; Wildling, L.; Johnson, W. T.; Ashcroft, B.; Nelson, J.; Lindsay, S. M.; Gruber, H. J.; Hinterdorfer, P. *ChemPhysChem* **2005**, *6*, 897–900.

65. Lo, Y. S.; Huefner, N. D.; Chan, W. S.; Stevens, F.; Harris, J. M.; Beebe, T. P. *Langmuir* **1999**, *15*, 1373–1382.

66. Crampton, N.; Bonass, W. A.; Kirkham, J.; Thomson, N. H. *Langmuir* **2005**, *21*, 7884–7891.

67. Gauthier, S.; Aimé, J. P.; Bouhacina, T.; Attias, A. J.; Desbat, B. *Langmuir* **1996**, *12*, 5126–5137.

68. Otsuka, H.; Arima, T.; Koga, T.; Takahara, A. *J. Phys. Org. Chem.* **2005**, *18*, 957–961.

69. Moon, J. H.; Shin, J. W.; Kim, S. Y.; Park, J. W. *Langmuir* **1996**, *12*, 4621–4624.

70. Jiang, Y.; Zhu, C.; Ling, L.; Wan, L.; Fang, X.; Bai, C. *Anal. Chem.* **2003**, *75*, 2112–2116.

71. Chen, X.; Davies, M. C.; Roberts, C. J.; Tendler, S. J. B.; Williams, P. M.; Davies, J.; Dawkes, A. C.; Edwards, J. C. *Langmuir* **1997**, *13*, 4106–4111.

72. Conti, M.; Donati, G.; Cianciolo, G.; Stefoni, S.; Samorì, B. *J. Biomed. Mater. Res.* **2002**, *61*, 370–379.

73. Karrasch, S.; Dolder, M.; Schabert, F.; Ramsden, J.; Engel, A. *Biophys. J.* **1993**, *65*, 2437–2446.

74. Yam, C.-M.; Xiao, Z.; Gu, J.; Boutet, S.; Cai, C. *J. Am. Chem. Soc.* **2003**, *125*, 7498–7499.

75. Schindler, H.; Badt, D.; Hinterdorfer, F. K. P.; Raab, A.; Wielert-Badt, S.; Pastushenko, V. P. *Ultramicroscopy* **2000**, *82*, 227–235.

76. Sungkanak, U.; Sappat, A.; Wisitsoraat, A.; Promptmas, C.; Tuantranont, A. *Biosens. Bioelectron.* **2010**, *26*, 784–789.

77. Chen, G.; Zhou, J.; Park, B.; Xu, B. *Appl. Phys. Lett.* **2009**, *95*.

78. Torres-Chavolla, E.; Alocilja, E. C. *Biosens. Bioelectron.* **2009**, *24*, 3175–3182.

79. Liu, W.; Montana, V.; Chapman, E. R.; Mohideen, U.; Parpura, V. *Proc. Natl. Acad. Sci. U.S.A.* **2003**, *100*, 13621–13625.

80. Negri, P.; Chen, G.; Kage, A.; Nitsche, A.; Naumann, D.; Xu, B.; Dluhy, R. A. *Anal. Chem.* **2012**, *84*, 5501–5508.

81. Boyce, M.; Bertozzi, C. R. *Nat. Methods* **2011**, *8*, 638–642.

82. Sletten, E. M.; Bertozzi, C. R. *Angew. Chem., Int. Ed.* **2009**, *48*, 6974–6998.

83. Best, M. D. *Biochemistry* **2009**, *48*, 6571–6584.

84. Baselt, D. R.; Lee, G. U.; Colton, R. J. *J. Vac. Sci. Technol., B* **1996**, *14*, 789–793.

85. Datar, R.; Kim, S.; Jeon, S.; Hesketh, P.; Manalis, S.; Boisen, A.; Thundat, T. *MRS Bull.* **2009**, *34*, 449–454.

86. Maraldo, D.; Mutharasan, R. *Anal. Chem.* **2007**, *79*, 7636–7643.

87. Campbell, G. A.; Medina, M. B.; Mutharasan, R. *Sens. Actuators, B* **2007**, *126*, 354–360.

88. Huang, X.; Li, M. F.; Xu, X. H.; Chen, H. J.; Ji, H. F.; Zhu, S. F. *Biosens. Bioelectron.* **2011**, *30*, 140–144.

89. Fu, L. L.; Zhang, K. W.; Li, S. Q.; Wang, Y. H.; Huang, T. S.; Zhang, A. X.; Cheng, Z. Y. *Sens. Actuators, B* **2010**, *150*, 220–225.

90. Ricciardi, C.; Castagna, R.; Ferrante, I.; Frascella, F.; Marasso, S. L.; Ricci, A.; Canavese, G.; Lore, A.; Prelle, A.; Gullino, M. L.; Spadaro, D. *Biosens. Bioelectron.* **2013**, *40*, 233–239.

91. Chen, G.; Ning, X.; Park, B.; Boons, G.-J.; Xu, B. *Langmuir* **2009**, *25*, 2860–2864.

92. Wang, B.; Guo, C.; Chen, G.; Park, B.; Xu, B. *Chem. Commun.* **2012**, *48*, 1644–1646.

93. Kirschner, C. M.; Brennan, A. B. *Annu. Rev. Mater. Res.* **2012**, *42*, 211–229.

94. Zhao, C.; Li, L.-Y.; Guo, M.-M.; Zheng, J. *Chem. Pap.* **2012**, *66*, 323–339.

Chapter 8

Nanoliter/Picoliter Scale Fluidic Systems for Food Safety

Morgan Hamon,[1] Omar A. Oyarzabal,[2] and Jong Wook Hong[*,1]

[1]Materials Research and Education Center,
Department of Mechanical Engineering,
Auburn University, Auburn, Alabama 36849, U.S.A.
[2]Institute for Environmental Health, Inc.,
Lake Forest Park, Washington 98155, U.S.A.
*E-mail: hongjon@auburn.edu.

Microfluidic technology is one of the latest platforms for detection of chemical and biological hazards. This technology may provide a higher sensitivity couple with the best specificity provided by targeting DNA molecules. This chapter introduces existing microfluidic systems and reviews some ongoing researches relevant to food safety. Microfluidic devices provide high throughput and large-scale analysis by multiplexing and parallelization of analyses on a single device.

Introduction

Food production is highly efficient and provides the convenience and variety of foods demanded by current lifestyles. To prevent food shortage and deliver low-cost and safe food and drinking water with low environmental impact, new sets of engineering and scientific processes, products and tools are being utilized. Microfluidic systems (also known as micro total analysis systems, μTAS, or lab-on-a-chip, LOC) are considered emerging technologies that provide high throughput and large-scale analysis via integration of multiple steps, multiplexing and parallelization on a single device (*1, 2*). Microfluidic systems is one of the latest platforms applied to the detection of chemical or biological contaminants in foods (Figure 1) (*3–10*).

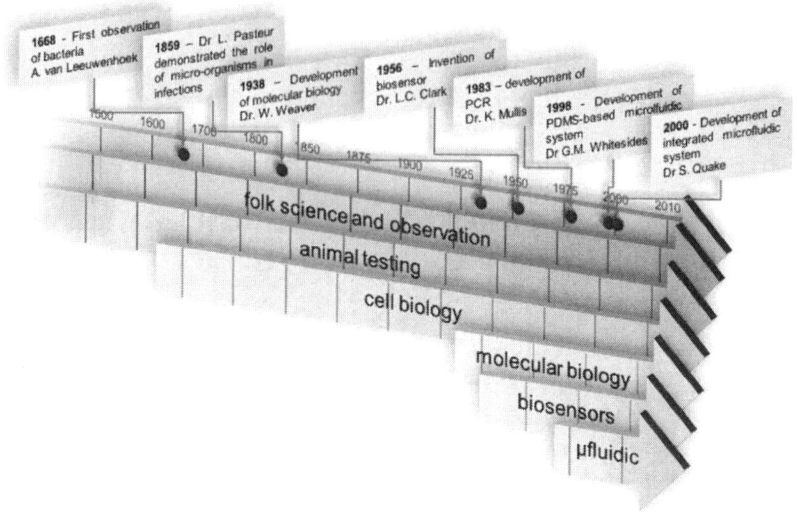

Figure 1. Timeline and cornerstones in the evolution of detection systems used for food safety.

Microfluidic System for Food Safety

Conventional Detection Method

Different methodologics have been developed to detect and identify biological and chemical agents in food and water. The main biological agents are bacterial pathogens, parasites and viruses. Chemicals agents include primarily natural toxins or intentional adulterations (*11–16*).

Pathogenic organisms have historicaly been detected by culture and colony counting method. The target bacterial pathogen present in food or water samples is cultured and isolated in a nutritive medium. Current molecular methods allow for the detection of bacterial pathogens in food before the actual isolation. Although the isolation of the pathogen is labor intensive and time consuming, the regulatory agencies in the USA require the actual isolation to determine if a sample is indeed positive (*17–20*). To decrease time and effort of detection of pathogens without compromising the reliability of the results, enzyme-linked immunosorbent assay (ELISA) and polymerase chain reaction (PCR) have been widely used to detect pathogen specific proteins and nucleic acid sequences, respectively. Although these techniques dramatically reduce the time for detection (*21–23*), key issues, such as cost, automation and simplicity, still remain a limitation.

To detect chemicals hazards in food samples, analytical chemistry methods such as gas chromatography (GC), high performance liquid chromatography (HPLC) (*24*) and mass spectrometry have been applied to get the best sensitivity

(*11*) and rapid detection (*25*) Although with advantages, these techniques remain slow, expensive, and require extensive sample preparation and trained personnel. In addition, the physical dimensions of the required equipment does not allow on-the-spot detection of chemical hazards (*21–23*). The need for cheap, high throughput and portable analytical systems has encouraged the development of new technologies and more suitable analytical methodologies (*26, 27*).

Microfluidic System-Based Method

Whole Cell Detection

Foodborne pathogenic bacteria in animals, plants and water have been found to be an important souce of contaminations for humas worldwide (*12–16, 28*). Strains of pathogenic Escherichia coli, *Campylobacter, Salmonella, Listeria, Staphylococcus, Clostridium* and *Bacillus*, the major foodborne pathogenic bacteria, are responsible for millions of foodborne illnesses, thousands of deaths, and several billion dollars in productivity losses, medical costs, and premature death (*29*). It is therefore of great importance to develop methods to efficiently detect pathogenic bacteria (*30*).

Microfluidic System for Cell Sorting and Capture

Microfluidics presents several characteristics that make it a suitable technique for isolating pathogens from suspended particle mixtures. Microfluidics is also an important platform to develop portable devices for accurate and rapid detection systems with limited amount of reagents (*31–36*). One main limitation to the use of whole cell detection in microfluidic system is the difficulty to detect low amounts of cells from nanoliter volumes. Among the methods used for whole cell detection, dielectrophoresis techniques have the capability to concentrate the target cells on a specific area and thus amplify the signal for efficient detection (*37, 38*). For example, Gagnon & Chang (2005) have developed a microfluidic device (Figure 2A) composed of microchannels align atop a thin continuous serpentine microwire of platinum. When alternating current passes through the wire, it generates an electro-osmotic flow that traps and concentrates the cells on the microwire area, and directs the cells towards a designated points of the device, while the suspended particles are swept towards the outlet along the fluidic flow (*37*). This system requires less than 100 µl of solution and can concentrate 10^3 particles/ml in few seconds.

Another microfluidic system for sorting and capturing pathogenic bacteria has been developed by Cheng *et al.* (2007) (*38*). This microfluidic system is composed of four different parts where different micro electrodes are used to filter, focus, sort and trap bacterial cells by creating a dielectrophoretic (DEP) force field cage, as shown in Figure 2B. In the first part (filtering section), parallel electrodes trap debris and particles with positive DEP (pDEP) while negative DEP (nDEP) particles, such as bacteria, pass to the focusing section where nDEP forces focus them toward the center of the channel. Particles are then sorted by three oblique

DEP gates in function of their size and their specific charges and are trapped and concentrated in three different channels where they are detected. Although this device is slower (~50 minutes) than Gagnon and Chang's device (*37*), it has a similar sensitivity as it can detect 500 cells within 100 μl, but has also the possibility to sort microorganisms in function of their size and electrical charges (*38*). All these microfluidic devices are more sensitive and faster (seconds vs. hours) than other optical biosensors, and therefore are suitable for the development of portable systems for rapid and accurate detection of microorganisms, especially when limited volume of sample (<100 μL) are available.

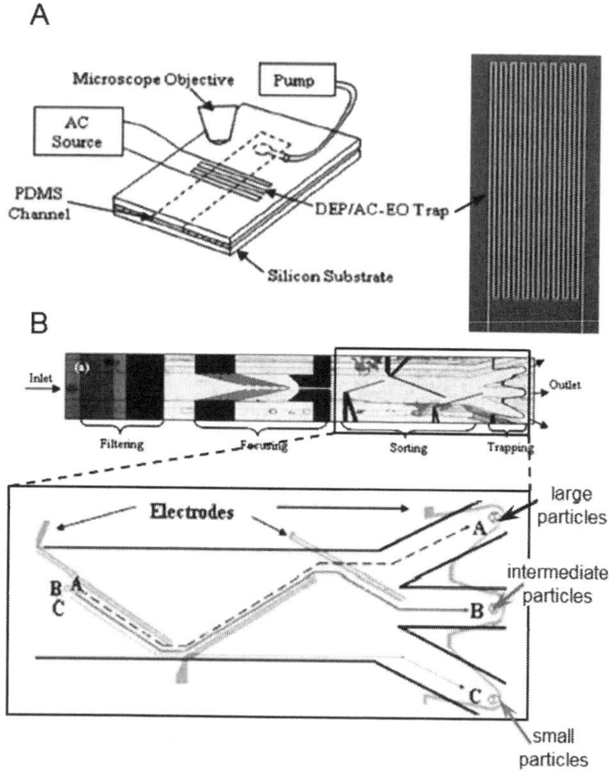

*Figure 2. Microfluidic chips for whole cell capture (**A**) and sorting (**B**) for harmful bacteria detection. Reproduced with permission from Reference (37) (Copyright 2007 Wiley-VCH Verlag GmbH & Co. KGaA) and from Reference (38) (Copyright 2007 American Institute of Physics), respectively.*

<u>*Impedimetric Biosensor*</u>

Impedimetric biosensors are other systems for concentration and detection of microorganisms present in food and water. They rely on the insulating properties of bacterial cell membranes. Impedimetric biosensors are composed of a solid electrode to which cells attach through specific or non-specific adsorption (Figure 3) (*39*). After attachment to the electrode, the natural capacitance (0.5-1.3 µF/cm^2) and resistance (10^2-10^5 Ω/cm^2) of the cell membrane (*40*) effectively reduce the current that reaches the electrode, and hence increase the impedance of the interface. The adsorption of the bacteria on the electrodeis is then revealed through the detection of either a shift in impedance or a change in capacitance or admittance at the bulk of the electrode interface (*41*).

Traditionally, macro-sized electrodes are used to measure impedance (*42–44*). To improve the sensitivity, microelectrodes, such as interdigitated array microelectrodes (IDAM), have been developed (*45–47*). IDAM consist of a series of parallel microband electrodes that are connected together, forming a set of interdigitating electrodes with low ohmic drop, fast establishment of steady-state, rapid reaction kinetics, and increased signal-to-noise ratio (*45–47*). In addition, their small dimensions (0.1–0.2 µm in height, 1–20 µm in width; 2–10 mm in length; 1–20 µm gap between the microbands) make them easy to combine with microfluidic systems to improve their performance (*33, 48–55*).

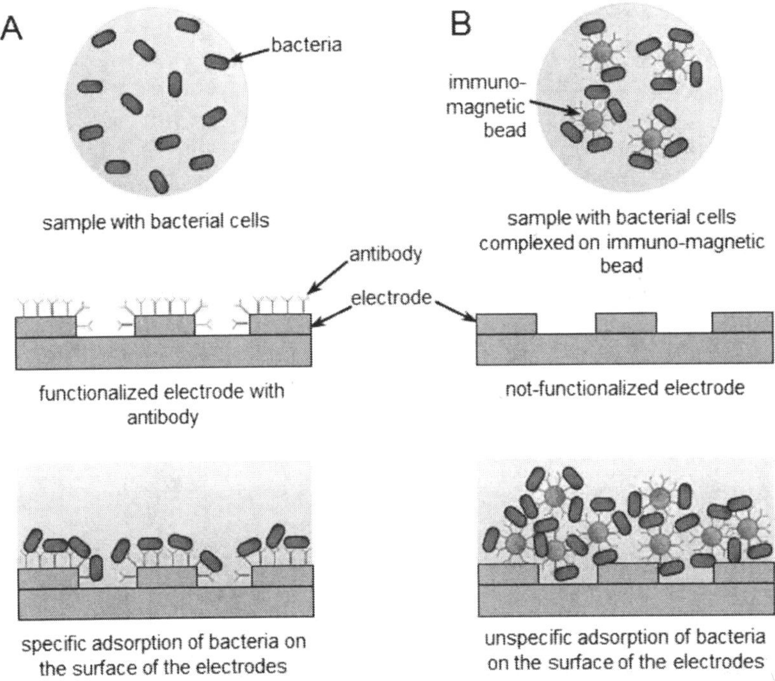

*Figure 3. Working principle of impedance detection technique with (**A**) or without (**B**) the use of bio-recognition elements for bacterial cell adherence on the electrodes. Adapted with permission from Reference (39). Copyright 2009 Elsevier.*

The detection techniques based on the specific adsorption of bacteria on the surface of the electrodes requires a functionalization of the electrode surface with bio-recognition elements (or bioreceptor), such as antibodies specific to pathogenic bacteria. When bacteria attach to the antibodies, the insulating properties of the cell membrane induce measurable changes in the electrical properties (*23*). The main advantage of the use of antibodies as bio-recognition elements is their selectivity. The critical part in the construction of antibody-based impedimetric biosensors is the immobilization method, because it can profoundly affect the analytical performance of impedance biosensors. Physical adsorption without the use of chemical linker is the simplest immobilization method. It depends on the non-specific interactions (such as non-covalent bridges, hydrophobic interactions, and van der Waals forces) of the antibody with the surface. However, the low stability and the random orientation of the antibodies immobilized on electrodes result in the decrease of the binding bacteria and a high detection limit (10^5 cfu/ml) (*56*), which restrict the use of this method to a limited number of applications. To improve the sensitivity of the biosensor, self-assembled monolayer (SAM) of chemical linkers have been use to immobilize antibodies on the electrodes (*57–59*). Recently, Tan *et al.* (2011) have integrated an electrochemical impedance spectroscopy (EIS)-based immunosensor, composed of an alumina nanoporous membrane with immobilized antibodies specific to foodborne pathogen, with a PDMS microfluidic system for the rapid detection of *E. coli* and *Staphylococcus aureus*, as shown in Figure 4A (*60*). This microfluidic immunosensor based on the nanoporous membrane impedance spectrum could achieve rapid bacteria detection within 2 hours with a sensitivity of 10^2 cell/ml in pure culture.

In 2007, Varshney et al (*61*) developed an impedimetric biosensors based on the non-specific absorption of bio-recognition element, based on gold IDAM for the detection of *E. coli* O157:H7 in food samples. In this method, cells are first separated, re-suspended and concentrated with the help of magnetic nanoparticle-antibody conjugates (MNAC), functionalized here with anti-*E. coli* antibodies (Figure 4B), and uniformly spread on the surface of IDAM. The impedance sensor detected a minimum of 7.4×10^4 and 8.0×10^5 cells/ml of *E. coli* O157:H7 in pure culture and ground beef samples, respectively (*39*). The integration of this technology within a microfluidic system increased the sensitivity of detection to 1.6×10^2 and 1.2×10^3 cells/ml of *E. coli* O157:H7 present in pure culture and ground beef samples, respectively with very low volume of sample (60 nl) (*61*).

When compared with impedance macrosystem (*62*), microfluidic slightly increase the sensitivity of the detection with (*60*) or without specific immobilization (*63*) (4.2×10^2 cells/ml, 10^2 cells/ml and 10^2 cells/ml in pure culture respectively) but is faster (about 2 hours, less than 2 hours and 35 minutes respectively) and requires lesser amounts of reagents (ml scale, μl scale and nl scale, respectively).

Fluorescence-based cell detection presents several advantages when compared with other techniques, including sensitivity, multicolor for simultaneous multi cell type detection, stability low hazard, availability and low cost (*64, 65*). It has been previously used for the detection of pathogenic bacteria. Recently, Yamaguchi *et al.* (2011) developed a simplified and rapid on-chip method using a microfluidic cytometry device for rapid quantification of bacteria cells, such as *E. coli*, in water (*66*). The microfluidic chip is composed of a T-shape introduction section, followed by a serpentine mixing channel, an alignment part composed of crossing channels, and a detecting part. Samples are injected and fluorescently labeled, and anti-*E. coli* O157:H7 antibodies are injected through the inlet of the T-shape introduction part. Bacterial cells in the sample are fluorescently stained during flow through the mixing part. After mixing, the cells are aligned by injection of a sheath fluid (PBS) through two microchannels, and fluorescently stained cells are detected in the detecting part. The authors demonstrated the capability of this simple fluorescent-based technique to detect a low number of bacterial cells (10^4 cells/ml). Although the sensitivity of the system is not significantly different from conventional epifluorescence microscopy (EFM), the use of microfluidic presents several advantages to EFM: it is simpler as it does not requires pretreatment apparatus, it is rapid and results can be obtain a limited amount of time (one hour) and it has the possibility to be integrated with small light source such as light-emitting diode (LED) for on-site analysis.

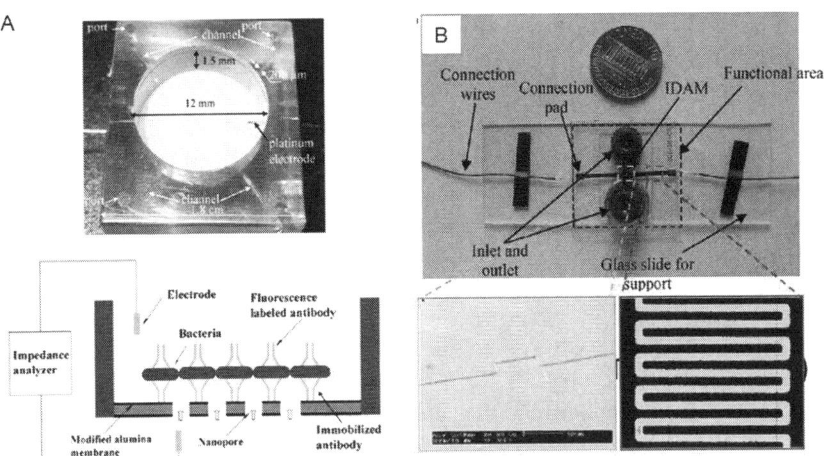

*Figure 4. Impedimetric biosensors for bacteria cell detection. **A**. Sensor based on the use of bio-recognition element. Adapted with permission from Reference (60). Copyright 2011 Elsevier. **B**. Sensor based on non-specific binding of bacteria. Adapted with permission from Reference (61). Copyright 2007 Elsevier.*

Detecting any pathogenic bacteria is a reliable method to prevent food poisoning. However, in some cases, although bacteria are killed during thermal process, bacterial toxins (such as staphylococcal enterotoxin B (SEB) or botulinum neurotoxin (BoNT) can remain active and be associated with food poisoning. To detect toxin, chemiluminescence methodologies such as ELISA-based and enzyme-substrate-based methods have been used for their sensitivity and reliability (*67–72*).

ELISA-Based Microfluidic System

ELISA is a reliable technique for the detection of low quantity of toxin (*67–70*). However, the large amount of reagent, space and time required to perform one ELISA are cost problems that limit the quantity of tests performed simultaneously outside a well-equipped laboratory. Microfluidics offers a cheap alternative to conventional ELISA methodology (*73–75*). The possibility to realize on a same chip the different step of the ELISA makes it a very attractive method to bring toxin detection systems on-site. For example, Yang *et al.* (2010) have developed a lab-on-a-chip system (Figure 5A) for a Carbon Nanotube (CNT) based immunoassay with optical detection of staphylococcal Enterotoxin B (SEB) for food safety applications (*73*). In this chip, four biosensing elements were combined: (1) CNT technology for primary antibody immobilization, (2) Enhanced Chemiluminescence (ECL) for light signal generation, (3) a cooled charge-coupled device (CCD) for detection and (4) polymer lamination technology for developing a point of care immunological assay for SEB detection. To combine those four elements, the authors have fabricated a six layer, 3D fluidic structure constructed with a rigid 3.2 mm PMMA (acrylic) core laminated with additional layers of thin polycarbonate (PC) film bonded with adhesive. The LOC contains two main analytical elements: an eight channel microfluidic design for the loading of the different samples and buffers and an interchangeable immunosensor surface where the ELISA reaction is done. Eight samples are introduced into the wells and go through the microchannels to the interchangeable immunosensor surface, which is composed of a polycarbonate strip with anti-SEB antibody immobilized on CNT. SEB contained in the sample first binds to the anti-SEB primary antibody of the interchangeable polycarbonate strip. After incubation and washing, horse radish peroxidase (HRP) labeled secondary antibodies are added. The HRP is then analyzed by ECL using a CCD detector. The combination of the four biosensing elements at a microscale level presents various advantages when compared with conventional ELISA technology as the assay can be performed without a laboratory, with a minimal amount of reagents (and consequently minimal exposure to users of toxins and other biohazards) and with a high sensitivity (approximately 0.1 ng/ml of SEB).

When the metabolite released by bacteria is an enzyme, an alternative detection method in the food or drink is to detect its enzymatic activity using, for example, fluorescent probes. Frisk *et al.* have developed a microfluidic device for the detection of BoNT/A with high sensitivity, on-site portability and multiplexing capabilities (Figure 5B) (*74, 75*). This device is composed of three parts: an input port, a reaction microchannel, and a detection port. In the reaction microchannel, the surface is coated with a fluorescently labeled peptide enzymatic substrate of BoNT/A. The sample is introduced into the input port and flows through the microchannel. There, BoNT/A catalyzes the cleavage reaction of the fluorescent labeled peptide that diffuses into the detection port where the evaporation of the solution preconcentrates the fluorescent-labeled fragments for detection. The evaporation-based concentration system leads to 3-fold signal amplification over 35 min and allows the detection of as little as 3 pg/ml of the toxin in buffer and 3nM in food (*75*). Because of the small size of the system, the platform could be easily modified and adjusted for detecting other serotypes of BoNT, for example by adding microchannels coated with fluorescent-labeled substrate specific to each serotypes. In addition, the integration of this device with a portable light source would enable real-time and on-site monitoring.

Nucleic Acid Sequence Detection

Molecular methods for the detection of nucleic acids (DNA and RNA) have been regularly used for detection of pathogenic bacteria in food because of their high sensitivity, selectivity and short time (*30*). With the use of smaller volumes and sample amounts, microfluidic systems increase the detection sensitivity and improve the experiment throughput. Several microfluidic platforms have been develop to perform molecular methods including nucleic acid hybridation (*76–79*), isolation (*80–82*), and amplification (*83, 84*).

Nucleic Acid Hybridation

Recently, various microfluidic DNA-based probes were coupled to different measurement techniques, including SPRi (*85*), conductance impedance (*86*), and fluorescence (*87*). Ben-Yoav *et al.* (2012) have developed a microfluidic-based electrochemical biochip shown in Figure 6A, which contains an array of individually addressable 25 nl reaction chambers, fabricated with micro-electromechanical systems (MEMS) technology (*88*). Each chamber contains a grid of 3×3 sensors and each row of 3 sensors also contains a counter and a reference electrode to complete the three-electrode system. Three unique single stranded DNA (ssDNA, 30-mers) probes were functionalized onto patterned electrodes of the chip to detect complementary DNA hybridization events using EIS analysis. This impedimetric biosensor-based technique is similar to those for whole cell detection but uses ssDNA as bio-recognition elements. This biosensor is able to detect ssDNA targets on the nM scale and with low cross-reactivity (13%).

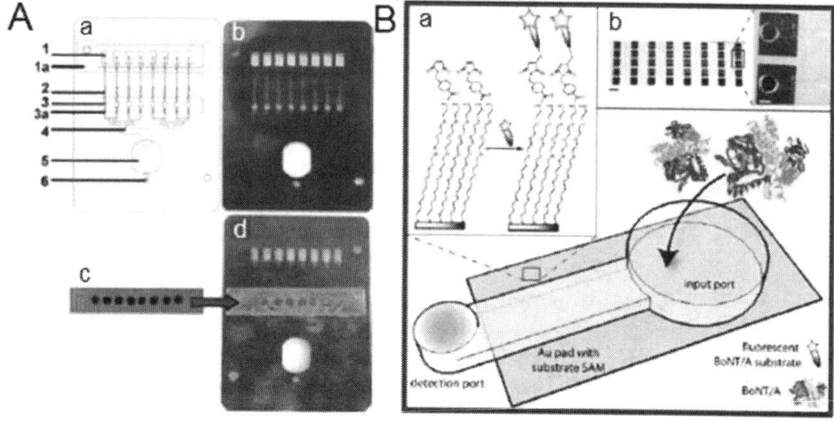

*Figure 5. Microfluidic chip for metabolite detection. **A**. ELISA-based microdevice. (a) schematic figure of the device with indication of the sample wells (1), rectangular loading frame to simplify reagent application (1a), channels (2), detection wells (3), interchangeable surface with the antibody ligand-CNT (3a), joining element for all eight channels (4), waste chamber (5) and the chip outlet (6), (b) top view of the chip, (c) immage of the immunosensor on polycarbonate strip and (d) bottom view. Adapted with permission from Reference (73). Copyright 2010 The Royal Society of Chemistry. **B**. Enzyme-substrate reaction-based system. (a) schematic representation of the microfluidic chip with the input port, the reaction microchannel and the detection por. Close up shows the formation of the self-assembled monolayer of the BoNT/A enzymatic substrate on Au surface. (b) 40 arrayed detection system for throughput screening. Adapted with permission from Reference (75). Copyright 2009 American Chemical Society.*

Nucleic Acid Amplification

Polymerase Chain Reaction (PCR) is a molecular technique for nucleic acid amplification that can replicate a specific fragment of a target nucleic acid and create several million DNA copies within a few hours. It is thus a key element in food safety to identify pathogens, allergens, and GMOs present in small quantity (*89, 90*). However, it remains difficult if not impossible to detect the presence of a single pathogen using this technique. Microfluidics has been integrated with

PCR to dramatically increase sensitivity. Ottesen *et al.* (2006) have developed a microfluidic digital PCR platform allowing single cell detection (*83*). The platform is composed of 1,176 individual independent 6.25 nl reaction chambers, allowing the realization of 1,176 independent PCR reactions. PCR was carried out on a conventional flat-block thermocycler. Amplification was monitored using 5′ nuclease probes to generate a fluorescent signal detected with a modified microarray scanner (Figure 6B).

In some devices, entire bacteria are introduced into the detection PCR platforms, lysed by thermal (*91*), chemical (*92*), physical (*93, 94*), and electrical methods (*95*), and their nucleic acid is then amplified on site. In 2008, Cheong *et al.* developed a one-step real-time PCR method for pathogen detection (*84*). In this design, Au nanorods were used to transform near-infrared energy into thermal energy and subsequently lyse the pathogens. Next, DNA was extracted and amplified in the PCR chamber. The entire process does not need to change or remove reagents, which improved the overall efficiency of the device.

Among all nucleic acid amplification techniques, isothermal techniques for amplification of DNA/RNA have recently drawn interest (*96, 97*) since they do not require high temperature and large variation of the temperature. This approach to a simpler and more energy efficient model holds with the goal of microfluidic systems to reduce cost and labor of routine experiments, making isothermal amplification techniques an excellent choice for nucleic acid detection LOC. Recently, we have developed, in our laboratory, a microfluidic chip for detection of pathogenic bacterial using the isothermal real-time helicase-dependent amplification (HDA) technique (unpublished data). The chip is a microfluidic platform composed of a 16×24 array of nanoliter-scale microchambers where real-time HDA for single cell analysis can be performed.

Sample Preparation

Sample preparation steps for PCR are of first importance to achieve high sensitivity and specificity. Hong *et al.* (2004) have developed nanoliter-scale nucleic acid processors for DNA and mRNA purification (*80*). Two different microfluidic chips were fabricated to isolate variable numbers of cells, lyse the cells, and purify their DNA or mRNA. All the steps were realized in one single microfluidic chip without pre- or post-treatment of the sample. For mRNA purification, oligo-dT polymer magnetic beads were introduced into the separation chamber to form an affinity column. After cell lysis in the cell chamber, lysate is flushed over the affinity column and mRNA was recovered from the column. For DNA purification, the chip was composed of three independent processors, as shown in Figure 6C. In each processor, lysis buffer, dilution buffer, and *E. coli* in culture medium are introduced into a metering section. The three solutions are then pushed into a microfluidic mixer where cells are lysed. Lysate is then flushed over the DNA affinity column, and the genomic DNA is then recovered from the chip for PCR amplification. This methodology allows DNA recovery with a reduced number of bacteria (< 28 bacteria), and thereby increases the sensitivity of this process three to four orders of magnitude over that of conventional methods.

155

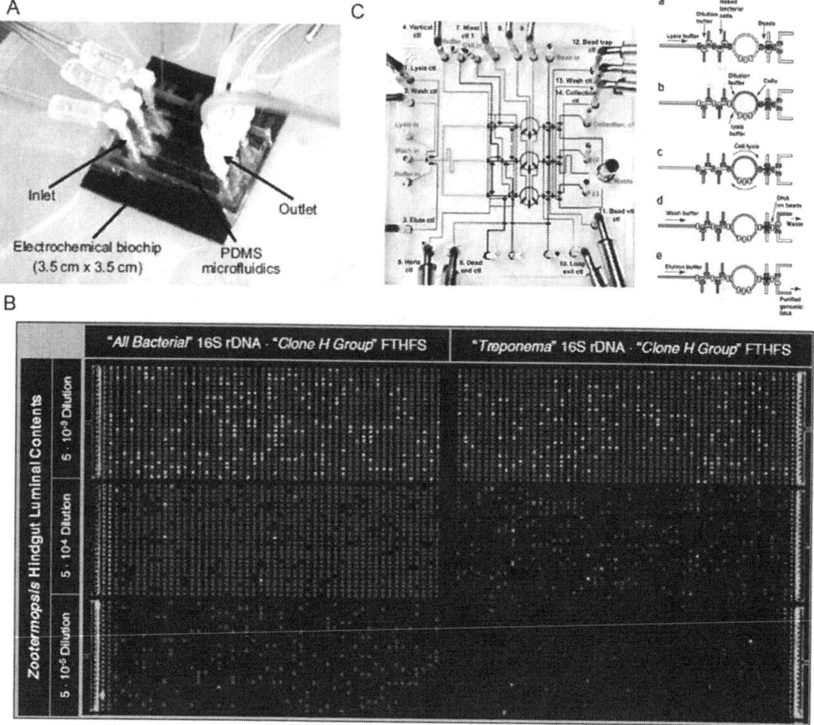

*Figure 6. Microfluidic chip for nucleic acid detection. **A**. Photography of a chip for nucleic acid hybridization-based detection. Reproduced with permission from Reference (88). Copyright 2012 Elsevier. **B**. Multiplex digital PCR platform for single cell analysis. Adapted with permission from Reference (83). Copyright 2006 HighWire Press. **C**. DNA purification chip and schematic diagram of one instance of the DNA isolation process (a-c). Adapted with permission from Reference (80). Copyright 2004 Macmillan Publishers Ltd: Nature Biotechnology.*

Microfluidic-Based Assay for Nutritional Hazard Detection from Chemical Origin

With the increased role of chemistry and chemical engineering in agriculture and food sciences, the presence of hazardous chemicals in the food, drink, and soil became a threat for human health and increased the importance of constant monitoring of ecosystems. A similar method to the one previously described for the detection of toxins with enzymatic activity can be used to detect harmful chemicals. With such a method, chemical reactions that provide detectable signal in function of the concentration of the target substrate can be used to monitor the presence of harmful chemicals. Several applications, based on a chemical detection concept, have been integrated in microfluidic systems (*98–100*). Recently, He *et al.* developed a microflow injection analysis (μFIA) system for the determination of nitrite in food, based on the reaction of ferrocyanide with nitrite in acidic medium (*99*). The microfluidic chip has three main parts: an introduction part composed of a Y-inlet for the introduction of the reagent (luminol) and the working solution (the sample in ferrocyanide and HCl solution), a mixing part, and a detection component. In the mixing part, nitrite from the sample metabolizes ferrocyanide to ferricyanide in acidic medium and chemoluminescence reaction of luminol with ferricyanide is sensed in the detection part.

Use of Microfluidic System To Improve Food Production

Microfluidic for Bacterial Plant Pathogen's Infection Studies

Microfluidic systems have the ability to recreate fluidic microenvironments that can be found in nature (*101–107*). This unique feature can be used to study cell behavior in a controlled environment, when it is difficult or impossible to do it *in vivo* or with conventional methods (*108, 109*). For example, microfluidic systems have been developed to recreate xylem vessels of plants and study the infection process and strategy of the plant pathogen bacteria *Xylella fastidiosa* (*108, 109*), which lives inside the xylem vessels of different plants, such as grapes, citrus, coffee, and almond (*110*). Using polydimethylsiloxane (PDMS) microfluidic chambers, researchers have mimicked the plant xylem vessel and demonstrated that the pili of *X. fastidiosa* are implicated in cells adherence (*108*) and migration against flowing currents of growth medium (*109*). With conventional methods, such as atomic force microscopy or laser tweezers, the molecular and biochemical aspects of the plant pathogen is difficult to study because of the large size of the cell chambers or the time required for the experiments. Scaling-down the experiment to the microscale reduces the size of the observation chamber and the time of the observation, overcoming the limitations of conventional apparatus.

Microfluidic System for Animal Food Production

Embryogenesis-based biotechnologies have various promising applications in the field of food production, from *in vitro* fecundation (IVF) for livestock breeding to embryos development studies (*111, 112*). Conventionally, experiments are performed manually, in 96 to 384 well plates (*113–116*), and monitored with imaging methodology (*117*), which keeps high-throughput analysis and imaging of embryos and juveniles challenging (*113, 115, 117, 118*). However, miniaturized LOC technologies overcome these limitations by providing an automated platform for the handling and immobilization of micron-size organisms (*113, 119–122*), , and for high-throughput screening (*123*) that simplify the different steps of fecundation and embryogenesis (*124, 125*). For example, Ma *et al.* (2011) have reported a novel microdevice that integrates each step of IVF, including oocyte positioning, sperm screening, fertilization, medium replacement, and embryo culture (*111*). The chip (Figure 7A) is composed of an oocyte positioning region and four symmetrical straight channels, crossing at the oocyte positioning region for sperm introduction, monitoring and selection. Oocytes are singly introduced in the 4 × 4 units. The fertilization process and early embryonic development of the individual zygote is traced with microscopic recording and analyzed by *in situ* fluorescent staining. In this system, the embryo growth rate and blastocyst formation are similar to the conventional Petri dish method, and healthy blastocysts developed in the microdevice can be easily retrieved through a routine pipetting operation and used for further embryo transfer. The present system presents the advantage to avoid the use of centrifugation that can increase the production of oxygen species and decrease the sperm motility, both negatively influencing the fecundation of the oocytes. In addition, healthy blastocysts can easily be collected and used for further embryo implantations or studies.

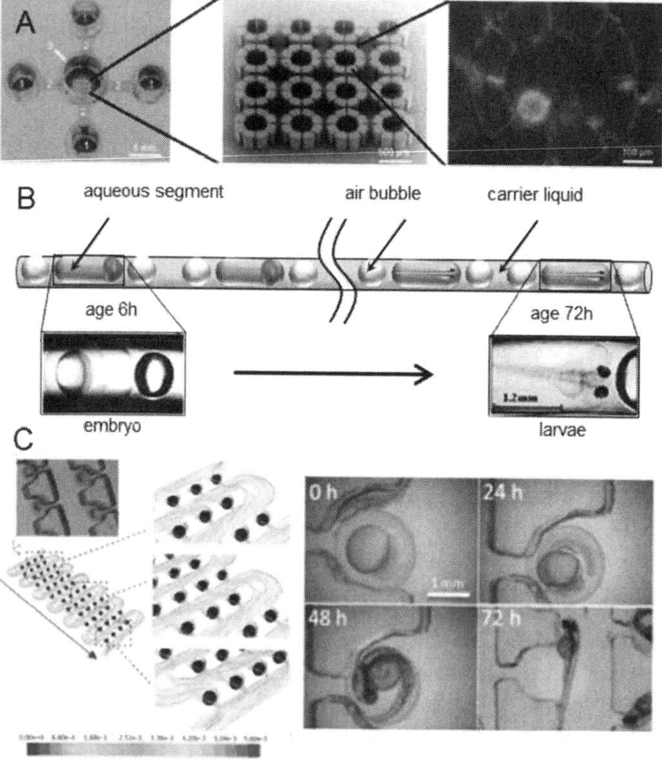

*Figure 7. Microfluidic system for food production. **A.** microfluidic chip for IVF with 4×4 array of individual fecondation chamber. Reproduced with permission from Reference (111). Copyright 2011 ACS Publication. **B and C.** Microfluidic system for on chip embryogenesis studies in droplet-based microfluidic system (**B**) or continuous-flow system (**C**). Adapted with permission from Reference (126) (Copyright 2007 The Royal Society of Chemistry) and Reference (131) (Copyright 2011 PLOS), respectively.*

To study embryogenesis, several microfluidic-based methodologies have been developed involving droplet-based microfluidic system (*126, 127*) or continuous-flow system (*128–132*) for screening and studies of fish embryos. Droplet-based microfluidic systems have been the first technology to be study fish embryogenesis in controlled environment. In 2006, Funfak *et al.* have developed a microfluidic system embryology based method (*126*). The system, shown in Figure 7B, is composed of a Teflon microtube in which fish egg-containing droplets are perfused with immiscible perfluoromethyldecalin (PP9). The development processes of fish embryos could be observed. After 5 days, larvae were collected from the droplets and transferred into breeding reservoirs for further experiments. More recently, Akagi *et al.* have introduced a miniaturized embryo array that automatically traps, immobilizes, and perfuses fish embryos (*131*). The devise, as it appears in the Figure 7C, is composed of a serpentine channels that contains embryo traps (each one designed for a single embryo), suction channels, and hydrodynamic deflectors. Embryos were loaded in the chip one-by-one, rolled on the bottom surface of the serpentine channels under the influence of drag force. Approaching an empty trap, the embryo was affected by the flow passing through the suction channel and directed toward the trap where it remained, without obstructing passage of other embryos in the main channel. The process was repeated with the second embryo and the following ones until all traps were occupied. After loading, the chip was perfused for up to 72 hours with normal and uniform development of all embryos observed. Although most of the microfluidic technology for fish embryological studies were performed on *Danio rerio*, one can easily transpose those technology to other fishes and other animals for food-purposed embryogenesis studies and engineering, such as fish spermatogonial transplantation for food production (*112*).

Summary

Microfluidic technologies are powerful tools that can revolutionize the entire food production system and introduce agriculture to a new era. In this chapter, we describe some of the systems and concepts that overcome a number of limitations imposed by conventional macroscale experimentations including the expensive and time-consuming methods, low throughput, and the need of large volume of reagents. With the miniaturization of the biomolecular, chemical, and biological processes on a single chip, the approach of the food production system is dramatically changed. This new set of tools for the supply of food aims to improve the production of low-cost and safe food and drink, and prevent a worldwide food shortage. However, a number of challenges still remain to achieve this new agricultural revolution. For example, although microfluidic systems dramatically downsize the space required in the lab to perform conventional experiments, the emptied space has been invaded by tubing, pumps, gas tanks, computers, and other components for chip handling making the "lab-on-a-chip" a "chip-in-the-lab". In order to substitute conventional methods, progress to make microfluidic devices more accessible to conventional laboratories has to be made. In addition, these efforts will render microfluidic chips mobile and bring this technolgy into fields and farms to improve food production in multiple areas. More importantly, mindsets need to change and participants of the food production system need to be convinced of the suitability of these systems by increasing the number of validation studies for the proposed processes. We are now at the door of a new era in agriculture, but a great deal of progress must be made to increase the utility of microfluidic systems.

Acknowledgments

The authors want to thank Lauren Bradley for her help in the redaction of the manuscript. This work was supported the USDA Nanotechnology Program (CSREES Grant No. 2008-01437).

References

1. Li, P. C. H. *Microfluidic Lab-on-a-chip for Chemical And Biological Analysis And Discovery*; Taylor & Francis Group: 2006.
2. Melin, J.; Quake, S. R. Microfluidic large-scale integration: the evolution of design rules for biological automation. *Ann. Rev. Biophys. Biomol. Struct.* **2007**, *36*, 213–31.
3. Brothwell, D. R.; Brothwell, P. *Food in Antiquity: A Survey of the Diet of Early Peoples*. Johns Hopkins University Press: Oxford, U.K., 1969.
4. Irwin, J. O. *J. Hyg.* **1950**, *48* (2), 215–38.
5. Eley, A. R. *Microbial Food Poisoning*; Springer: New York, 1996.
6. Morange, M.; Cobb, M. *A History of Molecular Biology*; Harvard University Press: Cambridge, MA, 2000.
7. Clark, L. C., Jr. *ASAIO J.* **1956**, *2* (1), 41–48.
8. Mullis, K.; Faloona, F.; Scharf, S.; Saiki, R.; Horn, G.; Erlich, H. *Cold Spring Harbor Symp. Quant. Biol.* **1986**, *51* (Pt 1), 263–73.
9. Duffy, D. C.; McDonald, J. C.; Schueller, O. J. A.; Whitesides, G. M. *Anal. Chem.* **1998**, *70* (23), 4974–4984.
10. Unger, M. A.; Chou, H.-P.; Thorsen, T.; Scherer, A.; Quake, S. R. *Science* **2000**, *288* (5463), 113–116.
11. Helferich, W.; Winter, C. K. *Food Toxicology*; CRC Press: Boca Raton, FL, 2001.
12. Park, S.; Worobo, R. W.; Durst, R. A. *Crit. Rev. Biotechnol.* **2001**, *21* (1), 27–48.
13. Gandhi, M.; Chikindas, M. L. *Int. J. Food Microbiol.* **2007**, *113* (1), 1–15.
14. Murphy, C.; Carroll, C.; Jordan, K. N. *J. Appl. Microbiol.* **2006**, *100* (4), 623–32.
15. Velusamy, V.; Arshak, K.; Korostynska, O.; Oliwa, K.; Adley, C. *Biotechnol. Adv.* **2010**, *28* (2), 232–54.
16. Wang, Y.; Ye, Z.; Si, C.; Ying, Y. *Sensors* **2011**, *11* (3), 2728–39.
17. Helrich, K. C. *Official Methods of Analysis of the AOAC. Volume 2*; Association of Official Analytical Chemists Inc.: Gaithersburg, MD, 1990.
18. Kaspar, C. W.; Tartera, C. *Methods Microbiol.* **1990**, *22*, 497–531.
19. Tietjen, M.; Fung, D. Y. C. *Crit. Rev. Microbiol.* **1995**, *21* (1), 53–83.

20. Hobson, N. S.; Tothill, I.; Turner, A. P. F. *Biosens. Bioelectron.* **1996**, *11* (5), 455–477.

21. Lazcka, O.; Del Campo, F. J.; Munoz, F. X. *Biosens. Bioelectron.* **2007**, *22* (7), 1205–17.

22. Sapsford, K. E.; Bradburne, C.; Delehanty, J. B.; Medintz, I. L. *Mater. Today* **2008**, *11* (3), 38–49.

23. Yang, L.; Bashir, R. *Biotechnol. Adv.* **2008**, *26* (2), 135–50.

24. Lehotay, S. J.; Hajslova, J. *TrAC, Trends Anal. Chem.* **2002**, *21* (9-10), 686–697.

25. Bagramyan, K.; Barash, J. R.; Arnon, S. S.; Kalkum, M. *PLoS One* **2008**, *3* (4), e2041.

26. Asensio-Ramos, M.; Hernandez-Borges, J.; Rocco, A.; Fanali, S. *J. Sep. Sci.* **2009**, *32* (21), 3764–800.

27. Escarpa, A.; González, M. C.; López Gil, M. A.; Crevillén, A. G.; Hervás, M.; García, M. *Electrophoresis* **2008**, *29* (24), 4852–4861.

28. Sidorowicz, S. V.; Whitmore, T. N. *J. Inst. Water Environ. Manage.* **1995**, *9* (1), 92–98.

29. USDA. *Foodborne Illness Cost Calculator*; United States Department of Agriculture Economic Research Service - USDA: Washington, DC, 2010.

30. Lauri, A.; Mariani, P. O. *Genes Nutr.* **2009**, *4* (1), 1–12.

31. Petersen, N. J.; Mogensen, K. B.; Kutter, J. r. P. *Electrophoresis* **2002**, *23* (20), 3528–3536.

32. Yi, C.; Zhang, Q.; Li, C.-W.; Yang, J.; Zhao, J.; Yang, M. *Anal. Bioanal. Chem.* **2006**, *384* (6), 1259–1268.

33. Boehm, D. A.; Gottlieb, P. A.; Hua, S. Z. *Sens. Actuators, B* **2007**, *126* (2), 508–514.

34. Boedicker, J. Q.; Li, L.; Kline, T. R.; Ismagilov, R. F. *Lab Chip* **2008**, *8* (8), 1265–1272.

35. Wu, J.; Ben, Y.; Chang, H.-C. *Microfluid. Nanofluid.* **2005**, *1* (2), 161–167.

36. Qiu, J.; Zhou, Y.; Chen, H.; Lin, J.-M. *Talanta* **2009**, *79* (3), 787–795.

37. Gagnon, Z.; Chang, H. C. *Electrophoresis* **2005**, *26* (19), 3725–37.

38. Cheng, I. F.; Chang, H. C.; Hou, D. *Biomicrofluidics* **2007**, *1* (2), 21503.

39. Varshney, M.; Li, Y. B. *Biosens. Bioelectron.* **2009**, *24* (10), 2951–2960.

40. Pethig, R. *Dielectric and Electronic Properties of Biological Materials*; Wiley: New York, 1979.

41. Sadik, O. A.; Aluoch, A. O.; Zhou, A. *Biosens. Bioelectron.* **2009**, *24* (9), 2749–2765.

42. Towe, B. C.; Pizziconi, V. B. *Biosens. Bioelectron.* **1997**, *12* (9-10), 893–899.

43. Berggren, C.; Bjarnason, B.; Johansson, G. *Biosens. Bioelectron.* **1998**, *13* (10), 1061–1068.

44. Mirsky, V. M.; Mass, M.; Krause, C.; Wolfbeis, O. S. *Anal. Chem.* **1998**, *70* (17), 3674–3678.

45. Amatore, C.; Saveant, J. M.; Tessier, D. *J. Electroanal. Chem.* **1983**, *146* (1), 37–45.

46. Ciszkowska, M.; Stojek, Z. *J. Electroanal. Chem.* **1999**, *466* (2), 129–143.

47. Maruyama, K.; Ohkawa, H.; Ogawa, S.; Ueda, A.; Niwa, O.; Suzuki, K. *Anal. Chem.* **2006**, *78* (6), 1904–12.

48. Mernier, G.; Hasenkamp, W.; Piacentini, N.; Renaud, P. *Proc. Eng.* **2010**, *5* (0), 37–40.

49. Gottschamel, J.; Richter, L.; Mak, A.; Jungreuthmayer, C.; Birnbaumer, G.; Milnera, M.; Bruckl, H.; Ertl, P. *Anal. Chem.* **2009**, *81* (20), 8503–12.

50. Richter, L.; Stepper, C.; Mak, A.; Reinthaler, A.; Heer, R.; Kast, M.; Bruckl, H.; Ertl, P. *Lab Chip* **2007**, *7* (12), 1723–31.

51. Yu, J.; Liu, Z.; Liu, Q.; Yuen, K. T.; Mak, A. F. T.; Yang, M.; Leung, P. *Sens. Actuators, A* **2009**, *154* (2), 288–294.

52. Zaytseva, N. V.; Goral, V. N.; Montagna, R. A.; Baeumner, A. J. *Lab Chip* **2005**, *5* (8), 805–811.

53. Zhu, T.; Pei, Z.; Huang, J.; Xiong, C.; Shi, S.; Fang, J. *Lab Chip* **2010**, *10* (12), 1557–1560.

54. Bayoudh, S.; Othmane, A.; Ponsonnet, L.; Ben Ouada, H. *Colloids Surf., A* **2008**, *318* (1–3), 291–300.

55. Liu, Y.-S.; Walter, T. M.; Chang, W.-J.; Lim, K.-S.; Yang, L.; Lee, S. W.; Aronson, A.; Bashir, R. *Lab Chip* **2007**, *7* (5), 603–610.

56. Mujika, M.; Arana, S.; Castano, E.; Tijero, M.; Vilares, R.; Ruano-Lopez, J. M.; Cruz, A.; Sainz, L.; Berganza, J. *Biosens. Bioelectron.* **2009**, *24* (5), 1253–8.

57. Radke, S. A.; Alocilja, E. C. *Biosens. Bioelectron.* **2005**, *20* (8), 1662–1667.

58. Radke, S. M.; Alocilja, E. C. *IEEE Sens. J.* **2005**, *5* (4), 744–750.

59. Radke, S. M.; Alocilja, E. C. *IEEE Sens. J.* **2004**, *4* (4), 434–440.

60. Tan, F.; Leung, P. H. M.; Liu, Z.-b.; Zhang, Y.; Xiao, L.; Ye, W.; Zhang, X.; Yi, L.; Yang, M. *Sens. Actuators, B* **2011**, *159* (1), 328–335.

61. Varshney, M.; Li, Y. B.; Srinivasan, B.; Tung, S. *Sens. Actuators, B* **2007**, *128* (1), 99–107.

62. Deng, L.; Bao, L. L.; Yang, Z. Y.; Nie, L. H.; Yao, S. Z. *J. Microbiol. Methods* **1996**, *26* (1-2), 197–203.

63. Varshney, M.; Li, Y. B. *Biosens. Bioelectron.* **2007**, *22* (11), 2408–2414.

64. Van Rood, J. J.; Van Leeuwen, A.; Ploem, J. S. *Nature* **1976**, *262* (5571), 795–797.

65. Zhao, J.-Y.; Chen, D.-Y.; Dovichi, N. J. *J. Chromatogr., A* **1992**, *608* (1), 117–120.

66. Yamaguchi, N.; Torii, M.; Uebayashi, Y.; Nasu, M. *Appl. Environ. Microbiol.* **2011**, *77* (4), 1536–1539.

67. Berdal, B. P.; Olsvik, O.; Omland, T. *Acta Pathol. Microbiol. Scand., Sect. B* **1981**, *89* (6), 411–5.

68. Crowther, J. S.; Holbrook, R. *Soc. Appl. Bacteriol. Symp. Ser.* **1980**, *8*, 339–57.

69. Lewis, G. E., Jr.; Kulinski, S. S.; Reichard, D. W.; Metzger, J. F. *Appl. Environ. Microbiol.* **1981**, *42* (6), 1018–22.

70. Notermans, S.; Hagenaars, A. M.; Kozaki, S. *Methods Enzymol.* **1982**, *84*, 223–38.

71. Parpura, V.; Chapman, E. R. *Croat. Med. J.* **2005**, *46* (4), 491–7.

72. Poras, H.; Ouimet, T.; Orng, S. V.; Fournie-Zaluski, M. C.; Popoff, M. R.; Roques, B. P. *Appl. Environ. Microbiol.* **2009**, *75* (13), 4382–90.

73. Yang, M. H.; Sun, S.; Kostov, Y.; Rasooly, A. *Lab Chip* **2010**, *10* (8), 1011–1017.

74. Frisk, M. L.; Berthier, E.; Tepp, W. H.; Johnson, E. A.; Beebe, D. J. *Lab Chip* **2008**, *8* (11), 1793–800.

75. Frisk, M. L.; Tepp, W. H.; Johnson, E. A.; Beebe, D. J. *Anal. Chem.* **2009**, *81* (7), 2760–7.

76. Berdat, D.; Rodriguez, A. C. M.; Herrera, F.; Gijs, M. A. M. *Lab Chip* **2008**, *8* (2), 302–308.

77. Schuler, T.; Kretschmer, R.; Jessing, S.; Urban, M.; Fritzsche, W.; Moller, R.; Popp, J. *Biosens. Bioelectron.* **2009**, *25* (1), 15–21.

78. Dutse, S. W.; Yusof, N. A. *Sensors* **2011**, *11* (6), 5754–5768.

79. Njoroge, S. K.; Chen, H. W.; Witek, M. A.; Soper, S. A. *Microfluid.: Technolog. Appl.* **2011**, *304*, 203–260.

80. Hong, J. W.; Studer, V.; Hang, G.; Anderson, W. F.; Quake, S. R. *Nat. Biotechnol.* **2004**, *22* (4), 435–9.

81. Bhattacharyya, A.; Klapperich, C. A. *Sens. Actuators, B* **2008**, *129* (2), 693–698.

82. Dimov, I. K.; Garcia-Cordero, J. L.; O'Grady, J.; Poulsen, C. R.; Viguier, C.; Kent, L.; Daly, P.; Lincoln, B.; Maher, M.; O'Kennedy, R.; Smith, T. J.; Ricco, A. J.; Lee, L. P. *Lab Chip* **2008**, *8* (12), 2071–2078.

83. Ottesen, E. A.; Hong, J. W.; Quake, S. R.; Leadbetter, J. R. *Science* **2006**, *314* (5804), 1464–7.

84. Cheong, K. H.; Yi, D. K.; Lee, J. G.; Park, J. M.; Kim, M. J.; Edel, J. B.; Ko, C. *Lab Chip* **2008**, *8* (5), 810–813.

85. Malic, L.; Sandros, M. G.; Tabrizian, M. *Anal. Chem.* **2011**, *83* (13), 5222–5229.

86. Javanmard, M.; Davis, R. W. *Sens. Actuators, B* **2011**, *154* (1), 22–27.

87. Chen, L.; Lee, S.; Lee, M.; Lim, C.; Choo, J.; Park, J. Y.; Lee, S.; Joo, S. W.; Lee, K. H.; Choi, Y. W. *Biosens. Bioelectron.* **2008**, *23* (12), 1878–1882.

88. Ben-Yoav, H.; Dykstra, P. H.; Bentley, W. E.; Ghodssi, R. *Biosens. Bioelectron.* **2012**, *38* (1), 114–120.

89. Heyries, K. A.; Loughran, M. G.; Hoffmann, D.; Homsy, A.; Blum, L. J.; Marquette, C. A. *Biosens. Bioelectron.* **2008**, *23* (12), 1812–1818.

90. Li, Y. Y.; Xing, D.; Zhang, C. S. *Anal. Biochem.* **2009**, *385* (1), 42–49.

91. Privorotskaya, N.; Liu, Y. S.; Lee, J. C.; Zeng, H. J.; Carlisle, J. A.; Radadia, A.; Millet, L.; Bashir, R.; King, W. P. *Lab Chip* **2010**, *10* (9), 1135–1141.

92. Grabski, A. C. *Methods in Enzymology: Guide to Protein Purification*, 2nd ed.; Burgess, R. R., Deutscher, M. P., Eds.; Elsevier, Inc.: London, 2009; Vol. 463, pp285−303.

93. Huh, Y. S.; Choi, J. H.; Park, T. J.; Hong, Y. K.; Hong, W. H.; Lee, S. Y. *Electrophoresis* **2007**, *28* (24), 4748–4757.

94. Di Carlo, D.; Jeong, K. H.; Lee, L. P. *Lab Chip* **2003**, *3* (4), 287–291.

95. Grahl, T.; Markl, H. *Appl. Microbiol. Biotechnol.* **1996**, *45* (1-2), 148–157.

96. Asiello, P. J.; Baeumner, A. J. *Lab Chip* **2011**, *11* (8), 1420–1430.

97. Craw, P.; Balachandran, W. *Lab Chip* **2012**, *12* (14), 2469–2486.

98. Suarez, G.; Jin, Y. H.; Auerswald, J.; Berchtold, S.; Knapp, H. F.; Diserens, J. M.; Leterrier, Y.; Manson, J. A.; Voirin, G. *Lab Chip* **2009**, *9* (11), 1625–30.

99. He, D. Y.; Zhang, Z. J.; Huang, Y.; Hu, Y. F. *Food Chem.* **2007**, *101* (2), 667–672.

100. Weng, X.; Chon, C. H.; Jiang, H.; Li, D. Q. *Food Chem.* **2009**, *114* (3), 1079–1082.

101. Tan, W.; Desai, T. A. *J. Biomed. Mater. Res., Part A* **2005**, *72A* (2), 146–160.

102. Vickerman, V.; Blundo, J.; Chung, S.; Kamm, R. *Lab Chip* **2008**, *8* (9), 1468–1477.

103. Young, E. W.; Beebe, D. J. *Chem. Soc. Rev.* **2010**, *39* (3), 1036–48.

104. Schaff, U. Y.; Xing, M. M. Q.; Lin, K. K.; Pan, N.; Jeon, N. L.; Simon, S. I. *Lab Chip* **2007**, *7* (4), 448–456.

105. Huh, D.; Matthews, B. D.; Mammoto, A.; Montoya-Zavala, M.; Hsin, H. Y.; Ingber, D. E. *Science* **2010**, *328* (5986), 1662–8.

106. Genes, L. I.; V. Tolan, N.; Hulvey, M. K.; Martin, R. S.; Spence, D. M. *Lab Chip* **2007**, *7* (10), 1256–1259.

107. Neeves, K. B.; Diamond, S. L. *Lab Chip* **2008**, *8* (5), 701–709.

108. De La Fuente, L.; Montanes, E.; Meng, Y.; Li, Y.; Burr, T. J.; Hoch, H. C.; Wu, M. *Appl. Environ. Microbiol.* **2007**, *73* (8), 2690–6.

109. Meng, Y.; Li, Y.; Galvani, C. D.; Hao, G.; Turner, J. N.; Burr, T. J.; Hoch, H. C. *J. Bacteriol.* **2005**, *187* (16), 5560–7.

110. Hopkins, D. L. *Ann. Rev. Phytopathol.* **1989**, *27*, 271–290.

111. Ma, R.; Xie, L.; Han, C.; Su, K.; Qiu, T. A.; Wang, L.; Huang, G. L.; Xing, W. L.; Qao, J.; Wang, J. D.; Cheng, J. *Anal. Chem.* **2011**, *83* (8), 2964–2970.

112. Yoshizaki, G.; Okutsu, T.; Morita, T.; Terasawa, M.; Yazawa, R.; Takeuchi, Y. *Reprod. Domest. Anim.* **2012**, *47*, 187–192.

113. Wlodkowic, D.; Khoshmanesh, K.; Akagi, J.; Williams, D. E.; Cooper, J. M. *Cytometry, Part A* **2011**, *79* (10), 799–813.

114. Lammer, E.; Kamp, H. G.; Hisgen, V.; Koch, M.; Reinhard, D.; Salinas, E. R.; Wendler, K.; Zok, S.; Braunbeck, T. *Toxicol. In Vitro* **2009**, *23* (7), 1436–42.

115. Giacomotto, J.; Segalat, L. *Br. J. Pharmacol.* **2010**, *160* (2), 204–16.

116. Pulak, R. *Methods Mol. Biol.* **2006**, *351*, 275–86.

117. Westerfield, M. *The Zebrafish Book: A Guide for the Laboratory Use of Zebrafish (Brachydanio Rerio)*; M. Westerfield: 1993.

118. Wheeler, G. N.; Brandli, A. W. *Dev. Dyn.* **2009**, *238* (6), 1287–308.

119. Chung, K.; Crane, M. M.; Lu, H. *Nat. Methods* **2008**, *5* (7), 637–43.

120. Hulme, S. E.; Shevkoplyas, S. S.; McGuigan, A. P.; Apfeld, J.; Fontana, W.; Whitesides, G. M. *Lab Chip* **2010**, *10* (5), 589–97.

121. Khoshmanesh, K.; Kiss, N.; Nahavandi, S.; Evans, C. W.; Cooper, J. M.; Williams, D. E.; Wlodkowic, D. *Electrophoresis* **2011**, *32* (22), 3129–32.

122. Lucchetta, E. M.; Munson, M. S.; Ismagilov, R. F. *Lab Chip* **2006**, *6* (2), 185–90.

123. Sundberg, S. A. *Curr. Opin. Biotechnol.* **2000**, *11* (1), 47–53.

124. Wheeler, M. B.; Rutledge, J. L.; Fischer-Brown, A.; VanEtten, T.; Malusky, S.; Beebe, D. J. *Theriogenology* **2006**, *65* (1), 219–227.

125. Wheeler, M. B.; Walters, E. M.; Beebe, D. J. *Theriogenology* **2007**, *68*, S178–S189.

126. Funfak, A.; Brosing, A.; Brand, M.; Kohler, J. M. *Lab Chip* **2007**, *7* (9), 1132–8.

127. Son, S. U.; Garrell, R. L. *Lab Chip* **2009**, *9* (16), 2398–401.

128. Wielhouwer, E. M.; Ali, S.; Al-Afandi, A.; Blom, M. T.; Riekerink, M. B.; Poelma, C.; Westerweel, J.; Oonk, J.; Vrouwe, E. X.; Buesink, W.; vanMil, H. G.; Chicken, J.; van't Oever, R.; Richardson, M. K. *Lab Chip* **2011**, *11* (10), 1815–24.

129. Yang, F.; Chen, Z.; Pan, J.; Li, X.; Feng, J.; Yang, H. *Biomicrofluidics* **2011**, *5* (2), 24115.

130. Pardo-Martin, C.; Chang, T. Y.; Koo, B. K.; Gilleland, C. L.; Wasserman, S. C.; Yanik, M. F. *Nat. Methods* **2010**, *7* (8), 634–6.

131. Akagi, J.; Khoshmanesh, K.; Evans, B.; Hall, C. J.; Crosier, K. E.; Cooper, J. M.; Crosier, P. S.; Wlodkowic, D. *PLoS One* **2011**, *7* (5), e36630.

132. Choudhury, D.; van Noort, D.; Iliescu, C.; Zheng, B.; Poon, K. L.; Korzh, S.; Korzh, V.; Yu, H. *Lab Chip* **2012**, *12* (5), 892–900.

Chapter 9

Applications of Nanoporous Materials in Agriculture

Michael Appell[*,1] and Michael A. Jackson[2]

**[1]Bacterial Foodborne Pathogens & Mycology Research, USDA-ARS, National Center for Agricultural Utilization Research, Peoria, Illinois 61604
[2]Renewable Product Technology Research, USDA-ARS, National Center for Agricultural Utilization Research, Peoria, Illinois 61604
*E-mail: michael.appell@ars.usda.gov.**

Nanoporous materials possess organized pore distributions and increased surface areas. Advances in the systematic design of nanoporous materials enable incorporation of functionality for better sensitivity in detection methods, increased capacity of sorbents, and improved selectivity and yield in catalyst-based synthesis. This chapter surveys the applications of nanoporous materials to improve food safety and utilization of agricultural products. Furthermore, agricultural commodities serve as feedstocks in the development of nanoporous materials.

Introduction

A limiting factor in utilizing solid surface molecular interactions is the size and spacing of molecular recognition and reaction sites. The high surface area and organized structure of nanoporous materials offer a means to increase the density of active sites to enable lower levels of detection and enhance reactivity and selectivity. This class of materials exhibits uniform pore size distributions, ordered pore networks, increased surface areas, special liquid permeability, and better material stability (*1*). These valuable properties benefit the fields of agriculture and food science with more sensitive and selective detection methods and materials with increased number of reaction and binding sites. The source of nanoporous materials is broad and range from metals, oxides, silicates, carbohydrates, polymers, and carbon-based materials with the potential for synthetic modification and rational design (*2*). Nevertheless, nanoporous materials share a fundamental purpose to provide an organized porous platform with pores between 1-100 nm.

Nanoporous materials have pore sizes in the nanometer range and include the class of mesoporous materials, a classification defined by pore size. Microporous materials possess pores from 0.2-2 nm, mesoporous materials have pores between 2-50 nm, and macroporous materials have pores in the range of 50-1000 nm (*3, 4*). Mesoporous materials are of recent interest as rational design materials with the benefits of nanoporous properties. It should be noted there is a history of practical use of materials with inherent micropores to improve agriculture, including activated carbon and zeolites to clean wastewater (*5, 6*).

This overview focuses on applications of nanoporous materials possessing uniform and specific porous architectures introduced during material synthesis, such as hexagonal mesoporous silicas. Figure 1 provides an example of synthetically derived functionalized nanoporous SBA-15 silica with a median pore diameter of 5.4 nm. Applications of nanoporous materials in agriculture include naturally microporous materials for separations and catalysis, nanoporous materials for detection, sorbents to bind toxicants, controlled release materials, solid supports for catalysis, and materials developed from agricultural commodities.

Figure 1. Surface transmission electron micrograph of mesoporous propylthiol functionalized SBA-15 at 100,000 x.

Microporous Materials in Agriculture

Materials naturally possessing micropores have a long history of applications in agriculture. Microporous silica, alumina, minerals, clays, and activated black carbon have been applied as size selective sorbents, non-selective sorbents, and catalysts. Generally, activated carbon materials possess affinities for organic molecules, while zeolites have affinities for gases, ions, metals, and small organic molecules. Furthermore, zeolites are used as popular scaffolds for catalysis. Early research in fuel development was based on catalytic cracking zeolites, with recent reports utilizing biomass for biofuels (*7*). Applications of microporous materials in agriculture have been extensively reviewed recently (*5, 6, 8*). Furthermore, microporous clays have been applied as sorbents, including as additives to sequester contaminants through in-feed applications as well as reduce human exposure to contaminants (*9, 10*). Synthetic modifications have led to expanding the uses of microporous materials through incorporating new functionality into the material structure. For example, mesoporous structure has been introduced into microporous zeolites to improve and expand the application of this material (*11*). In addition, current efforts focus on identifying green alternatives to activated carbon, including microporous biochar materials from the pyrolysis of agricultural waste (*8, 12*).

Nanoporous Materials for Food Safety

Interest in applying nanoporous materials to food safety can be attributed not only to the porous properties of the materials, but also the straightforward incorporation of useful components during synthesis, including components with magnetic, optical, and detection properties (*1*). Nanoporous materials have improved food safety methods by influencing data acquisition, detection, and molecular recognition (*13*). Moreover, materials can be modified through a variety of approaches, including conjugation with bio-based materials, such as antibodies. New synthetic approaches can be employed to incorporate functional groups not present in biological materials increasing the function of the materials.

Detection

Applications of nanoporous materials to improve detection in food safety address pathogenic biological organisms and viruses, including *Salmonella enteritidis*, bacteriophage virus MS2, *Escherichia coli* O157:H7 bacteria, and *Staphylococcus aureus*. Nanoporous membranes have been developed to detect the food allergen peanut. In addition, methods utilizing nanoporous materials have been developed to detect small organic molecules. These types of applications employ mesoporous materials.

Nanoporous silicon was utilized in biosensors capable of selective detection of *Salmonella enteritidis*, and these materials exhibit promise for high through-put screening applications. The porous material was synthesized electrochemically and functionalized with DNA probes selective for *Salmonella enteritidis* (*14*). The increased surface area associated with the nanoporous silicon permitted low levels of detection at 1 ng ml^{-1}, and outperformed nonporous silicon-based biosensors.

Films of nanoporous silicon bioconjugated with antibodies have enabled detection of the bacteriophage virus MS2, a difficult to detect and remove contaminant of drinking water. The high surface area associated with the nanoporous films increased the sensitivity of the assay allowing very low levels of detection of dye-labeled MS2 viruses (*15*). The films possessed an average pore dimension of 50 nm. The method developed using the nanoporous silicon outperformed the popular protein micro-array based detection methods.

Several polycarbonate membranes with nanopores of various sizes coated with gold were applied in an immunosensor based method to detect the food allergen peanut (*16*). The recognition component was an antibody selective for the Ara h1 peanut protein. A method based on detection through a change in conductivity was developed. Sensitivity and selectivity of the method was associated on the size of the pores. Membranes with the smallest pore size (15 nm) were the most sensitive and selective.

A polydimethylsiloxane microfluidic sensor with antibodies immobilized on an alumina nanoporous membrane was developed to detect *Escherichia coli* O157:H7 and *Staphylococcus aureus* with electrochemical impedance spectrum detection (*17*). The method could detect the bacteria pathogens selectively within 1-2 hours and had a sensitivity of 10 CFU ml^{-1}, which is a significant improvement over traditional microelectrode based impedance sensors. In addition, electrospun nanofibers utilizing magnetic nanoparticles were applied in a lateral-flow immuno-biosensor in a rapid method to detect *Escherichia coli* O157:H7 bacteria (*18*). The molecular recognition component was *Escherichia coli* O157:H7 selective antibodies bioconjugated to the electrospun nanofibers. The electrospun nanofibers provided capillary properties and high surface area to enable detection at levels less than 70 CFU ml^{-1} and the assay could be completed in 8 min.

An electrochemical method was developed for sensing streptomycin, **1**, using mesoporous silica with a mean size of 100 nm embedded with gold nanoparticles (see Figure 2) (*19*). The detection method employed was similar to competitive immunoassays and the limit of detection was 5 pg ml^{-1}. Validated streptomycin methods were developed in variety of commodities including meat, milk, and honey. Also, mesoporous nanoparticles were applied in an electrochemical immunosensor for the anabolic steroid norethisterone, **2**, a contaminant of concern in food from animal-based products (*20*). A sandwich type sensor was developed based on two types of antibodies for molecular recognition, and the limit of detection was less than 4 pg ml^{-1}. In addition, a method was developed for electrochemical detection of catechols in teas using mesoporous aluminum doped silica modified electrodes (*21*).

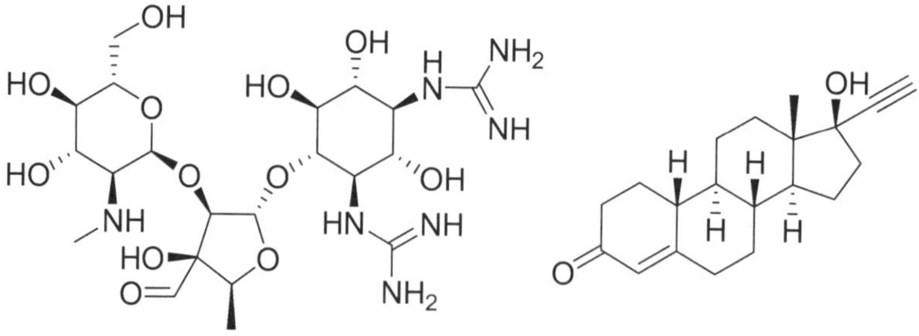

Figure 2. Streptomycin (1) and norethisterone (2).

Applications of nanoporous materials as sorbents have focused on sequestration of small organic molecules and metals with the size of less than 1 nm. Novel sorbents have been developed to aid in the nondestructive isolation of small organics from plant matrix. Gibberellic acid, **3**, (see Figure 3) regulates plant growth and is of great interest as a natural product bioactive. However, isolation and determination of levels of gibberellic acid is difficult because of its low concentrations in plant matter and lack of stability during isolation using traditional methods. A nanoporous silica-sucrose material was developed with tunable pore structure possessing the ability to bind gibberellic acid (*22*). The novel sorbent was capable of binding over three times more gibberellic acid compared to tradition protein and carbohydrate based sorbents. In addition, rational designed nanoporous materials have aided in the isolation of natural products. Mesoporous carbon nanocages created through green syntheses assisted isolation and detection of structurally diverse catechin and tannic acids from tea (*23*).

Functionalized SBA-15 silica was developed to remove the fungal toxin patulin from apple juice through covalent sequestration (*24*). Micelle template silicas were synthesized with propylthiol residues, a functional group known to react with patulin, **4**. Propylthiol functionalization of the SBA-15 reduced the pore size from 9.0 nm to 5.4 nm. It was demonstrated that patulin could be removed from apple juice and aqueous solutions. Computational chemistry studies using density functional methods at the B3LYP/6-31G(d,p) level predicted the adducts formed between the reaction of the thiol groups of the functionalized SBA-15 and patulin would be less toxic than patulin. This high capacity material offers a means to remove patulin from beverages without leaching of the patulin from the material. Moreover, since the material is insoluble it provides a convenient platform to remove the detoxified patulin adducts from beverages.

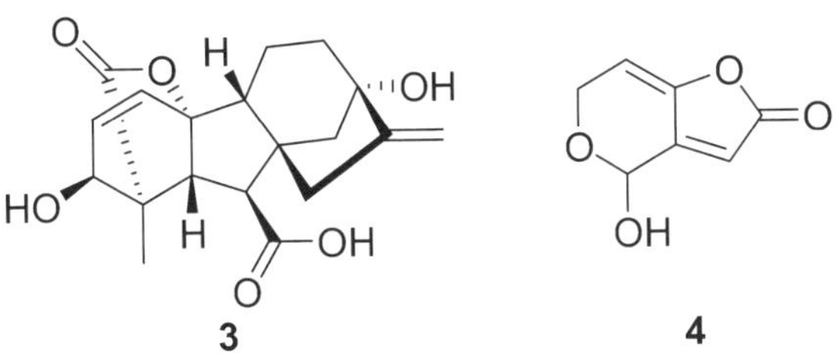

Figure 3. Gibberellic acid (3) and patulin (4).

Functionalized mesoporous silicas have been applied to reduce levels of heavy metals from water and beverages. The materials were capable of calcium fortification of water samples, while removing uranium (*25, 26*). Also, similar functionalized mesoporous silica materials are reported to remove a broad range of metal contaminants, including those resulting from nuclear disaster. Traditional and novel sorbents have been studied for their ability to remove a range of hard and soft metals from water, and offer promise for sequestration of radionuclides (*27*). Salt concentrations had a significant effect on adsorption of heavy metals, however, organic matrix did not influence adsorption.

Nanoporous Controlled Release Materials

Controlled release materials offer a means for direct delivery of bioactives to target release points, pulsed release, and extended release. Nanoporous materials have been investigated as controlled release materials to highly concentrate bioactives and reduce the amount of controlled release material needed in applications. In addition, the controlled release materials can be functionalized to tune the release rate. Theoretical studies report that surface barriers rather than diffusion influence release of bioactives from nanoporous materials (*28*). Nanoporous technology in the area of controlled release includes the use of agricultural commodities as controlled release matrix and applications of controlled release materials to improve agriculture.

A nanoporous silica was investigated for its ability to release of (-)-menthol, **5**, a flavor and bioactive (See Figure 4). Pore size (2.4 to 40.7 nm), structure, and wall thickness significantly influenced release of menthol and modulation of these parameters enabled controlled release (*29*). Functionalization with hydrophobic groups significantly reduced the release rate of menthol in aqueous solutions.

Chitin is an agricultural bioproduct from shellfish. Multilayer microcapsules of nanoporous poly(l-glutamic acid)/chitosan were developed and investigated for general drug release. Evaluations using 5-fluorouracil, **6**, as the model release compound revealed release rate can be modulated by ionic strength, specifically pH and salt concentration (*30*). Salt concentrations and pH influenced loading capacity and acidic conditions supported sustained release. The material exhibited favorable anticancer activity in the MTT cytotoxicity assay, and show promise for drug delivery applications.

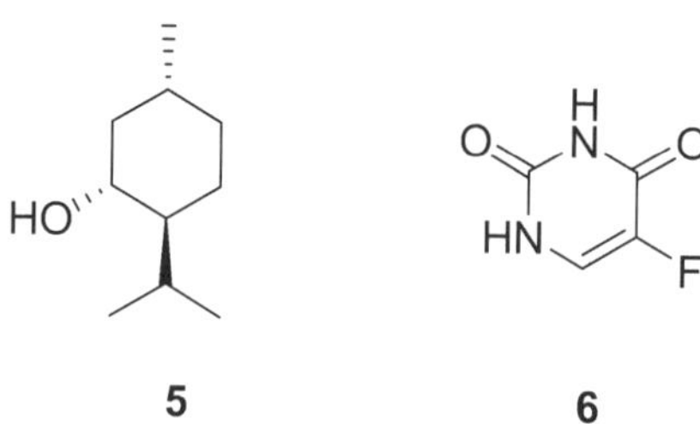

5   **6**

Figure 4. Materials investigated for the controlled release from nanoporous materials, (-)-menthol, 5, and 5-fluorouracil, 6.

Nanoporous Materials as Catalysis

Nanoporous materials have seen applications as high capacity supports to enable catalytic synthesis of fuels from food waste and algae. Mesoporous materials Al-MCM-41 and MF1 zeolites were investigated for their catalytic activity to produce energy materials from the oils of food waste leachate (*31*). The reaction produced oxygenated compounds and aromatics. The nanopores of MF1 zeolite increased acidity and provided volatiles, aromatics and hydrocarbons in the gasoline range. The weakly acidic Al-MCM-41 gave oxygenates and hydrocarbons in the diesel range. Bifunctional mesoporous catalysts containing acid functionality alongside hydrophobic regions have been used to make biodiesel from free fatty acids (*32*). In addition, agricultural waste has been converted into a fuel by a nanoporous aluminum coated with platinum, platinum-ruthenium, and platinum-ruthenium-carbon (*33*). This material was a key component in the development of a bioelectrochemical fuel based on the fermented rice husks. A molybdenum carbide catalyst supported on ordered mesoporous carbon was developed to produce hydrocarbons for diesel fuel from vegetable oil (*34*). The ordered mesoporous carbon material improved selectivity and reduced leaching during synthesis.

Lesquerella fendlerii is a robust mustard plant capable of growing in dry environments. Increasing the value of the oils of this plant can foster the use of this new crop dry lands with limited agricultural value. A mesoporous SBA-15 silica functionalized with sulfonic acid and chloroalkyl groups improved the catalytic synthesis of value-added products from oils of this plant (*35*). Specifically, the novel catalysts permitted the selective synthesis of phenoxy ether derivatives of methyl lesquerolate. The mechanism proposed is a selective decrease in Friedel-Crafts adducts. In addition, several mesoporous MCM-41 based catalysts were developed to synthesize *t*-butyl-*p*-cresol, a popular antioxidant used by the food industry (*36*). The best catalyst developed was based on MCM-41 functionalized with zinc and aluminum.

Attempts to esterify free ω-sulfhydryl fatty acid with sulfuric and *p*-toluene sulfuric acid produce a mixture of products, including reactions with the sulfhydryl groups. To address this issue, a novel SBA-15 functionalized with arenesulfonic acid and propyl sulfonic groups was developed to selectively catalyze ethanolic esterification of ω-sulfhydryl fatty acid while limiting side product formation (*37*). Scale-up to gram level production was possible by using nonporous Amberlyst-15.

Incorporating bioproteins onto mesoporous materials offers a means to highly concentrate active sites on a small insoluble surface and preserve the native and active conformation of the protein. The membrane protein chlorophyllase was entrapped in sol-gel silica to immobilize the protein (*38*). The effect of immobilization only reduced activity by 20%. Characterization of the sol-gel silica and sol-gel silica with chlorophyllase by thermoporometry indicate the chlorophyllase acts as a template during the sol-gel process and influences the morphology of the resulting polymer. In addition, *Candida Antarctica* lipase B has been immobilized on hydrophobic and hydrophilic SBA-15 and the catalytic activity has been maintained (*39*).

Nanoporous and nanoparticle electrodes based on Ti/TiO$_2$ films were prepared using the sol-gel method and evaluated for the ability to remove food dye from wastewater (*40*). The size of the nanoporous electrode was calculated to be 20 nm and the nanoparticle electrode was 10 nm. The nanoparticle electrode possessing the best activity also had the highest surface area and was capable of removing the dye in 15 minutes.

Agricultural Commodities in Nanoporous Materials

Agricultural materials have been applied as components in synthetic nanoporous materials. Polysaccharide aerogels have been developed as a promising class of nanoporous materials with potential uses as lightweight, highly porous bioactive delivery systems with high surface areas (*41*). Also, nanoporous materials have been developed using chitosan polysaccharide as the feed stock through electrospinning techniques and evaluated for antimicrobial activities (*42*). Several parameters of the chitosan and polylactic acid synthesis were investigated and optimized, including solvent, molecular weight, and spinning conditions.

Conclusion

There is a strong history of the use of microporous materials as sorbents in agriculture. Recent advances in mesoporous materials have improved food safety and utilization of agricultural products. Challenges for developing applications of nanoporous materials include matrix interference with the molecular recognition components of the materials and reducing the cost of the materials. Matrix interferences are not new and often complicate the practical application of technology in agriculture. Separation of nanomaterials from liquid matrices may be assisted through incorporation of components for magnetic separation or centrifugation. The cost of synthesizing nanoporous materials is expected to decrease as the materials are more broadly utilized in new applications. Overall, nanoporous materials offer a means to increase the selectivity and sensitivity of detection, isolate small molecules, and tune reactions.

Acknowledgments

Mention of trade names or commercial products in this article is solely for the purpose of providing specific information and does not imply recommendation or endorsement by the U.S. Department of Agriculture. USDA is an equal opportunity provider and employer.

References

1. Dai, Z.; Ju, H. *TrAC, Trends Anal. Chem.* **2012**, *39*, 149–162.
2. Wu, D.; Xu, F.; Sun, B.; Fu, R.; He, H.; Matyjaszewski, K. *Chem. Rev. (Washington, DC, U. S.)* **2012**, *112*, 3959–4015.

3. Rouquerol, J.; Avnir, D.; Fairbridge, W.; Everett, D. H.; Haynes, J. H.; Pernicone, N.; Ramsay, D. F.; Sing, K. S. W.; Unger, K. K. *Pure Appl. Chem.* **1994**, *66*, 1739–1758.

4. IUPAC. *Compendium of Chemical Terminology*, 2nd ed. (the "Gold Book"); Compiled by McNaught, A. D., Wilkinson, A.; Blackwell Scientific Publications: Oxford, 1997; XML online corrected version: http://goldbook.iupac.org (2006) created by M. Nic, J. Jirat, B. Kosata; updates compiled by A. Jenkins. ISBN 0-9678550-9-8. doi:10.1351/goldbook.

5. Jusoh, A.; Hartini, W. J. H.; Ali, N.; Endut, A. *Bioresour. Technol.* **2011**, *102*, 5312–5318.

6. Ramesh, K.; Reddy, D. D.; Biswas, A. K.; Rao, A. S. *Adv. Agron.* **2011**, *113*, 215–236.

7. Mortensen, P. M.; Grunwaldt, J.-D.; Jenson, P. A.; Knudsen, K. G.; Jensen, A. D. *Appl. Catal., A* **2011**, *407*, 1–19.

8. Manyà, J. J. *Environ. Sci. Technol.* **2012**, *46*, 7939–7954.

9. Slamova, R.; Trckova, M.; Vondruskova, H.; Zraly, Z.; Pavlik, I. *Appl. Clay Sci.* **2011**, *51*, 395–398.

10. Phillips, T. D.; Afriyie-Gyawu, E.; Williams, J.; Huebner, H.; Ankrah, N.-A.; Ofori-Adjei, D.; Jolly, P.; Johnson, N.; Taylor, J.; Marroquin-Cardona, A.; Xu, L.; Tang, L.; Wan, J.-S. *Food Addit. Contam., Part A* **2008**, *25*, 134–145.

11. Na, K.; Choi, M.; Ryoo, R. *Microporous Mesoporous Mater.* **2013**, *166*, 3–19.

12. Uchimiya, M.; Wartelle, L. H.; Klasson, K. T.; Fortier, C. A.; Lima, I. M. *J. Agric. Food Chem.* **2011**, *59*, 2501–2510.

13. Vaseashta, A. *Mater. Manuf. Process.* **2006**, *21*, 710–716.

14. Zhang, D.; Alocilja, E. C. *IEEE Sens. J.* **2008**, *8*, 775–780.

15. Rossi, A. M.; Wang, L.; Reipa, V.; Murphy, T. E. *Biosens. Bioelectron* **2007**, *23*, 741–745.

16. Singh, R.; Sharma, P. P.; Baltus, R. E.; Suni, I. I. *Sens. Actuators, B* **2010**, *145*, 98–103.

17. Tan, F.; Leung, P. H. M.; Liu, Z.-B.; Zhang, Y.; Xiao, L.; Ye, W.; Zhang, X.; Yi, L.; Yang, M. *Sens. Actuators, B* **2011**, *159*, 328–335.

18. Luo, Y.; Nartker, S.; Wiederoder, M.; Miller, H.; Hochhalter, D.; Drzal, L. T.; Alocilja, E. C. *IEEE Trans. Nanotechnol.* **2012**, *11*, 676–681.

19. Liu, B.; Zhang, B.; Cui, Y; Chen, H.; Gao, Z.; Tang, D. *ACS Appl. Mater. Interfaces* **2011**, *3*, 4668–4676.

20. Wei, Q.; Xin, X.; Du, B.; Wu, D.; Han, Y.; Zhao, Y.; Cai, Y; Li, R.; Yeng, M.; Li, H. *Biosens. Bioelectron.* **2010**, *26*, 723–729.

21. Lin, H.; Gan, T.; Wu, K. *Food Chem.* **2009**, *113*, 701–704.

22. Li, J.; Xia, J.-T.; Zhang, J.-H.; Liang, T. *J. Vac. Sci. Technol., A* **2010**, *28*, 658–661.

23. Ariga, K.; Vinu, A.; Miyahara, M.; Hill, J. P.; Mori, T. *J. Am. Chem. Soc.* **2007**, *129*, 11022–11023.

24. Appell, M.; Jackson, M. A.; Dombrink-Kurtzman, M. A. *J. Hazard. Mater.* **2011**, *187*, 150–156.

25. Alothman, Z. A.; Apblett, A. W. *Chem. Eng. J.* **2009**, *155*, 916–924.

26. Alothman, Z. A.; Apblett, A. W. *J. Hazard. Mater.* **2010**, *182*, 581–590.

27. Johnson, B. E.; Santschi, P. H.; Addleman, R. S.; Douglas, M.; Davidson, J. D.; Fryxell, G. E.; Schwantes, J. M. *Appl. Radiat. Isot.* **2011**, *69*, 205–216.

28. Heinke, L.; Kärger, J. *Phys. Rev. Lett.* **2011**, *106*, 074501.

29. Zhang, J.; Yu, M.; Yuan, P.; Lu, G.; Yu, C. *J. Inclusion Phenom. Macrocyclic Chem.* **2011**, *71*, 593–602.

30. Yan, S.; Rao, S.; Zhu, J.; Wang, Z.; Zhang, Y.; Duan, Y.; Chen, X.; Yin, J. *Int. J. Pharm. (Amsterdam, Neth.)* **2012**, *427*, 443–451.

31. Kim, S.-S.; Heo, H. S.; Kim, S. G.; Ryoo, R.; Kim, J.; Jeon, J.-K.; Park, S. H.; Park, Y.-K. *J. Nanosci. Nanotechnol.* **2011**, *11*, 6167–6171.

32. Mbaraka, I. K.; Shanks, B. H. *J. Catal.* **2005**, *229*, 365–373.

33. Paul, S. *J. Fuel Cell Sci. Technol.* **2012**, *9*, 1–9.

34. Han, J.; Duan, J.; Chen, P.; Lou, H.; Zheng, X.; Hong, H. *ChemSusChem.* **2012**, *5*, 727–733.

35. Jackson, M. A.; Appell, M. *Appl. Catal., A* **2010**, *373*, 90–97.

36. Selvaraj, M.; Sinha, P. K. *J. Mol. Catal. A: Chem.* **2007**, *264*, 44–49.

37. Compton, D. L.; Jackson, M. A. *J. Am. Oil Chem. Soc.* **2011**, *88*, 1799–1805.

38. Yi, Y.; Kermasha, S.; Neufeld, R. *Biotechnol. Bioeng.* **2006**, *95*, 841–849.

39. Laszlo, J. A.; Jackson, M.; Blanco, R. M. *J. Mol. Catal. B: Enzym.* **2011**, *69*, 60–65.

40. Guaraldo, T. T.; Pulcinelli, S. H.; Zanoni, M. V. B. *J. Photochem. Photobiol., A* **2011**, *217*, 259–266.

41. García-González, C. A.; Alnaief, M.; Smirnova, I. *Carbohydr. Polym.* **2011**, *86*, 1425–1438.

42. Torres-Giner, S.; Ocio, M. J.; Lagaron, J. M. *Eng. Life Sci.* **2008**, *8*, 303–314.

Chapter 10

Breaking up of Biofilms with *Moringa oleifera:* Insights into Mechanisms

Bishambar Dayal,[*,1,2] Vineela Reddy Yannamreddy,[2]
Chandran Ragunath,[3] Narayanan Ramasubbu,[3] Trinava Roy,[4]
Ritesh Amin,[2] Swayam. P. Nirujogi,[1,2] and Michael A. Lea[2]

[1]Department of Medicine, Rutgers University - New Jersey Medical School,
Newark, New Jersey 07103
[2]Department of Biochemistry and Molecular Biology,
Rutgers University-New Jersey Medical School, Newark, New Jersey 07103
[3]Department of Oral Biology, Rutgers University-New Jersey Dental School,
Newark, New Jersey 07103
[4]Undergraduate College of Natural Sciences-University of Massachusetts,
Amherst, Massachusetts 01003
*E-mail: dayalbi@rutgers.njms.edu.

The chemical properties of bioactives are important factors for nanoparticle drug delivery carriers. In this study, we systematically examine the antimicrobial and anticancer activities of different sections of the tropical fruit *Moringa oleifera*. This study showed that the seed cover (**seed coat**) had higher antimicrobial properties in comparison to the skin, the inner skin skeleton (pulp) and the seeds. Moringa seeds have been reported to act directly upon microorganisms resulting in growth inhibition but to our surprise seed cover had a dramatic effect and thus prevented the formation of biofilm itself. These experiments were carried out using strain NJ 9709-Staphylococcus epidermidis, a strain with antibiotic resistance. The parent glucosinolates do not possess significant biological activities and hydrolysis with the myrosinase enzyme transforms the glucosinolates to active chemopreventive isothiocyanate products that are more suitable for nanotechnology-based drug delivery systems.

Introduction

Moringa oleifera is a fast growing tree having multiple medicinal qualities (*1, 2*). It has been widely used in India, Pakistan, Bangladesh, Afghanistan and other parts of the world as a traditional medicine for thousands of years (*1–4*). Various parts of the *Moringa oleifera* tree have great use in Unani and Siddha medicine (*4, 5*). Recent data from World Health Organization shows that African and Asian countries use this and other herbal medicines extracted from natural plant products for primary health care (*1, 6, 7*).

In continuation of our research program we recently reported bioactive glucosinolates and their hydrolyzed products present in *Moringa oleifera* (Figure1). Their isolation and structure elucidation using thin-layer chromatography and electrospray ionization mass spectrometry was achieved (*1*). Further, a comparative inhibition of growth and induction of differentiation of colon cancer cells by extracts from okra (*Abelmoschus esculentus*) and drumstick (*Moringa oleifera*) was also elucidated (*8, 9*). The enzyme myrosinase (thioglucoside glucohydrolase) breaks down glucosinolates into bioactive compounds known as isothiocyanates, which have proven chemopreventive effects against different types of cancers including prostate, bladder, lung, colorectal and colon cancer (*10–13*). Posner and coworkers isolated and characterized sulforaphane, a major inducer of anticarcinogenic protective enzymes from broccoli (*14*). In a recent and important study, Ganin et al. (*15*) elucidated two natural isothiocyanates isolated from broccoli, namely sulforaphane (*13*) and erucin (*14*). These are potent selective antibacterials and possess anti-Helicobactor activity (*16*) acting as chemopreventive agents. Bassler and coworker have extensively studied quorum sensing, cell to cell communication in bacteria, a chemical discourse and bacterial small molecule signaling pathway (*17–21*). Ilana and colleagues used this small molecule signaling pathway approach to highlight that fours specific D-amino acids, tyrosine, lysine, methionine and tryptophan, trigger biofilm disassembly (*22*), thus fighting dangerous pathogens (*15–22*). Since *Staphylococcus epidermidis* is the primary cause of infections of indwelling medical devices such as intravascular catheters, alternative strategies are needed to inhibit both bacterial growth and *Staphylococcus epidermidis* biofilms (*23*). Present studies describe biofilm busting activities (*24*) of compounds present in the seed cover and the pulp tissue sections of the *Moringa oleifera*.

These bioactive compounds that have been elucidated in our studies from *Moringa oleifera* have the potential to be delivered via nanotechnology. An effective way to deliver such compounds is nanoencapsulation. The various properties of nanoparticles like their size similarity to biomolecules such as proteins allow them to be efficient drug delivery vehicles (*25*). The bioactive compounds can be encapsulated by nanoparticles composed of various cores, including silica and gold. These nanoparticle-encapsulated compounds can then be delivered to the system and released. The advantage of encapsulating the bioactive compounds with nanoparticles is a controlled release and the capsules rigidity (*26*).

Experimental (Materials and Methods)

Plant Materials

The *Moringa Oleifera* was dissected into four different sections: Skin, Pulp, seed coat, and seeds.

Figure 1. Fresh Moringa oleifera, dissected sections.

Extraction and Thin-Layer Chromatography

Extraction

The individual sections were treated with methanol and efficient extraction of the bioactives was carried out by microwave-assisted extraction (*24, 27–31*).

Chromatographic Materials

All TLC isolated products were detected as one spot when subjected to thin-layer chromatographic examination (precoated 0.25-mm silica plates from Analtech, Newark, DE) (*1*).

The compounds were resolved and analyzed by analytical and preparatory TLC and subjected to electrospray-ionization mass spectrometry as described previously (*8, 30*). Microwaved methanolic extracts of seed coat and pulp (10-20 mg) were evaporated under nitrogen and sequentially extracted with hexane and chloroform. The chloroform and hexane extracts from the seed coat and seed cover were then separated in the following TLC solvent system (chloroform: methanol: acetic acid 18:4:0.5 v/v/v Figure 2). Rf=0.18 was assigned to intact glucosinolate isolated from seed coat and pulp. The less polar spot with Rf=0.23/0.24 was assigned to its corresponding acetylated derivative. The extremely low polar spot (Rf=0.98) is a reference standard for sulforaphane and Rf=0.93/0.94 (Figure 2), is assigned to the hydrolyzed products of glucosinolates specifically the isothiocyanates. The spot with Rf=0.54 is an impurity in the reference standard sulforaphane.

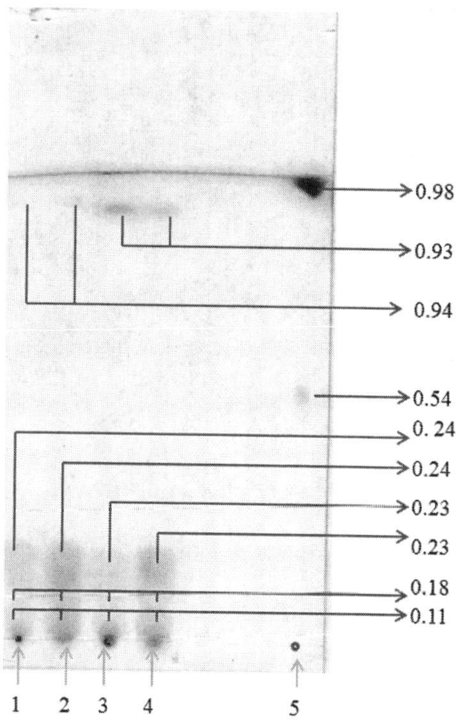

Figure 2. TLC profile of Moringa oleifera (CHCl₃ extract) in the solvent system chloroform: methanol: acetic Acid (18:4:0.5 v/v/v). Lane 1- skin, lane 2- pulp, lane 3- seed coat, lane 4- seed and lane 5- sulforaphane.

Electrospray-Ionization Mass Spectrometry

The Collision-induced (CID) mass spectra of glucosinolates and their break down products were recorded on a Micromass Quattro I tandem mass spectrometer equipped with an electrospray ion source as described previously (*1, 8, 32*). Samples were infused with acetonitrile–water–ammonia (9:1:0.00001, v/v/v) solutions at a flow rate of 6 µl/min and the source temperature was held at 80 °C.

Isolation and Structure Determination

As described previously the structure and stereochemistry of glucosinolates and isothiocyanates were elucidated via a combination of NMR and electrospray-ionization mass spectrometry (ESI-MS) (*1, 8, 24, 32*). Briefly, the hexane extraction provided low-molecular weight glucosinolates and their hydrolyzed products. Extraction of the seed coat with chloroform provided a mixture of intact glucosinolates, their acetyl derivatives, isothiocyanates and rhamnose carbamate as revealed by negative ion electrospray ionization mass spectrometry. Further MS/MS (MS2 scans) were performed for peaks at m/z 570 and 341, which demonstrated structural assignment for 4-($\pm\alpha$ L-rhamnopyranosyloxy) benzyl glucosinolates, and their myrosinase hydrolyzed low molecular weight isothiocyanates and rhamnose carbamate (Figure 3).

Bacterial Strains

Four *S. epidermidis* strains (designated NJ9709, NJ9710, NJ9711, and NJ9712) were isolated from the surfaces of infected intravenous catheters removed from patients at University Hospital, Newark, N.J. The strains were identified by using the Api-Staph biochemical identification kit (bioMérieux, Durham, N.C.).

All four strains contained the *ica* genetic locus (*23*) and produced black colonies on Congo red agar both of which were indicative of slime production as previously described (*23*).

Biofilm Formation

Media and Growth Conditions Protocol

The *Staphylococcus epidermidis* strain (designated NJ9707) was streaked onto a 5% sheep's blood agar plate and incubated for 24 h at 37°C in air. Three loops of bacteria from the agar plate were transferred into a microcentrifuge containing 200 µl of fresh medium Trypticase soy broth. The tube was vortexed for 1 min at high speeds, and the cells were allowed to settle for 5 min. 200 µl of the upper layer was transferred to a petri dish containing 25 ml of fresh medium, and the dish was incubated for 18 h. Biofilms were grown in Trypticase soy broth (Becton-Dickinson) supplemented with 6 g of yeast extract and 8 g of glucose per liter. The biofilm that formed on the surface of the dish was rinsed with phosphate-buffered saline (PBS) and subsequently scraped from the surface of the dish into 3 mL of PBS by using a cell scraper. The cells were then transferred to a centrifuge tube, vortexed for 1 min, and allowed to settle to the bottom of the tube for 10 min. The 150 µl bacteria and the *Moringa oleifera* extract aliquots were taken in the following manner (see Table 1).

Table 1. *Moringa oleifera* **and bacterial extract aliquots used in the experimental protocol as shown above**

Tube 1	Tube 2	Tube 3
300 µl PBS buffer and 150 µl bacteria	*Moringa oleifera* seed or pulp extract in methanol 150 µl and 150 µl PBS buffer 150 µl bacteria	*Moringa oleifera* seedcoat/pulp extract in methanol 300 µl and 150 µl bacteria

The tubes were incubated for 2 hrs in 37°C. A 0.5 ml aliquot of the top layer was transferred to a tube containing 5 ml of fresh broth and the tube was then vortexed for 30 s. The pellet at the bottom of the tube was broken up and allowed to settle to the bottom of the tube for 10 min. After the serial dilutions the biofilms on the 96 well plates were stained with crystal violet and observed.

The preparation of inocula was followed as described previously by Kaplan et. al (*23*).

Briefly, a loopful of colonies scraped from the surface of an agar plate was transferred to a microcentrifuge tube containing 200 µl of fresh medium. The tube was vortexed for 30 s at high speed, and the cells were allowed to settle for 5 min. Then, 100 µl of the upper layer were transferred to a 100-mm-diameter polystyrene Petri dish (model 3003; Falcon) containing 20 mL of fresh medium, and the dish was incubated for 16 h. The biofilm that formed on the surface of the dish was rinsed with phosphate-buffered saline (PBS) and then scraped from the surface of the dish into 3 ml of PBS by using a cell scraper. The cell aggregate was transferred to a 15 mL conical centrifuge tube, vortexed for 30 s, and allowed to settle to the bottom of the tube for 10 min. A 0.5 mL aliquot of the top layer was transferred to a tube containing 5 mL of fresh broth and the tube was vortexed briefly. The resulting inoculums contained 10^9 to 10^{10} CFU/mL. Serial decimal dilutions were made with fresh broth.

Growth of Biofilms in 96-Well Polystyrene Micro-Titer Plates

The wells of a 96-well polystyrene microtiter plate (model 3595; Corning) were filled with 100-µl aliquots of inoculum, and the plate was incubated for 16 h. The wells were rinsed with either three 200-µl aliquots of PBS or by submerging the entire plate in a tub of cold, running tap water. Biofilms were stained with crystal violet as previously described (*23, 33, 34*).

Results and Discussion

S. epidermidis is a major cause of hospital-based bloodstream, cardiovascular, eye, ear, nose, and throat infections. This pathogen leads to acute sepsis thus causing multiple organ failure in catheterized AIDS and HCV patients and premature death in newborns (*35*).

S. epidermidis grows on medical devices as an adherent biofilm consisting of communities of bacterial cells, a condition called quorum sensing (*17–21*), which are firmly attached to the underlying surface. Furthermore, *S. epidermidis* biofilms become highly resistant to antibiotics and thus cause infections in humans that are difficult to treat. In this context it is of paramount importance to develop alternative strategies to inhibit bacterial growth.

Recent preliminary studies from our laboratory indicated biofilm-busting activity is present in the compounds present in *Moringa oleifera*. These experiments were carried out using the strain NJ 9709-*Staphylococcus epidermidis*, a strain resistant to almost all antibiotics (*24*). We have now carried out a detailed evaluation of **in-vitro** antimicrobial activity of various tissues of *Moringa oleifera* (Drumsticks) against *Staph. epidermidis*. Specifically, the growth inhibition and prevention of biofilm formation by the bioactive compounds (isothiocyanates m/z 311, 354, 179, 175, 191, 129) present in the *Moringa oleifera* seed coat and pulp tissues was evaluated.

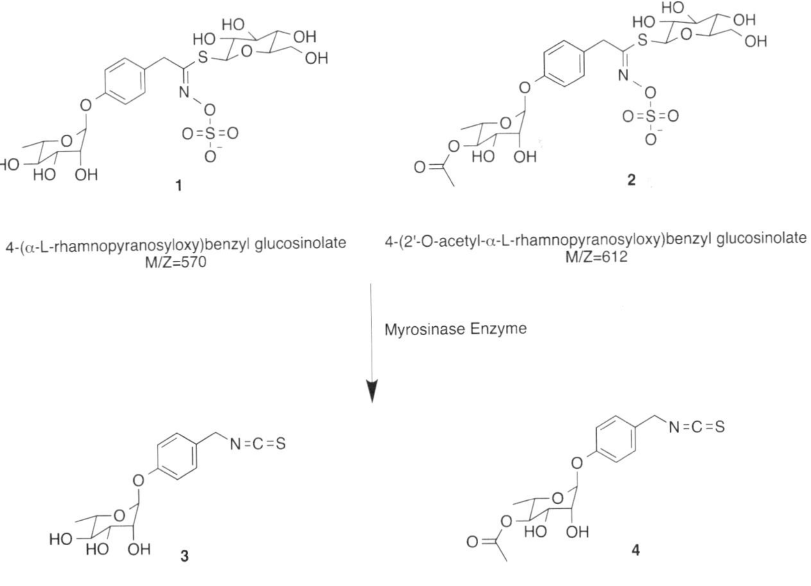

Figure 3. Myrosinase catalyzed hydrolysis products of 4-(±α-L-rhamnopyranosyloxy) benzyl glucosinolate and 4-[(2'-O- acetyl ± α-L-rhamnopyranosyloxy) benzyl glucosinolate.

The microwaved methanolic extract of bioactives present in *Moringa oleifera* were resolved and analyzed by analytical and preparatory TLC (Figure 2). Examination of the mass spectrometric studies of the seed cover and pulp sections of the *Moringa oleifera* showed an intact glucosinolate, 4-(±α-L rhamnopyranosyloxy) benzyl glucosinolate, (M/Z= 570, Rf=0.18) and its corresponding acetylated derivative (M/Z=612, Rf 0.24) (1, Figures 2,3) and their enzymatic hydrolysis products, 4-(±α-L rhamnopyranosyloxy) benzyl isothiocyanate (m/z=311) and its acetylated derivative m/z=354 as shown in (Figure 3). Further, hydrolyzed products of aliphatic glucosinolates namely 1). 5-methylsufinylpentylisothiocyanate m/z 191, 2). 3-methylsulfonylpropyl isothiocyanate m/z 179 and 3). progoitrin, (2R) -2-hydroxybut-3-enyl-isothiocyanate m/z 129, its corresponding glucosinolate m/z 388, 4-methylsufinyl-3-butenyl isothiocyanate m/z 175 were characterized by MS/ESI-MS studies (*1, 36*).

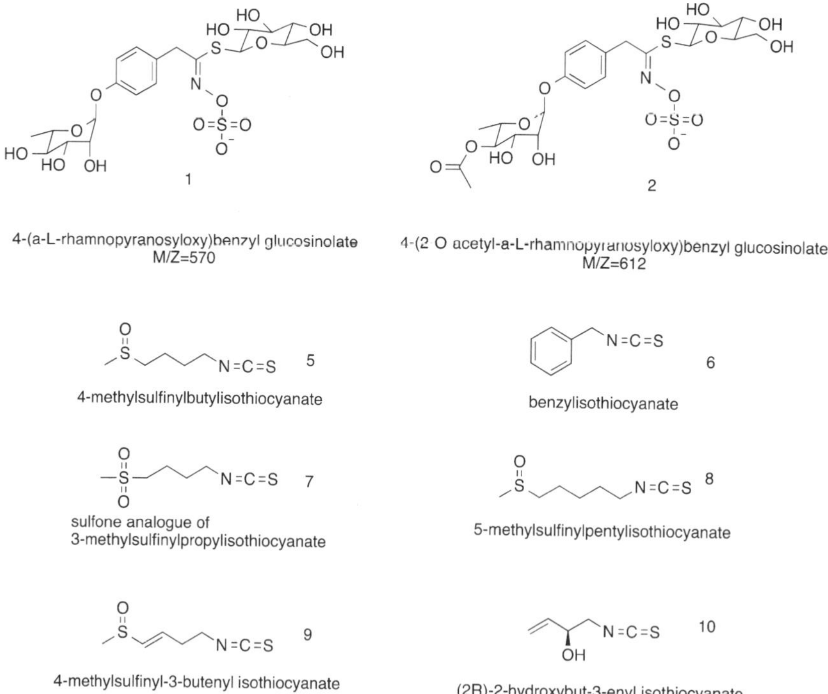

Figure 4. Structures of glucosinolates [1,2] and enzymatically hydrolyzed isothiocyanates [3-8].

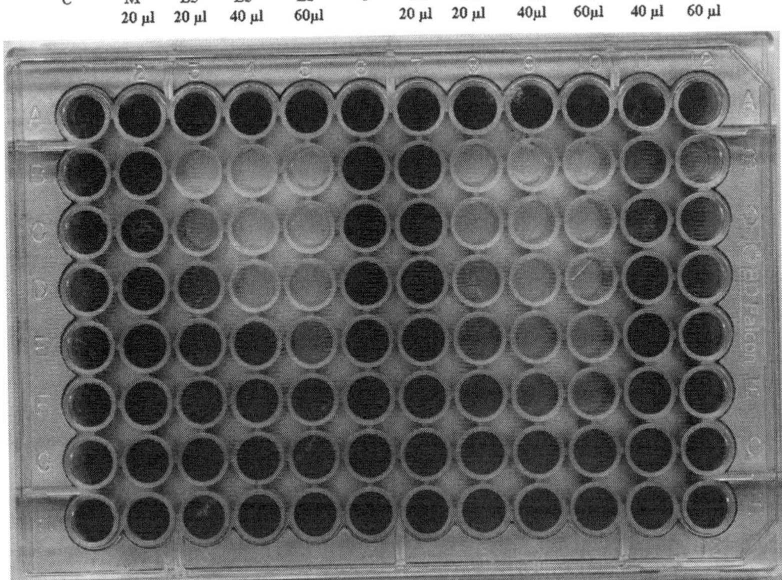

Figure 5. Growth inhibition and prevention of biofilm formation of Moringa oleifera seed coat and pulp tissues. Inhibition zones are exhibited on a 96 well-plate in a concentration dependent manner and are compared to control and methanol as shown (C- control, M-methanol, B5- pulp, C5- seed coat).

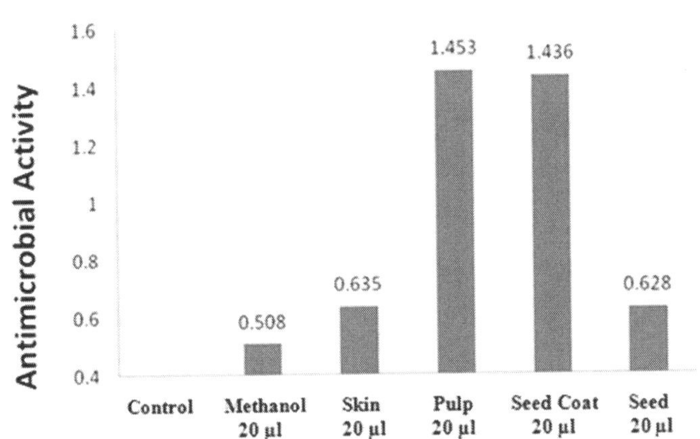

Figure 6. Antimicrobial activity of various tissues of Moringa oleifera (Drumsticks) against Staph. epidermidis.

185

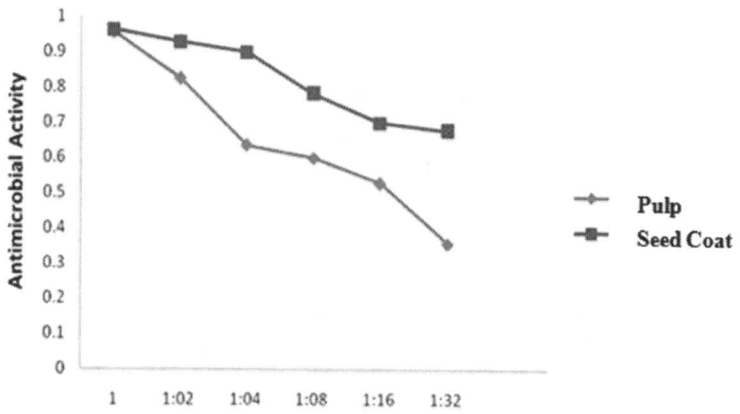

Figure 7. Comparison of antimicrobial activity and dilution ratio of pulp and seed coat.

As shown in Figures 5-7 the hydrolyzed products of glucosinolates (Figure 4) showed the highest antimicrobial activities in a concentration dependent manner. Inhibition zones as illustrated as shown on the 96 well plate as an example and are compared to control and methanol (Figure 5). We believe the antimicrobial effects of the seed cover are attributed to bioactives, 4-(±α-L rhamnopyranosyloxy) benzyl isothiocyanate (m/z=311), its acetylated derivative m/z=354 and its low molecular weight isothiocyanates as shown in (Figure 4). A recent study by Park et al. (*37*) elucidated inhibition of lipopolysaccharide-induced cyclooxygenase-2 and inducible Nitric Oxide Synthase (iNOS) expression by 4-[(2′-O-acetyl-α-L Rhamnosyloxy) benzyl]isothiocyanate from *Moringa oleifera*. Its novel bioactive inhibitory activity on NO production was found to be greater than the well-established isothiocyanate, sulforaphane, found in broccoli (*14, 15*).

More recent work by Drescher et al. (*38*) has described that biofilms cause total disruption of flow, which has serious consequences for environmental and medical catheter devices. Biofilm streamers cause catastrophic disruption of flow with consequences for environmental and medical systems. Another important study described in New Engl. Journal of Medicine, Spring 2012 issued a heart alert on the use of currently used antibiotics such as azithromycin which can cause changes in the electrical activity of the heart and may trigger a fatal irregular heart beat rhythm in heart disease patients (*39*). We believe the bioactive compounds present in *Moringa oleifera* seed coat and pulp are naturally occurring compounds and possess enhanced antibacterial activities (500 µg/ml) better than existing antibiotics including erythromycin, clarithromycin, amoxicillin and ciprofloxacin. Our results raise the possibility of new drugs that can dissolve biofilms across a variety of bacterial species. Mechanistic implications of dismantling biofilm communities in infectious diseases in wound healing have been highlighted by Losick and coworkers (*22*).

Potential of Nanotechnology on Isothiocyanates/HDAC Inhibitors as Chemotherapeutics and Biofilm Inhibitors

The enzymatically-hydrolyzed isothiocyanates (ITC) derived from the glucosinolates present in *Moringa oleifera* (Figure 4) can be used as chemotherapeutics (*40–44*) and in biofilm disassembly. Specifically, the ITCs act as Histone Deacetylase (HDAC) inhibitors, stimulating differentiation and/or apoptosis, thus inhibiting tumor growth in animals (*41, 42*). Nanotechnology can improve ITC delivery and interaction with the HDAC pocket (*40*). ITCs, as illustrated in Figure 4, react with Glutathione (GSH), Cysteine, and N-acetylcysteine becoming potent HDAC inhibitors (Figure 8). For example, Sulforaphane (SFN), isolated from broccoli, shows no HDAC inhibitory activity on its own. However, when SFN-GSH, SFN-Cysteine, and SFN-N-acetylcysteine adducts arc formed through catalyzed reactions in the body, the inhibitory activity increases tremendously. SFN-Cysteine shows the greatest inhibitory effect as illustrated by Rajendran and colleagues (*40–42*). Likewise, the bioactive compounds that we have isolated are expected to show similar inhibitory effects (Figures 8 and 9).

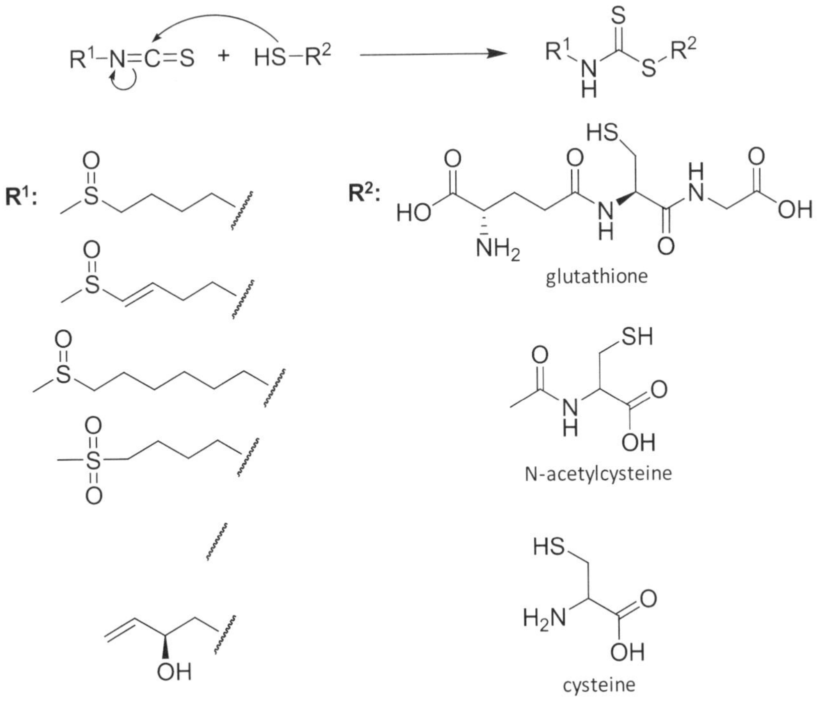

Figure 8. Mechanism of isothiocyanate conjugate formation.

Studies have shown that the cancer prevention activity of cruciferous vegetables is mainly derived from the isothiocyanates, which are hydrolyzed from glucosinolates (*42–44*). The ITC conjugates that are formed with cysteine, glutathione and N-acetylcysteine have similar effects. Nanoparticles are customizable and efficient delivery vehicles that can be used to deliver the formed adducts (*25*). Nanoparticles derived from various cores including gold and silica can bind to and deliver the ITC-GSH, ITC-cysteine, and ITC-N-acetylcysteine adducts to the tumor cells (*41, 42*). Subsequently, the conjugates would be able to bind to the HDAC binding pocket inhibiting histone deacetylation and causing the tumor cell to enter a phase of apoptosis and induce the release of p21[WAF1]-inhibiting CDK, which is required for cell proliferation (*40–43*). Nanotechnology can be applied as a drug delivery vehicle for the isothiocyanate adducts into the human body.

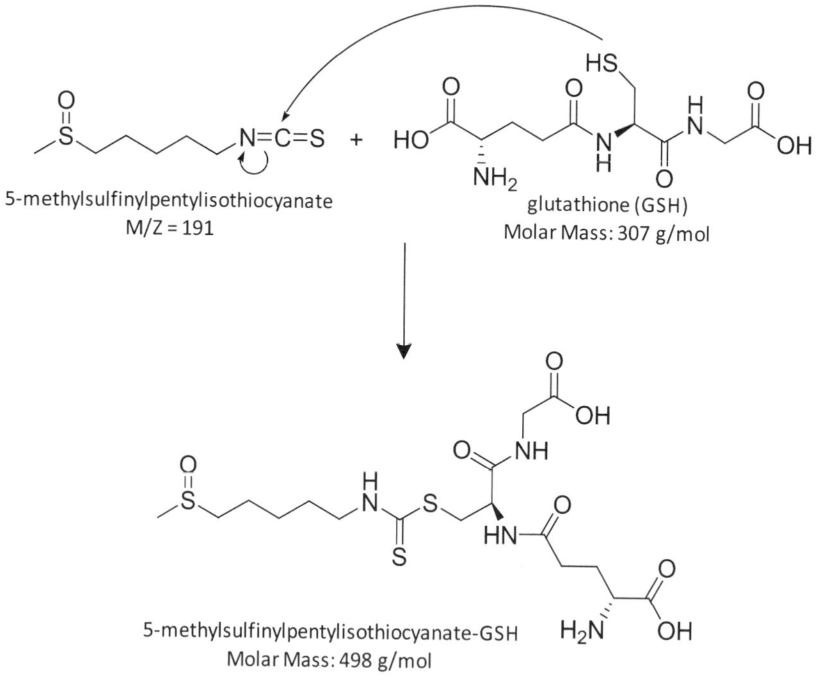

Figure 9. Mechanism of SFN-GSH adduct formation.

The bioactive compounds described in our studies from *Moringa Oleifera* can be potentially encapsulated by engineered nanoparticles. The subsequent nanoparticle-capsules can then be delivered to the various systems in which biofilms have formed. The encapsulated bioactive compounds, upon contact with the biofilm, will then be released to disassemble the biofilm and therefore inhibit bacterial growth and further biofilm formation. The importance of nanotechnology in terms of drug delivery through capsules and adducts is paramount because of its efficacy and the ability to control the release of the drug (*26*). We believe the present studies will play a significant role in future studies in order to maximize the effectiveness of treatments via nanoparticle-capsuled controlled drug delivery.

Summary

The *Moringa oleifera* compounds isolated in the present study demonstrate the prevention of biofilm formation and anti-cancer properties. This could benefit patients suffering from chronic/long term use of catheters. We believe that the application of these compounds as a drug will promote a healthy wound healing environment by preventing formation of biofilms from normal skin flora. Also, the antimicrobial and anti-proliferative activities of pure fractions of *Moringa Oleifera* pulp and seed coat may have potential beneficial effects for patients suffering from severe sepsis and trauma conditions. These compounds will eradicate pathogens and a condition called quorum sensing. The application of glucosinolates and their hydrolyzed derivatives will also be valuable in reducing cardiovascular risk, carcinogenesis and various neurodegenerative diseases caused by bacterial infections. The isothiocyanate derivatives of the glucosinolates found in *Moringa oleifera,* broccoli, and radish seeds are more desirable species for nanoparticle controlled release and exhibit their chemotherapeutic potential by way of Histone Deacetylation inhibition and biofilm and bacterial growth inhibition.

Acknowledgments

We are grateful to Shaheir Ali, Phani Paladugu, Bhaveshkumar Garsondiya, and Jacob Pourat summer research scholars in the Department of Medicine, Biochemistry and Molecular Biology for their excellent technical laboratory experimentation and skillful computer editing expertise in making this manuscript possible.

References

1. Dayal, B.; Yannareddy, V.; Amin, R.; Lea, M. A.; Attygale, A. B. In *Tropical and Subtropical Fruits: Flavors, Color, and Health Benefits*; Patil, B. S., Jayaprakasha, G. K., Roa, C. O., Mahattanatawee , K., Eds.; ACS Symposium Series 1129; American Chemical Society: Washington, DC, 2013; pp 203−219.
2. Fahey, J. W *Part 1. Trees for Life J.* **2005**, 1–5.
3. Anwar, F.; Latif, S.; Ashraf, M.; Gilani, A. H. *Phytother. Res.* **2007**, *21*, 17–25.
4. Amaglo, N. K.; Bennett, R. N.; Lo Curto, R. B.; Rosa, E. A. S.; Lo Turco, V.; Giuffrida, A.; Lo Curto, A.; Crea, F.; Timpo, G. M. *Food Chem.* **2010**, *122*, 1047–1054.

5. Posmontier, B. *Holistic Nurs. Pract.* **2011**, *25*, 80–87.
6. World Health Organization. *Traditional medicine fact sheet*; http://www.who.int/mediacentre/factsheets/fs134/en/. Accessed 8 Aug. 2012.
7. Fahey, J. W.; Zalcmann, A. T.; Talalay, P. *Phytochemistry* **2011**, *56*, 5–51.
8. Dayal, B.; Yannamreddy, V.; Singh, A. P.; Lea, M. A.; Ertel, N. H. In *Emerging Trends in Dietary Components for Preventing and Combatting Disease*; Patil, B., Jayaprakasha, G. K., Murthy, K. N. C., Seeram, N. P., Eds.; ACS Symposium Series 1093; Washington, DC: Washington, DC, 2012; pp 287–302.
9. Lea, M. A.; Akinpelu, T.; Amin, R.; Yarlagadda, K.; Enugala, R.; Dayal, B.; desBordes, C. *Proc. Am. Assoc. Cancer Res.* **2012**, *53*, 1314–1315.
10. Salah-Abbeès, J. B.; Abbeès, S.; Abdel-Wahhab, M. A.; Oueslati, R. *J. Pharm. Pharmacol.* **2010**, *62*, 231–239.
11. Guevara, A. P.; Vargas, C.; Sakurai, H.; Fujiwara, Y.; Hashimoto, K.; Maoka, T.; Kozuka, M.; Ito, Y.; Tokuda, H.; Nishino, H. *Mutat. Res.* **1999**, *440*, 181–188.
12. Abdull Razis, A. F.; De Nicola, G. R.; Pagnotta, E.; Iori, R.; Ioannides, C. *Arch. Toxicol.* **2012**, *86*, 183–194.
13. Juge, N.; Mithen, R. F.; Traka, M. *Cell. Mol. Life Sci.* **2007**, *64*, 1105–1127.
14. Zhang, Y.; Talalay, P.; Cho, C. G.; Posner, G. H. *Proc. Natl. Acad. Sci. U.S.A.* **1992**, *89*, 2399–2403.
15. Moon, J. K.; Kim, J. R.; Ahn, Y. J.; Shibamoto, T. *J. Agric. Food Chem.* **2010**, *58*, 6672–6677.
16. Ganin, J.; Rayo, J.; Amara, N.; Levy, N.; Krief, P.; Meijler, M. M. *MedChemComm* **2013**, *4*, 175–179.
17. Bassler, B. L. *Harvey Lect.* **2006**, *100*, 123–142.
18. Waters, C. M.; Bassler, B. L. *Annu. Rev. Cell. Dev. Biol.* **2005**, *21*, 319–346.
19. Bassler, B. L. *Cell* **2002**, *109*, 421–424.
20. Camilli, A.; Bassler, B. L. *Science* **2006**, *311*, 1113–1116.
21. Bassler, B. L.; Losick, R. *Cell* **2006**, *125*, 237–246.
22. Kolodkin-Gal, I.; Romero, D.; Cao, S.; Clardy, J.; Kolter, R.; Losick, R. *Science* **2010**, *328*, 627–629.
23. Kaplan, J. B.; Ragunath, C.; Velliyagounder, K.; Fine, D. H.; Ramasubbu, N. *Antimicrob. Agents Chemother.* **2004**, *48*, 2633–2636.
24. Dayal, B.; Amin R.; Yannamreddy, V.; Lea, M.; Attygalle, A. B. AGFD Abstract #58; Presented at the 243rd American Chemical Society National Meeting, San Diego, CA, March 2012.
25. De, M.; Ghosh, P. S.; Rotello, V. M. *Adv. Mater.* **2008**, *20*, 4225–4241.
26. Yang, X.-C.; Samanta, B.; Agasti, S. S.; Zhu, J.; Y., Z.-J.; Rana, S.; Miranda, O. R.; Rotello, V. M. *Angew. Chem. Int. Ed.* **2011**, *123*, 497–501.
27. Dayal, B.; Yannamreddy, V.; Racharla, L.; Desai J.; Lea, M. AGFD #95; Presented at the 241st American Chemical Society National Meeting, Anaheim, CA, March 2011.
28. Dayal, B.; Rao, K.; Salen, G. *Steroids* **1995**, *60*, 453–457.
29. Dayal, B.; Ertel, N. H.; Rapole, K. R.; Asgaonkar, A. *Steroids* **1997**, *62*, 451–454.

30. Dayal, B.; Rapole, K. R.; Salen, G. Orgn #46; Presented at the 244th American Chemical Society National Meeting, Philadelphia, PA, August 25-29, 2012.

31. Bose, A. K.; Banik, B. K.; Lavlinskaia, N.; Jayaraman, M.; Manhas, M. S. *CHEMTECH* **1997**, *27*, 18–23.

32. Bialecki, J. B.; Ruzicka, J.; Weisbecker, C. S.; Haribaland, M.; Attygalle, A. B. *J. Mass. Spectrom.* **2010**, *45*, 272–283.

33. Kaplan, J. B.; Fine, D. H. *Appl. Environ. Microbiol.* **2002**, *68*, 4943–4950.

34. Kaplan, J. B.; Meyenhofer, M. F.; Fine, D. H. *J. Bacteriol.* **2003**, *185*, 1399–1404.

35. Deitch, E. A. *Ann. Surg.* **1992**, *216*, 117–134.

36. Dayal, B.; Lea, M. A. AGFD Abstract #15; Presented at the 246th American Chemical Society National Meeting, Indianapolis, Indiana, Sept. 2013.

37. Park, E. J.; Cheenpracha, S.; Chang, L. C.; Kondratyuk, T. P.; Pezzuto, J. M. *Nutr. Cancer.* **2011**, *63*, 971–982.

38. Drescher, K.; Shen, Y.; Bassler, B. L.; Stone, H. A. *Proc. Natl. Acad. Sci. U.S.A.* **2013**, *110*, 4345–50.

39. Ray, W. A.; Murray, K. T.; Hall, K.; Arbogast, P. G.; Stein, C. M. *N. Engl. J. Med.* **2012**, *366*, 1881–1890.

40. Rajendran, P.; Williams, D. E.; Ho, E.; Dashwood, R. H. *Crit. Rev. Biochem. Mol. Biol.* **2011**, *46*, 181–199.

41. Richon, V. M.; Sandhoff, T. W.; Rifkind, R. A.; Marks, P. A. *Proc. Natl. Acad. Sci. U.S.A.* **2000**, *97*, 10014–10019.

42. Myzak, M. C.; Karplus, P. A.; Chung, F. L.; Dashwood, R. H. *Cancer Res.* **2004**, *64*, 5767–5774.

43. Song, D.; Liang, H.; Kuang, P.; Tang, P.; Hu, G; Yuan, Q. *J. Agric. Food Chem.* **2013**, *61*, 5097–5102.

44. Brinker, A. M.; Spencer, G. F. *J. Chem. Ecol.* **1993**, *19*, 2279–2284.

Chapter 11

Biochar: Sustainable and Versatile

Steven C. Peterson,*,[1] Michael A. Jackson,[2] and Michael Appell[3]

[1]Plant Polymer Research, USDA-ARS, National Center for Agricultural Utilization Research, Peoria, Illinois 61604
[2]Renewable Product Technology Research, USDA-ARS, National Center for Agricultural Utilization Research, Peoria, Illinois 61604
[3]Bacterial Foodborne Pathogens & Mycology Research, USDA-ARS, National Center for Agricultural Utilization Research, Peoria, Illinois 61604
*E-mail: Steve.Peterson@ars.usda.gov.

Biochar is a term that describes any charcoal that comes from biomass. It is a renewable, microporous, carbon-rich product that also contains nitrogen, hydrogen, oxygen, and ash. Several varieties of biochar are very porous with irregular surface areas. The pore size and distribution of a given biochar can affect its utility, and controlling the highest treatment temperature during the pyrolysis process is the most common way to affect this. Chemical and physical activation methods are another means to modify the pore size and/or distribution. Biochar can be produced from a very wide range of agricultural and wood-based waste streams and applications of biochar include carbon sequestration, soil amendment, and sorption of several classes of undesirable components from water, soil, or industrial processes. Additionally, recent research has shown potential in using biochar with high pore volume concentration as a filler for rubber composites.

Introduction

Figure 1 summarizes the differences between biomass combustion and pyrolysis. When biomass is combusted, heat and oxygen convert the biomass into gaseous steam and carbon dioxide, plus ash, which is primarily a mixture of solid metal oxides, along with smaller amounts of silica and other volatile inorganic elements. In order to make biochar, oxygen must be removed from the process so that the carbon in the biomass is not lost as carbon dioxide gas. This process, the thermochemical conversion of biomass with no oxygen present, is known as pyrolysis. Once oxygen is taken out of the equation, carbon dioxide cannot form, and solid carbon is left behind in the form of biochar. However, biochar is only one of the products of pyrolysis; liquid bio-oil may also be formed, and in many cases is a desirable alternative form of fuel (*1–3*). Syngas is the common term for the combustible gases that are produced during pyrolysis and typically consist of hydrogen, methane, and other gaseous hydrocarbons (*4*). The pyrolysis process can be altered in terms of the amount and rate of heat applied to the initial biomass.

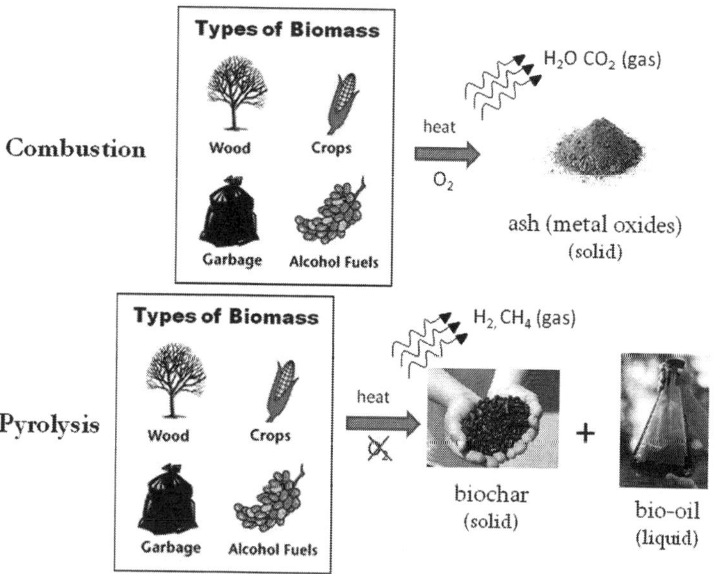

Figure 1. Schematic diagram of combustion versus pyrolysis.

Generally, pyrolysis processes are classified as fast or slow depending on the heating rate of the biomass. Heating rates less than around 100°C/min are classified as slow pyrolysis and favor the production of biochar and syngas, while fast pyrolysis can have heating rates exceeding 1000°C/min, and is used to maximize the production of bio-oil (*5*). The synthesis of biochar is green and carbon positive, which has spurred interest in identifying new applications of the biochars for soil enhancement, contaminant sequestration, and as materials additives. Furthermore, understanding the porous structure and surface area properties can aid in developing biochar-based applications.

Porous Structure of Biochar

Biochar that is produced from plant biomass will frequently retain its inherent phytotomy, or internal structure. This can be very advantageous in terms of surface area because the vascular systems of many plants have very ordered and tightly packed cell walls. When discussing biochar or other sorptive materials, pore sizes are grouped into three classes: macropores (diameter > 50 nm), mesopores (diameter 2-50 nm), and micropores (diameter < 2 nm). In Figure 2, two SEM images of biochar made from A) rubber tree feedstock, and B) a woody waste feedstock are shown. In both of these images one can see the inherent plant structure. Although not visible at this scale, macropores, mesopores, and micropores are also prevalent in many biochars. Micropores have been shown to be excellent sorptive sites for many metals and certain small organic molecules with sorption capacity being directly proportional to micropore surface area (*6–8*).

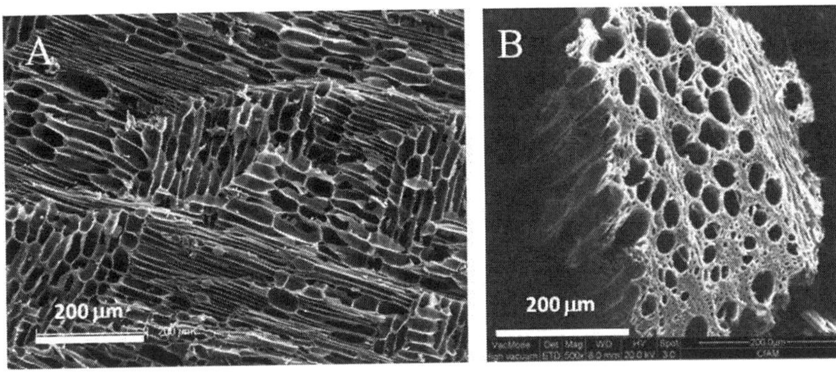

Figure 2. SEM images of biochar from (A) rubber tree feedstock and (B) woody waste feedstock. Image A was reproduced with permission from Dr. Nicola Cayzer, School of GeoSciences, University of Edinburgh. Image B was reproduced with permission from Dr. Peter Harris (Centre for Advanced Microscopy), Joanna Uglow, and Joanna Clark (both School of Human and Environmental Studies), University of Reading.

The pore structure of a given biochar is strongly influenced by its highest treatment temperature (HTT). The HTT and the duration of time the biomass is treated at the HTT is important because these factors determine the extent of carbonization; that is essentially the degree to which the volatile components in the biomass feedstock have been thermally driven off. As volatile components are driven off, micropores are formed as a result (*9–12*). In general, both total surface area and microporosity of a given biochar sample increase with HTT, but usually will plateau or even decrease around a local maximum. Local maxima in surface area are typically seen between 650 and 850°C, and there are many possible reasons as to why this occurs due to the widespread heterogeneity of biomass used as biochar feedstock. Pulido-Novicio and coworkers (*13*) postulated that tars present in wood during carbonization either prevent the formation of or impede the continuity of pores below 750°C, at which point they can finally be volatilized. Rodriguez-Marisol and coworkers (*14*) examined a lignin-based biochar and found that ash content of the starting material affected a fusion and swelling mechanism during carbonization, so they developed methods to reduce the ash content in the starting material and foster higher structure in the final biochar. Regardless of ash, in some biochar samples the cell structure itself can begin to melt and undergo plasticization that clogs any previously formed micropores during carbonization (*15*). The heating rate can also affect melting of cell structures; Cetin et al (*16*) studied pine dust char and showed that with low heating rates of 20°C/sec, carbonization led to biochar with fine structure that resembled the starting feedstock, while the same material heated at 500°C/sec destroyed the cell structure by melting and plastic transformations.

Additional treatments, both chemical and physical, can be applied to biochar in order to increase the internal surface area; this process is called activation and has been carried out in other forms of carbon for many years. In the past the carbon feedstock was generally a fossil fuel product such as coal or coke (*17, 18*), since these yielded the most pure sources of carbon. Today an emphasis towards biochar is being pursued in order to provide a green alternative and a sustainable feedstock for the future. Physical activation methods utilize oxidizing gases, typically steam, carbon dioxide, or oxygen applied to the biochar at high temperatures, typically from 700-1200°C (*19–22*). Chemical activation involves mixing the biochar thoroughly with acids (*23*), bases (*24*), or salts (*25*) prior to the carbonization step. Biochars that have been chemically activated can be carbonized at lower temperatures than those that have been physically activated with similar resulting surface areas (*19*).

Surface Area Measurements of Biochar

Many proposed uses of biochar, from soil enhancer to sorbent to filler, could benefit from high surface area. Unfortunately, "as prepared" biochars tend to have surface areas ranging from less than one to a few tens m^2/g depending on the biomaterial charred and the conditions under which the biochar is produced. Lower temperature charring, or torrefaction (200-320°C), gives lower surface area products, whereas biochars resulting from gasification (>700°C) tend to give higher surface-area products (*26–28*). Biochars typically have amorphous structure with heterogeneous surface textures consisting of limited pore structure and, often, higher ash content. These features can make measurement of the surface areas of biochars challenging. The surface area of biochar can be increased by treatments such as chemical activation (*28–30*) and ball milling (*31*). The methods described here are better applied to these activated, porous biochars.

The surface area of powders such as sorbents, catalysts, and pharmaceuticals are measured using the classical gas adsorption method of Brunauer, Emmet, and Teller (BET) (*32*), typically using N_2 isotherms collected on samples at 77K. The BET theory succeeds because it allows for the determination of the mass of molecules adsorbed on a surface despite the fact that the surface is never fully covered by a single layer of molecules, but is instead partially covered by stacked layers of molecules, the uppermost being in equilibrium with the vapor phase. Since these stacked layers are in equilibrium with the vapor and each other, the relative number of molecules in each layer is constant. This understanding leads to the BET equation which can be written in a linear form as:

$$\frac{1}{W\left(\frac{P}{P_o}-1\right)} = \left(\frac{C-1}{W_mC}\right)\left(\frac{P}{Po}\right) + \frac{1}{W_m} \qquad \frac{1}{W\left(\frac{P}{P_o}-1\right)} = \left(\frac{C-1}{W_mC}\right)\left(\frac{P}{Po}\right) + \frac{1}{W_m}$$

where W is the adsorbed weight of the gas, W_m is the weight of the gas in the monolayer, P/Po is the relative pressure, and C is the BET constant. A plot of P/Po verses $1/W(P/Po-1)$ gives a straight line of slope $(C-1)/W_mC$ which gives the weight of the adsorbed monolayer W_m. The molecular cross-sectional areas of the adsorptive gases are well established, so this weight can be converted to an area and then this can simply be divided by the weight of the sample to give the surface area of the sample. For common powders such as alumina and silica, the range of the linear portion of the BET plot is typically $0.05 \leq P/Po \leq 0.30$. However, this range is not definitive and some judgment must be used to find the linear portion of a BET plot. The only requirement in choosing the pressure range is to include the inflection point, known as Point B, from the original isotherm. Point B is regarded as the pressure at which the monolayer begins to develop into stacked layers. For biochar samples, we find the linear range to most frequently fall at lower P/Po values, often $0.025 \leq P/Po \leq 0.10$. It is best that the linear range be given when the BET surface area is reported.

The BET C constant is a dimensionless measure of the interaction between the solid surface and the gas. It also reflects the level of coverage by the monolayer. That is, the portion of the solid not covered by the monolayer, $(\theta)0$, can by defined as

$$(\theta)_0 = \frac{\sqrt{C}-1}{C-1}$$

From this equation it is found that only as C becomes very large, does the monolayer actually cover the entire surface and this reflects more chemisorption occurring rather than physisorption. In biochar samples, we find C values in the wide range of 50-1200. There is also a strong correlation between C and the amount of surface area of the biochars that is in micropores.

Unlike the larger pores (> 2 nm), adsorbate behavior in micropores is controlled by gas-solid interaction which occurs at very low relative pressures. Biochars tend to be largely microporous and the ash content contributes to the gas-solid interaction. Characterization of the pores can, therefore, be tricky. The classical method for micropore analysis is known as the t-Method of Lippens and de Boer (*33*). This method is based on the assumption that BET results are valid for microporous materials. Results from the method are micropore volume and area and are complementary to the BET surface area measurement. The t-Method calculation uses a comparison of the isotherm of the sample of interest to that of a sample without porosity. The comparison requires knowing the statistical thickness, t, for the adsorbate layer. Lippens, Linsen, and de Boer (*34*) put forth the following equation for t

$$t(nm) = 0.1\left[\frac{13.99}{log(P_0/P) + 0.034}\right]^{\frac{1}{2}}$$

This equation was developed for oxidic, nonporous alumina and should be used against material with similar values of C and, therefore, surface chemistry. Its use in the analysis of biochar is questionable since the chars are mostly carbon although they do contain some, and often considerable amounts of, oxides. Other definitions of t have been devised for other gases and solids and current instrumentation allows for easy calculation of results from each. Table 1 shows the results of the t-method analysis of an activated biochar using different calculations for the statistical thickness. Despite the wide variability in micropore area, an attractive feature of the t-method is the quick collection of data, with as few as eight points needed.

Table 1. Micropore surface areas from the t-Method using different calculations for the statistical thickness

Calculation Method	Micropore Surface Area (m²/g)
De Boer	70
Carbon Black	74
Halsey	54
Generalized Halsey	54

An alternative approach to measuring the pore size distribution in microporous carbon is to use CO_2 as the adsorptive gas (*35–37*). Carbon dioxide measurements taken at 273K can be made at higher pressures and shorter times compared to nitrogen measurements at 77K. The high saturation pressure of CO_2 at 273K (26141 torr) allows for rapid equilibration at each point and penetration of the smallest pores by the gas. Whereas a nitrogen isotherm at 77K covering the micropore region requires P/P_0 values on the order of 10^{-6} and as long as 30 hours to measure, the same measurement with CO_2 which generates the same, or at least complementary, results can be done with P/P_0 values on the order of 10^{-3} and be completed in around three hours. The data are interpreted by using the Dubinin-Radushkevich (DR) equation

$$\log V = \log(V_{mp}) - 2.303\left(\frac{RT}{\beta E_0}\right)^2\left[\log\left(\frac{P_0}{P}\right)\right]^2$$

where V_{mp} is the micropore volume, V the total volume of adsorbate adsorbed, β is the "affinity coefficient" (*38*) which is a property of the adsorbate, E_0 is a property of the adsorbent and reflects the energetic heterogeneity of the adsorbent due to the range of micropore sizes. A plot of $[\log(P_0/P)]^2$ verses log V yields a straight line with log V_{mp} at the y-intercept. β is defined as the affinity of the adsorbate toward the adsorbent relative to benzene which was assigned the value of one by Dubinin. βCO_2 has earned a classical value of 0.35 although the origin of this value is unclear; some uncertainty exists as to the value of βCO_2. Bickford et. al (*39*) surveyed the literature for reports on pore size in activated carbons and calculated a range of β values of 0.26-0.69 which is wide enough to create an uncertainty in pore diameter of ± 0.5 nm. Some of this uncertainty can be sidestepped by using simulations of individual pore filling based on statistical mechanical model calculations known as non-localized Density Function Theory and the Grand Canonical Monte Carlo simulation (*40, 41*). Unfortunately, these models are based on a slit pore orientation for the carbons and may not be entirely suitable for biochar.

The determination of surface area and pore size composition of biochar is an important step in evaluating its value in different applications. However, care must be taken in using appropriate models and data points must be selected judiciously.

Biochar for Soil Amendment

The most common application of biochar is as a soil additive to enhance the uptake of nutrients and water by plants, mitigate the release of components, and store carbon in soil. Sequestering carbon in soil can have a significant impact on the environment by reducing the amount of carbon dioxide released to the atmosphere as well as providing opportunities for distributed energy production. Matovic has shown (*42*) that carbon sequestration using biochar is viable on a global scale, although it is very important to minimize the physical distance between the biomass feedstock and biochar dispersal in order to maintain cost and efficiency advantages. Biochar is most frequently used in gardening and agricultural commodity production (*43*). Biochar can be used to alter poor soil qualities with respect to pH and sorption release of undesirable components such as pollution. For example, soil enriched with biochar has been shown to mitigate the release of nitrous oxide and ammonia runoff from the soil, and the release is related to the residence time of the biochar in the soil (*44*). Wetting and drying the biochar several times significanctly helped retain nitrous oxide and ammonium. However, use of biochar influences soil microbes and the effects are unclear because many studies neglect the changes of biochar while in the soil (*45*). A further complication is that the effectiveness of biochar depends on the type and properties of the biochar, including source, method of preparation, pore structure, surface area, and stability (*46–48*). It has been shown that pyrolysis temperature and soil type can influence the sequestration and release of heavy metals in soil (*49, 50*). A more detailed review of biochar and its impact on soil microbiology has been done by Ennis and coworkers (*51*); they believe the key to utilizing biochar as a soil amendment effectively is to develop robust methods that can characterize the wide range of mechanisms possible for biochar in targeted soils.

Biochar as Sorbent Medium for Contaminants

Due to its inherent structure discussed above, biochar has been investigated as a sorptive medium to remove undesirable contaminants such as small organic toxins and metals from water, food, or industrial wastewater (*52–54*). Biochar sorption can be modeled using Freundlich and Langmuir isotherm models, and as dual mode sorption (*12, 55, 56*). Dual mode sorption has two different mechanisms that can occur: a) linear, non-competitive partitioning, and b) nonlinear, extensive and competitive surface adsorption. Furthermore, biochar sorption can be more complex than these models and influenced by competitive constituents in the media, such as minerals, metals, organic contaminants, nutrients, herbicides, pesticides and fertilizers (*50, 52, 57–59*).

Partitioning occurs mainly in the non-carbonized domains of the biochar, so this mechanism is seen more in chars that have been produced using lower HTTs (*60*). In this process the contaminant is absorbed into the amorphous, non-carbonized part of the biochar. This could be a myriad of different materials, such as, but not limited to, polysaccharides, lignins, cellulose, amino acids, or lipids, depending on the biochar feedstock (*61*). In these non-carbonized domains of biochar, there are oxygen, hydrogen, and nitrogen atoms present in various functional groups that, depending on the pH, can affect the partitioning of contaminants (*62, 63*). Polar functional groups present on biochar surfaces tend to adsorb water and other polar adsorbates (*64, 65*) that form water clusters (*66, 67*); these water clusters may then reduce the accessible surface area of the biochar and therefore lower its efficacy as a sorbent (*62, 68*). Due to the vast range of possible surface characteristics of non-carbonized biochar, specific characterization of the char along with the targeted contaminants must take place to help elucidate at what rate partitioning between the contaminant and biochar can take place.

The other sorption mechanism, surface adsorption, occurs on the carbonized domains of the biochar as a pore-filling mechanism (*69*). This mechanism is competitive, since there are only a finite number of pores to fill in the biochar. This mechanism happens much more quickly than the partitioning mechanism; so much so that it can be used to determine the carbonized vs non-carbonized ratio of the biochar, which is in turn a function of the pyrolytic temperature. This was shown in work done by Chen and coworkers (*12*), where the model solute naphthalene was sorbed to a series of pine wood biochars prepared at different pyrolysis temperatures.

Biochar as Potential Filler in Rubber Composites

Carbon black (CB) has been by far the most popular reinforcing filler for rubber composites for decades, due to its purity and availability in a large range of particle/aggregate sizes. However, since it is a fossil-fuel product, partially or fully substituting CB with renewable biochar would be highly desirable. However, research up to this point has been limited; Adeosun and Olaofe studied the use of charcoal along with other inorganic materials such as clays and silica sand as filler for natural rubber (*70, 71*), and previous work by Peterson has evaluated biochars from corn stover (*72*) and woody waste feedstock (*73*) blended with CB in order to reinforce styrene-butadiene rubber (SBR). In all of these cases, it was found that biochar did indeed reinforce rubber to a certain extent, but only at low loadings. The difference in particle size between biochar and CB is thought to be the main reason for this behavior. Biochars used in the aforementioned studies to reinforce SBR had an average number-distribution size of 0.74 μm compared to the control CB which was ~30 nm (25 times smaller). Also, the particle size distribution for that particular biochar was much more heterogeneous than CB, and had a small percentage of larger particles with diameters up to 100 microns. Large particle size in filler material is a significant problem for two reasons: a) larger filler particles have less surface area to interact with (reinforce) the polymer matrix, and b) fracture, tearing, or other types of defects result from the polymer matrix containing localized stresses (*74, 75*). Tensile properties of SBR composites made from CB/biochar blended filler surpass 100% CB in terms of percent elongation, tensile strength, and toughness, but only at 10% (weight) total filler concentration. Above 10% total filler these advantages are lost relative to 100% CB (*73*). Hence any practical applications that would utilize this technology to reduce CB usage would require softer, flexible, lightly-filled components.

A recent patent claims that rubber belt composites filled with 100% biochar can perform as well as or better than analogous composites filled with CB (*76*). In this work the inventors state that one of the key factors for successful use of biochar as filler is a high pore volume of both meso- and macropores, in order to provide the polymer matrix accessibility to the biochar surface and foster reinforcement interactions between polymer and filler. Ideally the inventors are striving for filler material with a pore volume of at least 0.5 cm^3/g.

Other interesting morphologies of carbon used in conjunction with biochar such as graphene and nanowires may also provide advantages as filler, but research to this point has focused on these materials' sorptive properties (*77–79*). It remains to be seen whether these morphologies can provide either the surface area or surface chemistry to match or surpass carbon black as reinforcing filler.

Conclusions

Biochar is a renewable resource that can be made from practically any carbonaceous material, including low-value agricultural and forest waste. Its inherent fine structure and high surface area can be advantageous in sorbent applications. The pyrolysis conditions have a major effect on the resulting biochar; in general surface area and sorptive quality increases directly with the highest treatment temperature. The increased surface area associated with the microporous nature of biochars is important for certain sorbent applications. Recent studies also suggest that biochars with a high meso- and macropore concentration may be effective as filler for rubber composites. Larger pore structure facilitates more interactions between the polymer matrix and the biochar acting as filler. In order for biochar to make its way into the rubber composite filler industry, cost-effective methods of production must be established. Large scale characterization and standardization of biochar is also needed, since it can come from a virtually infinite array of source feedstocks.

Acknowledgments

Mention of trade names or commercial products in this article is solely for the purpose of providing specific information and does not imply recommendation or endorsement by the U.S. Department of Agriculture. USDA is an equal opportunity provider and employer.

References

1. Wright, M. M.; Daugaard, D. E.; Satrio, J. A.; Brown, R. C. *Fuel* **2010**, *89*, S2–S10.

2. Jones, S. B.; Valkenburg, C.; Walton, C. W.; Elliott, D. C.; Holladay, J. E.; Stevens, D. *Production of gasoline and diesel from biomass via fast pyrolysis, hydrotreating and hydrocracking: A design case*; PNNL-18284, Rev. 1; Pacific Northwest National Laboratory: Richland, Washington, 2009.

3. Mullen, C. A.; Boateng, A. A.; Goldberg, N. M.; Lima, I. M.; Laird, D. A.; Hicks, K. B. *Biomass Bioenergy* **2010**, *34*, 67–74.

4. Balat, M.; Kirtay, E.; Balat, H. *Energ. Convers. Manage.* **2009**, *50*, 3158–3168.

5. Brown, T. R.; Wright, M. M.; Brown, R. C. *Biofuels, Bioprod. Biorefin.* **2011**, *5*, 54–68.

6. Kasozi, G. N.; Zimmerman, A. R.; Nkedi-Kizza, P.; Gao, B. *Environ. Sci. Technol.* **2010**, *44*, 6189–6195.

7. Yu, X.-Y.; Ying, G.-G.; Kookana, R. S. *J. Agric. Food Chem.* **2006**, *54*, 8545–8550.

8. Peterson, S. C.; Appell, M.; Jackson, M. A.; Boateng, A. A. *Can. J. Agric. Sci.* **2013**, *5*, 1–8.

9. Raveendran, K.; Ganesh, A. *Fuel* **1998**, *77*, 769–781.

10. Lua, A. C.; Yang, T.; Guo, J. *J. Anal. Appl. Pyrolysis* **2004**, *72*, 279–287.

11. Downie, A. Ph.D. Thesis, University of New South Wales, Kensington, Australia, 2011.

12. Chen, Z.; Chen, B.; Chiou, C. T. *Environ. Sci. Technol.* **2012**, *46*, 11104–11111.

13. Pulido-Novicio, L.; Hata, T.; Kurimoto, Y.; Doi, S.; Ishihara, S.; Imamura, Y. *J. Wood Sci.* **2001**, *47*, 48–57.

14. Rodriguez-Mirasol, J.; Cordero, T.; Rodriguez, J. J. *Carbon* **1993**, *31*, 87–95.

15. Biagini, E.; Tognotti, L. In *Characterization of biomass chars*; Seventh International Conference on Energy for Clean Environment, Lisbon, Portugal, 2003.

16. Cetin, E.; Moghtaderi, B.; Gupta, R.; Wall, T. F. *Fuel* **2004**, *83*, 2139–2150.

17. *Chemistry and Physics of Carbon. A Series of Advances*; Walker, P. L., Jr., Ed.; Dekker: New York, 1968; Vol. 4, p 399.

18. Mills, G. A. *CHEMTECH* **1977**, *7*, 418–423.

19. Azargohar, R.; Dalai, A. K. *Microporous Mesoporous Mater* **2008**, *110*, 413–421.

20. Avelar, F. F.; Bianchi, M. L.; Goncalves, M.; da, M. E. G. *Bioresour. Technol.* **2010**, *101*, 4639–4645.

21. Valix, M.; Cheung, W. H.; McKay, G. *Chemosphere* **2004**, *56*, 493–501.

22. Wang, S.-Y.; Tsai, M.-H.; Lo, S.-F.; Tsai, M.-J. *Bioresour. Technol.* **2008**, *99*, 7027–7033.

23. Azargohar, R.; Dalai, A. K. *Can. J. Chem. Eng.* **2011**, *89*, 844–853.

24. Hina, K.; Bishop, P.; Arbestain, M. C.; Calvelo-Pereira, R.; Macia-Agullo, J. A.; Hindmarsh, J.; Hanly, J. A.; Macias, F.; Hedley, M. J. *Aust. J. Soil Res.* **2010**, *48*, 606–617.

25. Tay, T.; Ucar, S.; Karagöz, S. *J. Hazard. Mater.* **2009**, *165*, 481–485.

26. Brewer, C. E.; Schmidt-Rohr, K.; Satrio, J. A.; Brown, R. C. *Environ. Prog. Sustainable Energy* **2009**, *28*, 386–396.

27. Ghani, W. A. W. A. K.; Mohd, A.; da Silva, G.; Bachmann, R. T.; Taufiq-Yap, Y. H.; Rashid, U.; Al-Muhtaseb, A. H. *Ind. Crop. Prod.* **2013**, *44*, 18–24.

28. Lima, I. M.; Boateng, A. A.; Klasson, K. T. *Ind. Eng. Chem. Res.* **2009**, *48*, 1292–1297.

29. Kastner, J. R.; Miller, J.; Geller, D. P.; Locklin, J.; Keith, L. H.; Johnson, T. *Catal. Today* **2012**, *190*, 122–132.

30. Ormsby, R.; Kastner, J. R.; Miller, J. *Catal. Today* **2012**, *190*, 89–97.

31. Peterson, S. C.; Jackson, M. A.; Kim, S.; Palmquist, D. E. *Powder Technol.* **2012**, *228*, 115–120.

32. Brunauer, S.; Emmett, P. H.; Teller, E. *J. Am. Chem. Soc.* **1938**, *60*, 309–319.

33. Lippens, B. C.; de Boer, J. H. *J. Catal.* **1965**, *4*, 319.

34. de Boer, J. H.; Lippens, B. C.; Linsen, B. G.; Broekhoff, J. C. P.; van, d. H. A.; Osinga, T. J. *J. Colloid Interface Sci.* **1966**, *21*, 405–414.

35. Garrido, J.; Linares-Solano, A.; Martin-Martinez, J. M.; Molina-Sabio, M.; Rodriguez-Reinoso, F.; Torregrosa, R. *Langmuir* **1987**, *3*, 76–81.

36. Yao, Y.; Gao, B.; Inyang, M.; Zimmerman, A. R.; Cao, X.; Pullammanappallil, P.; Yang, L. *Bioresour. Technol.* **2011**, *102*, 6273–6278.

37. Mukherjee, A.; Zimmerman, A. R.; Harris, W. *Geoderma* **2011**, *163*, 247–255.

38. Dubinin, M. M. *Chem. Rev. (Washington, DC, U. S.)* **1960**, *60*, 235–241.

39. Bickford, E. S.; Clemons, J.; Escallon, M. M.; Goins, K.; Lu, Z.; Miyawaki, J.; Pan, W.; Rangel-Mendez, R.; Senger, B.; Zhang, Y.; Radovic, L. R. *Carbon* **2004**, *42*, 1867–1871.

40. Lastoskie, C.; Gubbins, K. E.; Quirke, N. *Langmuir* **1993**, *9*, 2693–2702.

41. Ravikovitch, P. I.; Vishnyakov, A.; Russo, R.; Neimark, A. V. *Langmuir* **2000**, *16*, 2311–2320.

42. Matovic, D. *Energy (Oxford, U. K.)* **2011**, *36*, 2011–2016.

43. Schnell, R. W.; Vietor, D. M.; Provin, T. L.; Munster, C. L.; Capareda, S. *J. Environ. Qual.* **2012**, *41*, 1044–1051.

44. Singh, B. P.; Hatton, B. J.; Singh, B.; Cowie, A. L.; Kathuria, A. *J. Environ. Qual.* **2010**, *39*, 1224–1235.

45. Lehmann, J.; Rillig, M. C.; Thies, J.; Masiello, C. A.; Hockaday, W. C.; Crowley, D. *Soil Biol. Biochem.* **2011**, *43*, 1812–1836.

46. Keiluweit, M.; Nico, P. S.; Johnson, M. G.; Kleber, M. *Environ. Sci. Technol.* **2010**, *44*, 1247–1253.

47. Singh, B.; Singh, B. P.; Cowie, A. L. *Aust. J. Soil Res.* **2010**, *48*, 516–525.

48. Roberts, K. G.; Gloy, B. A.; Joseph, S.; Scott, N. R.; Lehmann, J. *Environ. Sci. Technol.* **2010**, *44*, 827–833.

49. Uchimiya, M.; Wartelle, L. H.; Klasson, K. T.; Fortier, C. A.; Lima, I. M. *J. Agric. Food Chem.* **2011**, *59*, 2501–2510.

50. Uchimiya, M.; Klasson, K. T.; Wartelle, L. H.; Lima, I. M. *Chemosphere* **2011**, *82*, 1431–1437.

51. Ennis, C. J.; Evans, A. G.; Islam, M.; Ralebitso-Senior, T. K.; Senior, E. *Crit. Rev. Environ. Sci. Technol.* **2012**, *42*, 2311–2364.

52. Kolodynska, D.; Wnetrzak, R.; Leahy, J. J.; Hayes, M. H. B.; Kwapinski, W.; Hubicki, Z. *Chem. Eng. J. (Amsterdam, Neth.)* **2012**, *197*, 295–305.

53. Kong, H.; He, J.; Gao, Y.; Wu, H.; Zhu, X. *J. Agric. Food Chem.* **2011**, *59*, 12116–12123.

54. Sun, K.; Ro, K.; Guo, M.; Novak, J.; Mashayekhi, H.; Xing, B. *Bioresour. Technol.* **2011**, *102*, 5757–5763.

55. Pignatello, J. J.; Xing, B. *Environ. Sci. Technol.* **1996**, *30*, 1–11.

56. Xing, B.; Pignatello, J. J. *Environ. Sci. Technol.* **1997**, *31*, 792–799.

57. Zhang, G.; Liu, X.; Sun, K.; He, F.; Zhao, Y.; Lin, C. *J. Environ. Qual.* **2012**, *41*, 1906–1915.

58. Spokas, K. A.; Novak, J. M.; Venterea, R. T. *Plant Soil* **2012**, *350*, 35–42.

59. Martin, S. M.; Kookana, R. S.; Van, Z. L.; Krull, E. *J. Hazard. Mater.* **2012**, *231-232*, 70–78.

60. Chun, Y.; Sheng, G.; Chiou, C. T.; Xing, B. *Environ. Sci. Technol.* **2004**, *38*, 4649–4655.

61. Cornelissen, G.; Gustafsson, O.; Bucheli, T. D.; Jonker, M. T. O.; Koelmans, A. A.; van, N. P. C. M. *Environ. Sci. Technol.* **2005**, *39*, 6881–6895.

62. Boehm, H. P. *Carbon* **1994**, *32*, 759–769.

63. Braida, W. J.; Pignatello, J. J.; Lu, Y.; Ravikovitch, P. I.; Neimark, A. V.; Xing, B. *Environ. Sci. Technol.* **2003**, *37*, 409–417.

64. Bandosz, T. J.; Jagiello, J.; Schwarz, J. A. *Langmuir* **1993**, *9*, 2518–2522.

65. Jia, Y. F.; Thomas, K. M. *Langmuir* **2000**, *16*, 1114–1122.

66. Leboda, R.; Skubiszewska-Zieba, J.; Bogillo, V. I. *Langmuir* **1997**, *13*, 1211–1217.

67. Boehm, H. P. *Carbon* **2002**, *40*, 145–149.

68. Jiang, Z.; Liu, Y.; Sun, X.; Tian, F.; Sun, F.; Liang, C.; You, W.; Han, C.; Li, C. *Langmuir* **2003**, *19*, 731–736.

69. Nguyen, T. H.; Cho, H.-H.; Poster, D. L.; Ball, W. P. *Environ. Sci. Technol.* **2007**, *41*, 1212–1217.

70. Adeosun, B. F.; Olaofe, O. *Pak. J. Sci. Ind. Res.* **2005**, *48*, 63–67.

71. Adeosun, B. F.; Olaofe, O. *Pak. J. Sci. Ind. Res.* **2003**, *46*, 20–26.

72. Peterson, S. C. *J. Elastomers Plast.* **2011**, *44*, 43–54.

73. Peterson, S. C. *J. Elastomers Plast.* **2012** in press.

74. Gent, A. N.; Pulford, C. T. R. *J. Mater. Sci.* **1984**, *19*, 3612–3619.

75. Skelhorn, D. In *Particulate-filled polymer composites*, 2nd ed.; Rothon, R. N., Ed.; Rapra Technology: Shrewsbury, U.K., 2003; pp 326–331.

76. Sealey, J. H.; Sedlacek, D. R. U.S. Patent 20110028257A1, 2011.

77. Zhang, M.; Gao, B.; Yao, Y.; Xue, Y.; Inyang, M. *Sci. Total Environ.* **2012**, *435-436*, 567–572.

78. Ghaedi, M. *Spectrochim. Acta, Part A* **2012**, *94*, 346–351.

79. Ghaedi, M.; Khajesharifi, H.; Hemmati, Y. A.; Roosta, M.; Sahraei, R.; Daneshfar, A. *Spectrochim. Acta, Part A* **2012**, *86*, 62–68.

Chapter 12

Bioinspired Layered Nanoclays for Nutraceutical Delivery System

Soo-Jin Choi*[*,1] and Young-Rok Kim[2]

[1]Department of Food Science and Technology, Seoul Women's University, Seoul 139-774, Korea
[2]Department of Food Science and Biotechnology, Kyung Hee University, Yongin 446-701, Korea
*E-mail: sjchoi@swu.ac.kr. Tel.: +82-2-970-5634/+82-31-201-3830.
Fax.: +82-2-970-5977/+82-31-204-8116.

This review focused on the promising potential of nanoclays as nutraceutical delivery systems. In nature, two forms of nanoclays, anionic and cationic clays, are present depending on the surface layered charge and the types of interlayer ions. Nanoclays including anionic and cationic clays have unique layered structure which enables intercalation of nutraceuticals and bioactive agents into the gallery spaces without any changes in their chemical and functional properties. They also have efficient cell permeation, high mucoadhesive, and low toxic properties that offer a new route to enhance the bioavailability of bioactive agents through oral administration. The preparation, physical/chemical characterization and bioavailability of nanoclay based delivery systems as well as their applications in food system as a nanocarrier for vitamins, antioxidants, linoleic acid, and other neutraceuticals are discussed.

Keywords: Nanoclays; layered double hydroxides (LDHs); nanotechnology; nanomaterials; food; nutraceuticals; delivery system

Introduction

Nanotechnology is the science and technology that focused on the understanding and control of matters at scale ranging from few atoms to submicron dimension. Unique phenomena arisen from the nanomaterials and nanostructures have been applied to produce variety of physical, chemical, and biomolecular systems at nanometer scale, as well as integrating them into larger scale. The technologies in nanoscience have provided promising breakthroughs in areas such as materials science, electronics, medicine, food, healthcare, biotechnology, and thus having consequences for the improved quality of human life. It is also widely accepted that nanotechnology will be the next revolution in food industry (*1*). Control of food materials in nanometer scale could lead to the modification of macromolecular characteristics of food, such as taste, flavor, texture, sensory property, processibility and stability during the process and storage. The applications of nanotechnology in food and ingredients sector have been growing fast.

Food safety is among the fast growing fields to which nanotechnology provides useful tools. Advanced nanoelectronics in combination with appropriate functional nanomaterials and smart biological components enable us to develop highly specific and selective sensing devices for the detection of hazardous agents including viruses, pathogenic microorganisms as well as undesirable chemical and physical contaminants in food (*2–4*). Micro/nano fluidic technology accelerates miniaturization of those sensing devices to the level that is suitable for field application (*5–7*). In addition to ensuring the safety of food, the potential of nanotechnology to monitor the quality and status of food would significantly improve our food life. Food packaging is another area that has employed various nanomaterials to enhance the shelf life of food by improving the barrier property of container (*8, 9*). Nylon based nanocomposites developed by Honeywell and Nanocor have already been employed to produce commercial beer bottles in Korea and United States of America (*10*). Polymer nanocomposites are made by dispersing inorganic filling nanomaterials into polymer (*11*). For example, nylon layered silicate nanocomposite containing 2% inorganic nanofiller has two-fold higher tensile strength and thermal stability over 100 °C compared with pure nylon (*11*). Nylon based nanocomposites are formed by dispersing silicate layers into a continuous polymer matrix. This structure significantly reduces the diffusion rate of oxygen or carbon dioxide across the packaging film since those gases should take tortuous path through the space between dispersed nanosilicate layers (*12*). The notable improvements in strength, heat stability and barrier property made this nanocomposite ideal for food packaging materials.

On the other hand, there is growing interest in nutraceutical benefits of food. There have been numerous efforts and trials to deliver various forms of antioxidants, nutrients, minerals, drugs and other functional agents through foods (*13*). To increase the stability and absorption rate of the active compounds, a number of encapsulation materials have been developed including liposome, micelle, polymersome, dendrimer as well as diverse biodegradable polymeric materials. Reducing the size to nanometer scale is another way of improving the bioavailability of bioactive compounds (*14*). Currently, most of efforts for the development of delivery systems focus on drugs since their high production costs are still acceptable in medicine. Meanwhile, the cost and safety concerns to near zero-toxicity level are the most crucial factors that determine the choice of food for customer. Therefore, food grade or "generally recognized as safe" (GRAS) materials such as polysaccharides, food protein, emulsifier, and inert inorganic materials, with controlled release behavior, are attractive candidates for effective and safe encapsulation system.

Nanoclays belonging to clay minerals have attracted a great deal of attention for their biological application due to their abundance in nature, convenience for preparation, and biocompatibility (*15–18*). They also have great potential for food application because nanoclays have been traditionally used for curative and protective purpose as a folk medicine since the earliest days of civilization (*19–21*). For example, administered clay minerals have been used as laxatives, anti-diarrheal, and anti-inflammatory agents to purify the blood, reduce infection, and even heal ulcers (*22*). Moreover, biocompatible clay minerals have been currently applied as oral antacid or antipepsin agents (*23, 24*). Clays have also been ingested as trace mineral supplements which supports their low or null toxicity.

Nanoclays have unique layered structures consisting of stacking nanosheets with metal ions and interlayer counter ions for charge-balancing (*25*). The 2-dimensional structure represents interesting strategies to develop novel nano-bio hybrid systems by intercalating bioactive molecules into the gallery spaces (Figure 1). Potentially unstable and labile agents can be protected against harsh environmental and processing conditions, and eventually released at desired site with controlled manner (*26, 27*). Layered nanoclays have been applied as host materials to encapsulate various biomolecules because flexible interlayer spaces allow accommodation of the biomolecules depending on the volume and molecular size of intercalated molecules. Many researches demonstrated that layered nanomaterials can encapsulate DNA (*28–30*), nucleotides (*31, 32*), drugs (*33, 34*), proteins (*35*), and even viruses (*36*). These nano-bio hybrid systems were determined to enhance the delivery efficiency into cells and the efficacy of bioactive molecules inside cells as well (*37–39*). However, application of nanoclays for delivering nutrients or nutraceuticals is extremely limited.

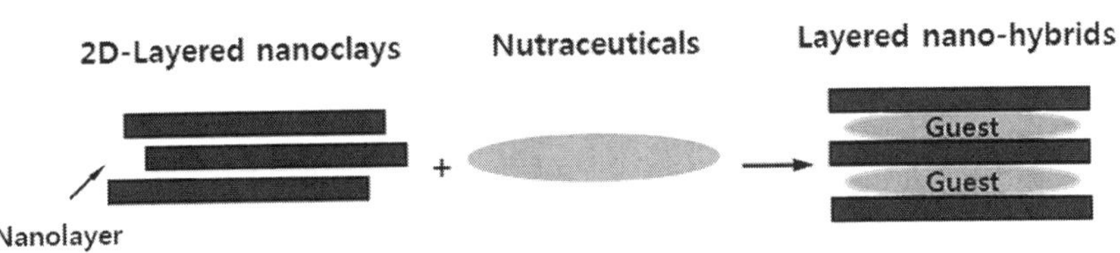

Figure 1. Scheme for nutraceuticals-loaded layered nanoclay hybrid system.

In this chapter, we describe nanoclay based delivery systems which have shown enhanced stability, bioavailability and biological efficacies for diverse nutraceutical compounds. Preparation, characterization and bioavailability of nanoclay based delivery system as well as examples of the biologically active encapsulation system are provided.

Delivery Efficiency and Biocompatibility of Nanoclays

Structural Features

Layered nanoclays are divided into two different classes, anionic and cationic clays, depending on the surface layered charge and the types of interlayer ions. Anionic layered nanomaterials typically stand for layered double hydroxides (LDHs) with exchangeable anions in the interlayer spaces. LDHs consist of a wide range of chemical compositions and their layer structure can have a variety of stacking faults to generate many different polytypes. The general chemical formula of LDHs is expressed as $[M^{2+}_{1-x}M^{3+}_{x}(OH)_2]\cdot(A^{m-})_{x/m}\cdot nH_2O$, where the M^{n+} represents a metal cation ($M^{2+} = Mg^{2+}$, Zn^{2+}, Ni^{2+}, Cu^{2+} and so on; $M^{3+} = Al^{3+}$, Fe^{3+}, and so on) and A^{m-} is a interlayer anion ($A^{m-} = CO_3^{2-}$, NO_3^{-}, Cl^{-}, SO_4^{2-} and other anionic species) located in the hydrated gallery spaces, giving rise to layered structure (Figure 2A) (*18*). The structure of LDHs is based on brucite $(Mg(OH)_2)$-like layers, where the partial isomorphous substitution of divalent cation, Mg^{2+}, with a trivalent cation such as Al^{3+} yields positively charged sheets in stacked position. Thus, the interlayer space is balanced by charge-compensating hydrated anions, stabilized through electrostatic interaction together with water molecules. This unique structural feature of LDHs results in anion exchange capacity, thereby contributing to intercalation ability of anionic molecules into the layers and controlled release property as well. The exchangeability of interlayered anions differs with their electrostatic interaction with positively charged sheets; divalent or trivalent anions are more strongly stabilized into LDH layers than monovalent anions, associated with less anion exchangeability of the former than the latter. They also have high layer charge density (2-5 meq/g), leading to strong electrostatic forces between the sheets and interlayer anions, thus swelling is more difficult than cationic clays. On the other hand, the solubility of LDHs is highly dependent on pH, namely, they are stable in an alkaline environment, but can be dissolved into ions under acidic conditions. These characteristics of LDHs are fascinating feature as delivery carriers since they can be gradually and easily decomposed inside cells where pH is slightly acidic. Meanwhile, surface coating of LDH layers with acid-resistant polymers will protect them from dissolution, which is essential for oral application.

Cationic layered nanoclays, for example, aluminosilicates such as montmorillonite (MMT), consist of octahedral and tetrahedral sheets, having high internal surface area. The basic structure of cationic clays is based on a mica framework (Figure 2B) (*17*), where the unit layer is composed of one octahedral sheet sandwiched between two tetrahedral sheets. The cations in tetrahedral sheets are typically Si^{4+} and Al^{3+}, while those in the octahedral sheet are Al^{3+}, Fe^{3+}, Mg^{2+} and Fe^{2+}. Cations of tetrahedral and/or octahedral sheets can be isomorphically substituted by lower-valent cations, resulting in negatively charged layer framework. This negatively charged layer is generally delocalized over the layer surface, thereby maintaining permanent negative charge, while not affected by environmental conditions such as pH and ion activity. On the contrary, additional negative charge could be also generated by deprotonation of the exposed hydroxyl groups in the cleaved edge or structurally deformed region in the sheets, which is highly influenced by experimental conditions. In the same manner as LDHs, cationic clays can also accommodate exchangeable counter ions, cations in their gallery space, endowing cation exchange capacity (CEC) as well. Cationic clays including smectite group and vermiculite group have weak negative charge and low charge density (0.2-0.9 meq/g) in their layers, which allows swelling property in the presence of water. This structural swelling feature is also related to CEC and intercalation ability of cationic molecules.

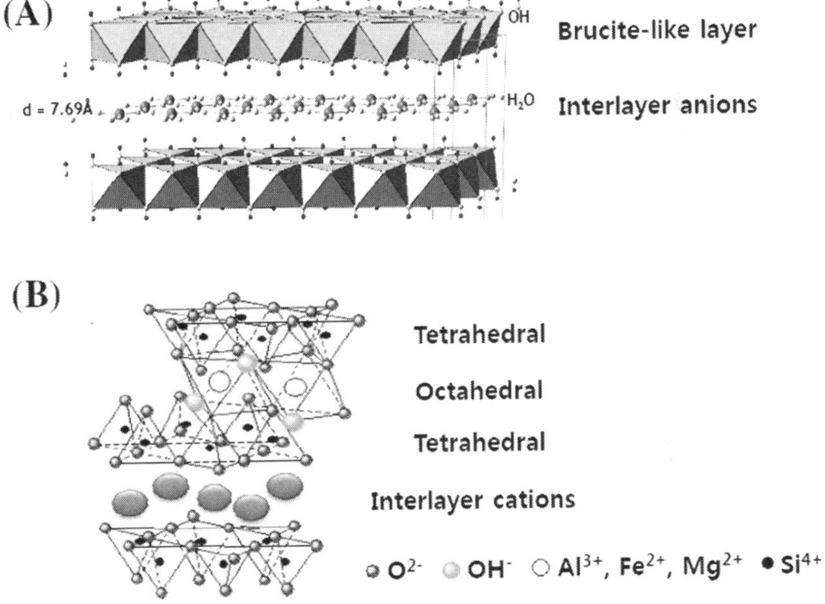

Figure 2. Schematic structure of (A) a layered double hydroxide and (B) a cationic clay, montmorillonite.

Anionic or cationic nanoclays can therefore be applied as delivery carriers, depending on the charge of intercalated molecules, fundamentally based on their unique layered structure. Of course, the size and volume of biomolecules which will be stabilized into the layers should be also carefully considered for successful intercalation with high stability. The fact that bioactive molecules are stabilized into the interlayer spaces through electrostatic interaction also provide controlled release ability, by exchanging them with other ions massively present in biological environment. Thus, sustainable bioactivity could be obtained by using nanoclays-bio hybrid systems. Moreover, labile nutraceuticals intercalated into the layers could be effectively protected against both low gastric pH and various enzymatic attacks, which is crucial to improve their activity by oral intake at the systemic level.

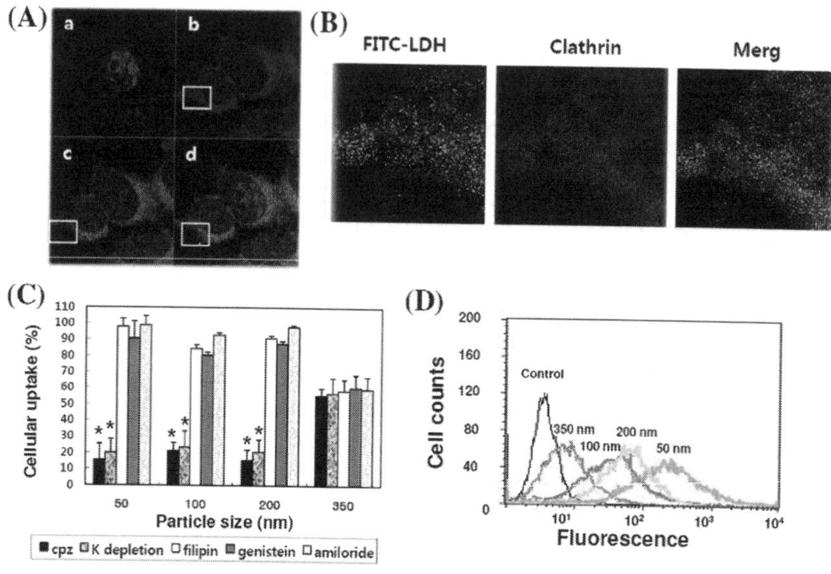

Figure 3. (A) Confocal microscopic images of LDH conjugated to fluorescein isothiocyanate (FITC) in cells (a, nuclei stained by Dapi; b, clathrin stained by Texas-Red; c, LDH-FITC; d, merged image) and (B) magnified image of the white box in part (A) showing high colocalization of LDH-FITC with clathrin protein. Reproduced with permission from Ref. (71). © 2006 American Chemical Society. (C) Size-dependent cellular uptake mechanism (chlorpromazine (cpz) and K depletion, inhibitor for clathrin-mediated endocytosis; filipin and geninstein, inhibitor for caveolae-mediated endocytosis inhibitor; amiloride, inhibitor for macropinocytosis). Reproduced with permission from Ref. (41). © 2009 Wiley Interscience. (D) Size-dependent cellular uptake amount. Reproduced with permission from Ref. (41). © 2009 Wiley Interscience.

Cellular Uptake

One of the most important factors as nutraceutical delivery systems is that nanomaterial-based delivery vehicles should be easily and efficiently taken up by cells and tissues (40). In this context, LDHs are effective because they enter cells through clathrin-mediated endocytosis, the most common endocytic pathway in all mammalian cells (Figure 3A and 3B) (37). especially when the particle size is controlled in the range from 50 to 200 nm (Figure 3C) (41). Particle size of LDHs is a key factor affecting uptake efficiency (41); the smaller the particle size, the more massive amount can be internalized (Figure 3D). Recent study demonstrated that intracellular trafficking pathway of LDHs after internalization into cells via clathrin-coated vesicles is highly dependent on particle size (*42*). Fifty nm particles follow a typical endo-lysosomal pathway as well as direct exocytic route at the same degree, while most of larger particles of 100 nm escape endosome-lysosome trafficking mechanism. They rather follows exocytic pathway represented by endosome-golgi apparatus-endoplasmic reticulum. Delivery efficiency of various biomolecules-loaded LDHs was demonstrated in many researches, showing high uptake of drugs (*43, 44*), genes (*28, 45, 46*), siRNA (*47*) and bioactive molecules (*48–50*) delivered through LDH carriers, in comparison with that delivered without LDH nanovehicles.

Meanwhile, the size of cationic clays varies ranging from nano-scale to micro-sized. In case of micro-sized clays, the main purpose of their biological application is not to enhance cellular delivery efficiency but rather to stabilize and protect molecules against unfriendly physiological environment. Bulk particles like aluminosilicates have, however, other fascinating characteristics to be used as oral delivery carriers. They possess highly mucoadhesive property (*51*), which is essential for molecules to across the GI barrier. Some aluminosilicates such as MMT also act as a potent detoxifier in the intestine because they have high adsorption ability for protein and microorganisms. Thus, they can adsorb not only dietary, microbial, metabolic toxic, and xenobiotics but also abnormally increased hydrogen ions found in acidosis (*52*).

Biocompatibility

In general, clays have been considered as bioinspired layered nanomaterials. Indeed, cationic clay minerals like smectite group have been traditionally applied in diverse fields including cutaneous chemotherapy, laxatives, anti-diarrheal and anti-inflammatory as well as anti-microbial agents (*53–55*). Recently, they are used as excipients, lubricants and dispersants in pharmacological applications to improve organoleptic, physical and chemical properties (*56–58*). Another cationic clay, MMT is also generally recognized as a biocompatible material (*59*), since it is commonly applied in many pharmaceutical formulations as both excipient and active substance (*60*). Moreover, anionic nanoclay, LDH in carbonate form, is applied as an antacid or antipepsin agent to neutralize stomach acidity, which is related to its alkali characteristics under physiological pH (*61*). However, nanomaterials-based nutraceutical delivery system could exhibit different toxicological effects induced by pharmaceutical formulations because nutraceuticals could be daily taken up without any prescription. Therefore, the safety issue of nanoclays must be addressed when applied for delivering nutraceuticals or functional nutrients, in particular, to the point of view of long-term and chronic toxicity.

In this regard, the toxicity of LDHs was evaluated *in vitro* and *in vivo*. Cytotoxicity of LDH was assessed in four different cell lines, in comparison with other inorganic nanoparticles extensively applied for biological purpose (*62*). The results demonstrated that LDH showed little cytotoxicity compared to silica, iron oxide and single walled carbon nanotubes (Figure 4A). LDH nanoparticles of about 200 nm did not significantly cause inhibition of cell proliferation, cell death, apoptosis and membrane damage up to 250 µg/ml with 48 h incubation. They only induced oxidative stress and pro-inflammatory mediator like interleukin-8 (IL-8) at 500 µg/ml for 72 h exposure. Interestingly, their cytotoxicities differ with tested cell lines (Figure 4B); the most remarkable cytotoxicity was found in lung cancer A549 cells, while they were not toxic at all to normal lung L-132 cells. It is worthy to note here that LDHs as delivery carriers are normally applied at concentration less than 100 µg/ml. It seems that LDH nanoparticles do not cause acute cytotoxicity at dose level used for practical biological application. On the other hand, size-dependent cytotoxicity of LDH nanoparticles was reported in A549 cells, showing higher induction of IL-8 by 50 nm LDH nanoparticles than by larger-sized one (100 to 350 nm) (*63*).

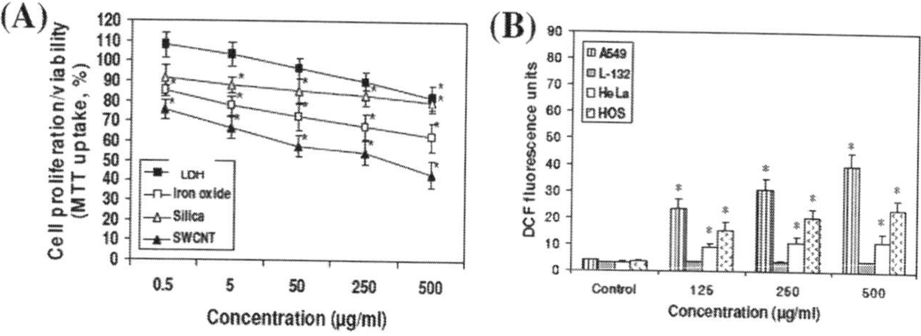

Figure 4. (A) Effect of LDH nanoparticles on cell proliferation of A549 cells after 72 h. (B) Induction of reactive oxygen species in different cell lines treated with 500 µg/ml of LDH nanoparticles for 72 h. Reproduced with permission from Ref. (62). © 2009 Elsevier.

Toxicity of LDH nanoparticles was also investigated in mice, with respect to particle size after repeated intraperitoneal injection (once a week) for 4 weeks (*64*). They did not significantly affect body weight gain or survival rate up to the highest dose tested, 600 mg/kg, in mice. However, when 50 nm particles were treated, some typical inflammatory responses such as alveolar sac filled with collagen or infiltration of neutrophils were found in about 10% of total treated mice (*27, 65*). It is likely that small-sized LDH nanoparticles could induce inflammation response, the most critical toxicological effect caused by LDH nanoparticles (*66*). However, the particle size of LDH nanoparticles for oral delivery should not be necessarily less than 100 nm. Larger size may be more efficient to protect nutraceuticals against GI environment. According to the results of many studies described above, LDHs are considered very attractive and safe delivery system for oral application.

Nanoclays for Nutraceutical Delivery System

As described earlier, nanoclays are biocompatible materials with high internal space area, high ion exchange capacity and low toxicity, which make it ideal as a delivery system for bioactive compounds. Nanoclay encapsulation system can be prepared by relatively simple and inexpensive process compared with other delivery system. Various inorganic and organic bioactive compounds could readily be introduced into the hydroxide interlayer space by simple ion exchange reaction or co-precipitation method. In addition to improving the solubility of drug or active compounds, nanoclay encapsulation system is particularly attractive for oral application due to its high mucoadhesive property which is useful for delivered molecules to across the GI barrier (*51*). Choi research group reported a facile way of encapsulating glutathione, a most powerful antioxidant, into MMT by cation exchange reaction (Figure 5A) (*67*). Glutathione (γ-L-glutamyl-L-cysteinylglycine, GSH) is an endogenous antioxidant protecting cells from reactive oxygen species by neutralizing free radicals. GSH is also essential for maintaining the normal function of immune system, and its deficiency occurs commonly in immunocompromised patients. But oral intake of GSH results in low bioavailability because of the chemical and enzymatic hydrolysis reaction in intestine and liver. The stability of GSH could be increased by encapsulating GSH into MMT. Oral administration of GSH-MMT hybrid to mice resulted in significantly increased bioavailability and high antioxidant activity in plasma (Figure 5B). The stability of GSH was further increased by coating GSH-MMT hybrid with polyvinylacetal diethylaminoacetate (AEA) (Figure 5B).

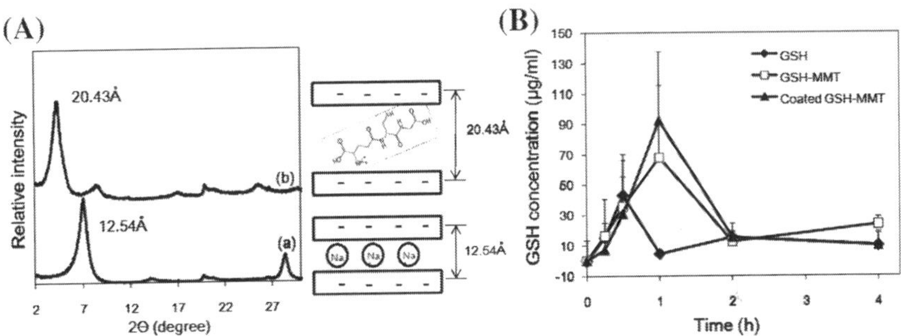

Figure 5. (A) Powered X-ray pattern and schematic diagram of (a) MMT and (b) GSH-MMT hybrid. (B) Plasma concentration of GSH in mice after oral administration of free GSH, GSH-MMT, and polymer-coated GSH-MMT, respectively. Reproduced with permission from Ref. (67). © 2012 Elsevier.

Conjugated linoleic acid (CLA) is another attractive functional ingredient with its health benefits including anticancer, antioxidant and anti-arteriosclerotic activities. However, the application of CLA as a food ingredient is limited because CLA oxidizes easily to volatile compound and impart off-flavor to the food. Oh research group encapsulated CLA into zinc basic salt by co-precipitation method (*68*). CLA molecules were shown to be packed with zigzag form between the intracrystalline spaces nanoparticles. The thermal stability of CLA at 180 °C increased by about 25% through nanoencapsulation, compared to that of pristine CLA.

Gasser research groups also encapsulated vitamin C into LDH by ion exchange process between anionic derivatives in LDH and the anionic ascorbic acid (*69*). Vitamin C is sensitive to oxidation in the presence of oxygen of which the reaction is assisted by trace amount of metal ions under neutral and alkaline condition. Intercalation of vitamin C into gallery space of LDH is expected to offers a new way of stabilizing the sensitive compound as well as controlling the release at given condition. The release of encapsulated vitamin C out from the LDH lattice occurred by ion exchange and diffusion reaction in carbonated aqueous solution. Choy group was also successful to intercalate vitamin C and E, tocopherol acid succinate, into LDH by coprecipitation reaction, clearly demonstrating controlled release property of both vitamin C-LDH and vitamin E-LDH hybrids (*48*). Figure 6 shows powdered X-ray patterns and schematic illustration of vitamin-loaded LDH nanohybrid systems.

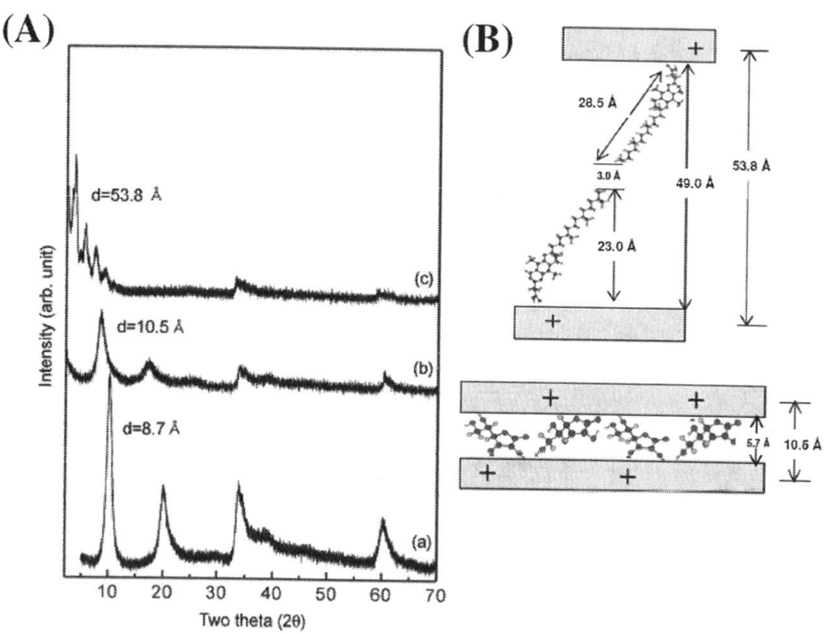

Figure 6. (A) Powder X-ray diffraction patterns of (a) ZnAl(NO₃)-LDH, (b) ZnAl(vitamin C)-LDH and (c) ZnAl(vitamin E)-LDH. (B) Schematic illustration of (a) ZnAl(vitamin C)-LDH and (b) ZnAl(vitamin E)-LDH. Reproduced with permission from Ref. (48). © 2004 KCSNET.

Edible dyes are widely used in food industry, however their stability is often challenged by thermal treatment which commonly takes place during food processing. Thus, enhancing the thermal stability of those edible dyes is essential for the production of high quality and more acceptable food. Choy group has employed ZnAl layered double hydroxyde as an encapsulation material for several edible dyes, such as Allura® Red AC, Sunset Yellow FCF, and Brilliant

Blue FCF (*70*). The dye molecules were intercalated into hydroxide layers during co-precipitation process without any chemical and structural changes. The thermal stability and absorbance intensity of dye-LDH were found to be clearly enhanced compared to those of the corresponding dye salt because of the charge transfer between the dye molecules and LDH lattices.

Nanoclays have received a great deal of attention as a good candidate to encapsulate diverse nutraceuticals and clearly exhibited enhanced stability and bioactivities of encapsulated functional molecules. But most of studies so far have been conducted *in vitro*. More oral administration research using animal model will be necessary for actual application of the nanoclay encapsulation systems in food industry.

Conclusion

Nanoclays have been traditionally used in many fields such as pelotherapy, cutaneous chemotherapy, and medicine to improve human health and life. Moreover, they have been recently developed as excipients, lubricants, and active substances in diverse pharmacological formulations. Although their application in food science is relatively limited, they have promising potential as oral carriers for effectively delivery of nutraceuticals and nutrients; 1) their unique structure enables to intercalate bioactive molecules into the interlayer spaces, 2) nutraceuticals-loaded nanoclays have controlled release property and protection ability against GI environments, 3) they can be easily internalized into cells, 4) they have high mucoadhesive property which facilitates transportation of molecules across the GI barrier, 5) they exhibit low or null toxicity. Research on the development of nanoclays based oral delivery system for nutraceuticals or functional ingredients showed that they are effective in protective and controlled delivery of intercalated molecules, contributing to enhance bioavailability as well. The promising potential of nanoclays provides novel perspectives to develop nanomaterials-based oral delivery systems.

References

1. Weiss, J.; Takhistov, P.; McClements, D. J. *J. Food Sci.* **2006**, *71*, R107–R116.
2. Lim, M. C.; Shin, Y. J.; Jeon, T. J.; Kim, H. Y.; Kim, Y. R. *Anal. Bioanal. Chem.* **2011**, *400*, 777–785.
3. Joung, C. K.; Kim, H. N.; Im, H. C.; Kim, H. Y.; Oh, M. H.; Kim, Y. R. *Sens. Actuators, B* **2012**, *161*, 824–831.
4. Huang, Y.; Dong, X.; Liu, Y.; Li, L. J.; Chen, P. *J. Mater. Chem.* **2011**, *21*, 12358–12362.
5. Batt, C. A. *Science* **2007**, *316*, 1579–1580.
6. Min, J.; Kim, J. H.; Y., L.; Namkoong, K.; Im, H. C.; Kim, H. N.; H.Y., K.; Huh, N.; Y.R., K. *Lab Chip* **2011**, *11*, 259–265.
7. Yager, P.; Edwards, T.; Fu, E.; Helton, K.; Nelson, K.; Tam, M. R.; Weigl, B. H. *Nature* **2006**, *442*, 412–418.

8. Sorrentino, A.; Gorrasi, G.; Vittoria, V. *Trends Food Sci. Technol.* **2007**, *18*, 84–95.

9. de Azeredo, H. M. C. *Food Res. Inter.* **2009**, *42*, 1240–1253.

10. Blasco, C.; Pico, Y. *Trends Anal. Chem.* **2011**, *30*, 84–99.

11. Giannelis, E. P. *Adv. Mater.* **1996**, *8*, 29–35.

12. Grunlan, J. C.; Grigorian, A.; Hamilton, C. B.; Mehrabi, A. R. *J. Appl. Polym. Sci.* **2004**, *93*, 1102–1109.

13. Huang, Q.; Yu, H.; Ru, Q. *J. Food Sci.* **2010**, *75*, R50–R57.

14. Hilty, F. M.; Arnold, M.; Hilbe, M.; Teleki, A.; Knijnenburg, J. T. N.; Ehrensperger, F.; Hurrell, R. F.; Pratsinis, S. E.; Langhans, W.; Zimmermann, M. B. *Nat. Nanotechnol.* **2010**, *5*, 374–380.

15. Theng, K. G. *The Chemistry of Clay-Organic Reactions*; Adam Hilger: London, U.K., 1974.

16. Nemecz, E. *Clay Minerals*; Akademiai Kiado: Budapest, Hungary, 1981.

17. Ogawa, M.; Kuroda, K. *Chem. Rev.* **1995**, *95*, 399–438.

18. Vaccari, A. *Catal Today* **1998**, *41*, 53–71.

19. Robertson, R. H. S. *Br. Miner Soc. Bull.* **1996**, *113*, 3–7.

20. Carretero, M. I. *Appl. Clay Sci.* **2002**, *21*, 155–163.

21. Veniale, F.; Barberis, E.; Carcangiu, G.; Morandi, N.; Setti, M.; Tamanini, M.; Tessier, D. *Appl. Clay Sci.* **2004**, *25*, 135–148.

22. Gorchakov, V. N.; Dragun, G. N.; Kolmogorov, Y. P.; Smelova, V. A.; Tikhonova, L. I.; Tysjachnova, Y. V. *Nucl. Instrum. Methods Phys. Res., Sect. A* **2001**, *470*, 437–440.

23. Burzlaff, A.; Brethauer, S.; Kasper, C.; Jackisch, B. O.; Scheper, T. *Cytometry, Part A* **2004**, *62*, 65–9.

24. Peterson, C. L.; Perry, D. L.; Masood, H.; Lin, H.; White, J. L.; Hem, S. L.; Fritsch, C.; Haeusler, F. *Pharm. Res.* **1993**, *10*, 998–1004.

25. Baes, C. F.; Mesmer, R. E. *The hydrolysis of cations*; A Wiley-Interscience Publication: New York, 1976; Vols. 95-98, pp 112–122.

26. Choy, J. H.; Choi, S. J.; Oh, J. M.; Park, T. *Appl. Clay Sci.* **2007**, *36*, 122–132.

27. Choi, S. J.; Choy, J. H. *Nanomedicine* **2011**, *6*, 803–814.

28. Choy, J. H.; Kwak, S. Y.; Park, J. S.; Jeong, Y. J.; Portier, J. *J. Am. Chem. Soc.* **1999**, *121*, 1399–1400.

29. Nakayama, H.; Hatakeyama, A.; Tsuhako, M. *Int. J. Pharm.* **2010**, *393*, 104–11.

30. Desigaux, L.; Belkacem, M. B.; Richard, P.; Cellier, J.; Leone, P.; Cario, L.; Leroux, F.; Taviot-Gueho, C.; Pitard, B. *Nano Lett.* **2006**, *6*, 199–204.

31. Choy, J. H.; Kwak, S. Y.; Jeong, Y. J.; Park, J. S. *Angew. Chem., Int. Ed.* **2000**, *39*, 4042–4045.

32. Wong, Y.; Cooper, H. M.; Zhang, K.; Chen, M.; Bartlett, P.; Xu, Z. P. *J. Colloid Interface Sci.*, *369*, 453–9.

33. Choy, J. H.; Jung, J. S.; Oh, J. M.; Park, M.; Jeong, J.; Kang, Y. K.; Han, O. J. *Biomaterials* **2004**, *25*, 3059–3064.

34. Choi, S. J.; Oh, J. M.; Choy, J. H. *J. Phys. Chem. Solids* **2008**, *69*, 1528–1532.

35. Rahman, M. B.; Basri, M.; Hussein, M. Z.; Rahman, R. N.; Zainol, D. H.; Salleh, A. B. *Appl. Biochem. Biotechnol.* **2004**, *118*, 313–20.

36. Jin, S.; Fallgren, P. H.; Morris, J. M.; Chen, Q. *Sci. Technol. Adv. Mater.* **2007**, *8*, 67–70.

37. Oh, J. M.; Park, M.; Kim, S. T.; Jung, J. Y.; Kang, Y. K.; Choy, J. H. *J. Phys. Chem. Solids* **2006**, *67*, 1024–1027.

38. Kim, J. Y.; Choi, S. J.; Oh, J. M.; Park, T.; Choy, J. H. *J. Nanosci. Nanotechnol.* **2007**, *7*, 3700–5.

39. Oh, J. M.; Choi, S. J.; Lee, G. E.; Han, S. H.; Choy, J. H. *Adv. Funct. Mater.* **2009**, *19*, 1–8.

40. Xu, Z. P.; Zeng, Q. H.; Lu, G. Q.; Yu, A. B. *Chem. Eng. Sci.* **2006**, *61*, 1027–1040.

41. Oh, J. M.; Choi, S. J.; Lee, G. E.; Kim, J. E.; Choy, J. H. *Chem. Asian J.* **2009**, *4*, 67–73.

42. Chung, H. E.; Park, D. H.; Choy, J. H.; Choi, S. J. *Appl. Clay Sci.* **2012**, *65-66*, 24–30.

43. Minagawa, K.; Berber, M. R.; Hafez, I. H.; Mori, T.; Tanaka, M. *J. Mater. Sci. Mater. Med.*, *23*, 973–81.

44. Choi, S. J.; Choi, G.; Oh, J. M.; Oh, Y. J.; Park, M. C.; Choy, J. H. *J. Mater. Chem.* **2010**, *20*, 9463–9469.

45. Ladewig, K.; Xu, Z. P.; Lu, G. Q. *Expert Opin. Drug Delivery* **2009**, *6*, 907–22.

46. Xu, Z. P.; Walker, T. L.; Liu, K. L.; Cooper, H. M.; Lu, G. Q.; Bartlett, P. F. *Int. J. Nanomed.* **2007**, *2*, 163–74.

47. Wong, Y.; Markham, K.; Xu, Z. P.; Chen, M.; Max Lu, G. Q.; Bartlett, P. F.; Cooper, H. M. *Biomaterials* **2010**, *31*, 8770–9.

48. Choy, J. H.; Son, Y. H. *Bull. Korean Chem. Soc.* **2004**, *25*, 122–126.

49. Yang, J. H.; Lee, S. Y.; Han, Y. S.; Park, K. C.; Choy, J. H. *Bull. Kor. Chem. Soc.* **2003**, *24*, 499–503.

50. Tagaya, H.; Sasaki, N.; Morioka, H.; Kadokawa, J. *Mol. Cryst. Liq. Cryst.* **2000**, *341*, 1217–1222.

51. Doborozsi, D. *Oral liquid mucoadhesive compositions.* U.S. Patent 6.638.521, 2003.

52. Sun, B.; Ranganathan, B.; Feng, S. S. *Biomaterials* **2008**, *29*, 475–86.

53. Cornejo, J.; Hermosin, M. C.; White, J. L.; Barnes, J. R.; Hem, S. L. *Clays Clay Miner* **1983**, *31*, 109–112.

54. Bolger, R. *Ind. Min.* **1995** August, 52–63.

55. Ferrand, T.; Yvon, J. *Appl. Clay Sci.* **1991**, *6*, 21–38.

56. Poensin, D.; Carpentier, P. H.; Fechoz, C.; Gasparini, S. *Jt., Bone, Spine* **2003**, *70*, 367–370.

57. Summa, V.; Tateo, F. *Appl. Clay Sci.* **1998**, *12*, 403–417.

58. Cara, S.; Carcangiu, G.; Padalino, G.; Palomba, M.; Tamanini, M. *Appl. Clay Sci.* **2000**, *16*, 125–132.

59. Gupta, G.; Gardner, W. *J. Hazard Mater.* **2005**, *118*, 81–3.

60. Wang, X.; Du, Y.; Luo, J. *Nanotechnology* **2008**, *19*, 065707.

61. Tarnawski, A.; Pai, R.; Itani, R.; Wyle, F. A. *Digestion* **1999**, *60*, 449–455.

62. Choi, S. J.; Oh, J. M.; Choy, J. H. *J. Inorg. Biochem.* **2009**, *103*, 463–71.

63. Choi, S. J.; Oh, J. M.; Choy, J. H. *J. Nanosci. Nanotechnol.* **2008**, *8*, 5297–5301.

64. Choi, S. J. O. J. M.; Choy, J. H. *J. Ceram. Soc. Jpn.* **2009**, *117*, 543–549.

65. Choi, S. J. O. J. M.; Choy, J. H. *J. Ceram. Soc. Jpn.* **2009**, *117*, 543–549.

66. Choi, S. J.; Choy, J. H. *J. Mater. Chem.* **2011**, *21*, 5547–5554.

67. Baek, M.; Choy, J. H.; Choi, S. J. *Int. J. Pharm.* **2012**, *425*, 29–34.

68. Choy, J. H.; Shin, J.; Lim, S. Y.; Oh, J. M.; Oh, M. H.; Oh, S. *J. Food Sci.* **2010**, *75*, N63–N68.

69. Gasser, M. S. *Colloid Surf., B* **2009**, *73*, 103–109.

70. Choy, J. H.; Kim, Y. K.; Son, Y. H.; Choy, Y. B.; Oh, J. M.; Jung, H.; Hwang, S. J. *J. Phys. Chem. Solids* **2008**, *69*, 1547–1551.

71. Oh, J. M.; Choi, S. J.; Kim, S. T.; Choy, J. H. *Bioconjugate Chem.* **2006**, *17*, 1411–7.

Chapter 13

Nano/Microencapsulation of Functional Ingredients and Drugs into Biopolymer Matrices: A Study of Stability and Controlled Release

G. K. Kouassi,[*,1] V. Gogineni, T. Ahmad, N. M. Gowda,[1] M. S. Boley,[2,1] and N. Koissi[3]

[1]Department of Chemistry, Western Illinois University, Macomb, Illinois
[2]Department of Physics, Western Illinois University, Macomb, Illinois
[3]Department of Chemistry, University of Maryland at Baltimore County, Baltimore, Maryland
*E-mail: gk-kouassi@wiu.edu.

Nano/microencapsulation of Vitamin K into whey protein and κ-carrageenan and the microencapsulation of Piroxicam into whey protein, κ-carrageenan, and chitosan are described. Power ultrasound was used as the coating technique. The sizes of the particles were characterized using AFM and FlowCam imaging. The turbidity study suggested that in the presence of whey protein, coating was successful in both acidic and basic conditions. Furthermore, controlled-released of vitamins K and Piroxican was water activity-dependent. Optimal release of both compounds was found at water activity values of 0.662 and 0.769 due to water plasticization of the glassy coating matrix. The models studied show that a proper control of pH, temperature, and water activity could be useful for target delivery of encapsulated lipophilic substances.

Introduction

Encapsulation is one of nature's techniques used to protect essential materials from degradation or from attacks from the surrounding conditions (*1*). Thus, most biological systems are protected with capsules or layers. For example, at a molecular level, the cell, known as the basic structural and functional unit of all living things is protected by the cellular membrane. At a microscopic level, the essential components of seeds in the embryo are surrounded by the seed coat.

Encapsulation can provide a physical barrier between the core compound and the other components of the product (2). It can also be used to control the release of the core substance, or to regulate the interactions between the core components and the surroundings, and/or to mask or halt undesirable properties of the core component. Encapsulation of drugs and nutraceuticals within colloidal-sized polymer particles can provide sustained release, reduce the side effects of the drugs, and improve delivery and bioavailability (1). When structures have at least one dimension with a size below 100 nm, the process of encapsulation is termed nanoencapsulation. The term microencapsulation is used when the structures have sizes comprised between a few micrometers and a few millimeters (2). Nano-and/or microencapsuation are increasingly used in formulation of food or pharmaceuticals. Liposoluble food ingredients such as vitamin K have many benefits to health (3–5). A vast majority of lipophilic ingredients and drugs are prone to oxidation and degradation under the effects of various environmental conditions, or due to their interactions with other components (1, 3). Thus, nano/microsize formulation processes have been adapted to limit their degradation. The structure, size, shape, and functionality of the nano/microcapsules depend on the materials and the techniques used for the encapsulation, as well as the types of core/shell interactions occurring during processing and storage. Biopolymers, including polysaccharides, biodegradable polymers, and proteins, present a wide array of potential applications, particularly for the food industry (6–10). Carrageenans are hydrocolloids obtained by extraction from the red sea weeds (members of the class Rhodophyceae). They consist of potassium, sodium, calcium, magnesium, and ammonium sulfate esters of galactose and 3,6-anhydrogalactose copolymers (11). They are mainly divided into three types based on the position of sulfate groups and presence or absence of anhydrogalactose. *Kappa(κ), Iota(ι),* and *lambda(τ)* carrageenan have wide applications in the world of encapsulation (12). Chitosan is a cationic polysaccharide obtained by partial deacetylation of chitin, the major component of crustacean shells. It is a hydrophilic polymer with positive charges that come from weak basic groups, useful for special applications (13). Proteins are used as emulsifiers during the encapsulation processes owing to their amphotheric character, and their ability to form complexes with both anionic and cationic substances (1, 14, 15).

In general, encapsulation is conducted in a continuous fluidic phase and the core material is embedded into a coating material, in the form of emulsions. Emulsions may exhibit limited thermodynamic stability due to poor miscibility of their components. As a consequence, the emulsion may be dispersed if energy is applied for mixing (9). Dispersion of the components of an emulsion often leads to coalescence or phase separation. Therefore, thermodynamic compatibility between the coating materials and the core component/s is critical in encapsulation processes that are conducted through emulsions. Nano/microcapsules prepared in the form of emulsions are made amorphous through vitrification. For example, spray-drying, freeze-drying, and complex coarcevation are techniques that are used to prepare dry capsules (*10*). Thus, like most amorphous materials containing polymeric materials, the stability of the coating matrix is dependent on change in the physical state (*9*). In the dry amorphous state, encapsulated materials are glassy. At temperatures above the glass transition temperature, the solid glassy material can become a supercooled liquid, referred to as a plasticized material. For example, water is a plasticizer. Increasing the water content of a glassy material leads to significant water plasticization (*10, 16, 17*). Plasticization is necessary for the release of encapsulated substances, because the molten supercooled liquid state resulting from plasticization is often associated with increase in molecular mobility at temperature above the glass transition (*17*). Therefore, the controlled release of encapsulated substances may be achieved through appropriate management of water content, water activity, and temperature. Vitamin K is a fat soluble substance that exists in the two natural forms of Vitamin K_1, (phylloquinone) and Vitamin K_2 (menaquinone). Vitamin K_1 consists of four isoprenoid residues in its side chain, and is an obligatory cofactor in the post-translational carboxylase modification of calcium-binding proteins involved in antihemorrhagic activity (*18*). It has a great physiological role in blood coagulation and bone metabolism. Due to the presence of the hydrophilic functional group naphthoquinone, all forms of vitamin K are active in the human body. The sensitivity of vitamin K_1 to external heat and light and its high degree of lipophilicity, makes it highly unstable. Therefore, encapsulation of vitamin K could improve stability, biological efficiency, better characteristics, and good solubility profile (*19*).

Piroxicam is a derivative of the Oxicam group of non-steroidal anti-inflammatory drugs (NSAID). It is a 4-hydroxy-2-methyl-N-(pyridin-2-yl)-2H-1,2-benzothiazine-3-carboxamide-1,1-dioxide that exhibits weakly acidic 4-hydroxy proton at pKa 5.1 and weakly basic pyridyl nitrogen at pKa 1.8 (*20*). It is used to relieve the symptoms of rheumatoid and osteoarthritis, primary dysmenorrhea, and postoperative pain (*20*) Pixoricam is poorly soluble in both hydrophilic and hydrophobic media. These limitations reduce its absorption into the intestinal mucosa because, the cell membrane is lipophilic and limits the diffusion of compounds that are ionized. In spite of the important efforts that are still being made to develop suitable systems of Piroxicam that allow its delivery, very few reports on the physical state of Piroxicam containing amorphous systems are available in the literature (*20*).

The objective of this study is two-fold. First, encapsulate vitamin K and Piroxicam into nanosize and microsize biopolymer matrices. Secondly, investigate the effects of the change in the glass transition temperature (T_g) and water activity on the stability of the matrices and the release profile.

Material and Methods

Carrageenan and Chitosan were received as gifts from Ingredient Solutions, Inc, Waldo, and whey protein was purchased from Bulkfood, Toledo, Ohio. All reagents and chemicals used in this study were purchased from Sigma Aldrich, St Louis, Missouri.

Nano/Microencapsulation of Vitamin K and Piroxicam

Figure 1 shows the procedure for nano/microencapsulation of vitamin K and Piroxicam. Droplets of Vitamin K_1 were prepared in whey protein emulsion and coated with κ-carrageenan as follows: 1.25g of whey protein was dissolved in 15 mL of distilled water using a magnetic stirrer. A solution of κ-carrageenan was prepared by dissolving 0.1 g of κ-carrageenan into 10 mL of distilled water. About 100 mg of vitamin K_1 was dissolved in 2 mL of n-hexane and then dropped into the whey-protein solution and mixed using a magnetic stirrer. The prepared protein-vitamin mixture was homogenized using a Power Ultrasonicator. Emulsification was achieved by adding 0.125 mg of Tween 20. Tween 20 was added as a surfactant to improve the miscibility of the lipophilic vitamin K and the hydrophilic whey protein.

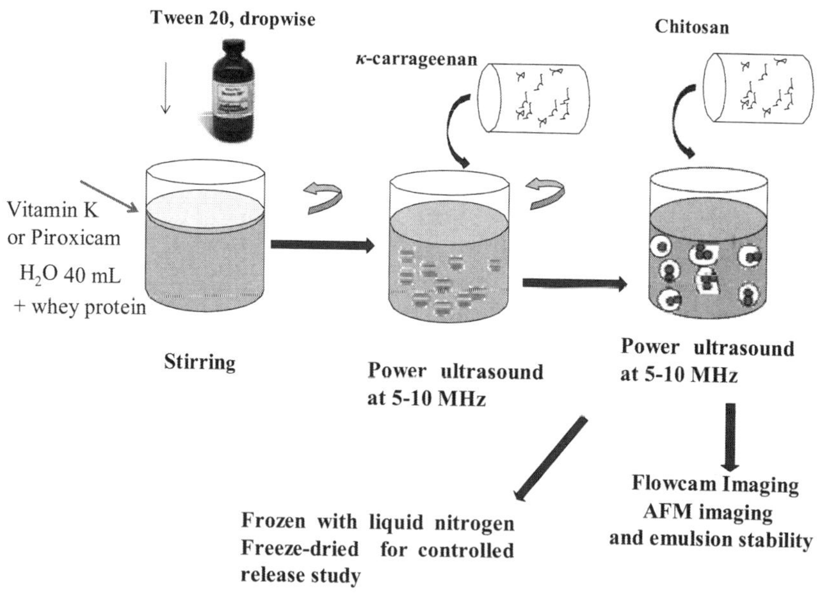

Figure 1. Nano-microencapsulation of vitamin K and Piroxicam by controlled emulsification.

Microencapsulation of Piroxicam into a dual polymer system of κ-carrageenan and chitosan was similar to the nanoencapsulation of vitamin K, except that in the later case, a solution of chitosan was added as a second coating biopolymer during ultrasonication, and in each phase of the ultrasonication, the frequency was kept at 10 MHz for 10 min instead of 5 MHz in the former case. The stability of the encapsulated vitamin K was evaluated at various pH levels to assess the extent of the interactions between the layers surrounding vitamin K droplets. Drops of the emulsions were taken for AFM and FlowCam imaging. The turbidity of the emulsion was evaluated at various pH values, and the remainder of the emulsion was freeze-dried. The freeze-dried amorphous microparticles were stored in evacuated desiccators containing salt solutions of $MgCl_2$, K_2CO_3, $Mg(NO_3)_2$, $NaNO_3$, and $NaCl$, corresponding to water activity (a_w) values of 0.333, 0.444, 0.538, 0.663, and 0.763, respectively. Water sorption properties of the samples, and the amounts of vitamin K released was determined gravimetrically.

HPLC Analysis of Release of Vitamin K

Duplicate samples of 100 mg of freeze-dried encapsulated vitamin K were stored in desiccators at various water activity conditions described above. After 6 hours of storage, the samples were added to 5 ml of petroleum ether contained in a glass vial and gently mixed to extract the surface oil. Evaporation of petroleum ether was done by flashing petroleum ether with nitrogen gas for about 5 min. Reverse phase HPLC with a UV-Visible detector and a Grace Smart C18 column was used to detect the amount of Vitamin K in the sample Methanol 90% HPLC grade and 10% dichloromethane (DCM) were used as a mobile phase for HPLC analysis. About 2ml of Methanol/DCM (90:10) was added as a mobile phase to vitamin K and filtered with a 0.2 µm filter. The injection volume of the sample was 10 µl, and the flow rate was maintained at 0.9 ml/min. The detection was conducted at a wavelength of 294 nm. Samples were analyzed for 90 min. The amount of vitamin K was determined using a least square analysis using external standard method. The vitamin K released was calculated by subtracting the amount of total vitamin K extracted from the sample of an equal amount of freeze-dried encapsulated vitamin K.

HPLC Analysis of Piroxicam

The extraction of Piroxicam was similar to that of vitamin K. The mobile phase for HPLC analysis of released Piroxicam was a mixture of buffer and methanol in the ratio of 65:35. About 7.72g of anhydrous citric acid was dissolved in 400ml of water and separately 5.35g of dibasic sodium phosphate was dissolved in 100 ml of water. The buffer was prepared by adding phosphate solution to the citric acid solution, then diluted with water to make up to 1000 ml. The columns were equilibrated for at least one hour with each mobile phase. The column and HPLC conditions were similar to those used for vitamin K analysis, except that the wavelength of the UV detector was set at 254 nm. The amount of Piroxicam in the encapsulated sample was calculated using a least square analysis using external standard method. In this method, the concentration of unknown Piroxicam extract was determined by extrapolating the graph of absorbance versus concentration of standard samples of Piroxicam. The amount of released Piroxicam was calculated by subtracting the amount of Piroxicam extracted from equal amount of freeze-dried encapsulated structures which has not been stored at water activity conditions. The encapsulation efficiency (EE) was calculated by dividing the amount of surface vitamin K or Piroxicam extracted using petroleum ether by the amount of vitamin K or Piroxicam extracted using hexane. The following equation was used:

$$EE = \frac{TP-SP}{TP} * 100$$

where SP represents the surface product (non-encapsulated product) extracted with hexane and TP stands for the total contents of the product in a given amount of freeze-dried sample.

Glass Transition Temperature and Water Sorption

In general, the state of vitrification can be achieved after freeze-drying. Prepared emulsions were pre-frozen with liquid nitrogen to entrap the nano/microcapsules. Then, the pre-frozen preparation was frozen overnight at -20°C, and finally freeze-dried at -40 °C using a freeze-dryer model 4.5 originated from Labconco, Cambrdige MA. To remove any residual water, the samples were stored for two days in desiccators in the presence of phosphorous pentoxide (P_2O_5). A Differential Scanning Calorimeter (DSC) was used to determine the glass transition temperature of freeze-dried amorphous nano/microcapsules. The glass transition is a well known change in the state of amorphous materials (*10*). About 20 mg of dried capsules were placed in aluminum pans and stored in evacuated desiccators containing the saturated salt solutions described above. The samples in aluminum pan were sealed after 24h of storage over the saturated salt solutions and analyzed using a DSC. The amorphous nano/microcapsules in aluminum pans were scanned from 2 °C to 120 °C, to measure the change in heat capacity, as a function of temperature. The glass transition temperature (T_g) of the particles was measured. T_g serves as a reference temperature above which possible alteration of the polymeric coating material mayoccur. The effects of water content and water activity (a_w) on the release profile of vitamin K and Piroxicam were investigated after exposure of the encapsulated materials to various salt solutions in evacuated desiccators corresponding to the same water activities described above.

Sizes Characterization of Capsules by AFM and FlowCam Imaging

Size and structure characterization of the particles were conducted using atomic force microscopy (AFM) and FlowCam imaging. A drop of the emulsion was deposited on a gold plate and dried overnight. The dry capsules were scanned with atomic force microscopy model Q-250 (c.1999) from Quesant Instruments, Agoura, CA, and was operated in the non-contact wave mode in order to obtain 3D topography images. Size, structure, and dispersity of capsules in emulsion were analyzed using the FlowCam imaging technique from Fluidic imaging Technology, Yarmouth, Maine, as described by Kouassi et al. (*1*).

Measurement of Turbidity in Whey Protein, κ-Carrageenan, and Encapsulated Vitamin K

The stability of the nano/microcapsules depends to some extent on the strength of the interactions between the layers surrounding the core material. To assess the strength of these interactions, the turbidity of encapsulated vitamins K, κ-carrageenan, and whey protein solutions were measured using a turbidimeter. The pH of the emulsions was adjusted with either 1 M HCl or 0.5 M NaOH solution and nephelometric readings were made in the pH range from 2 to 12 pH.

Results and Discussions

Protein-Biopolymer Interactions

Changes in turbidity of whey protein κ-carrageenan interactions at various pH are presented in Figure 2.

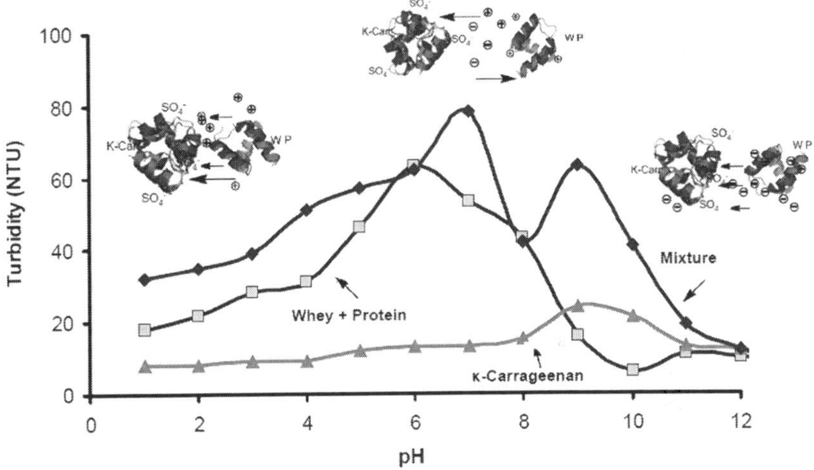

Figure 2. Turbidity of whey protein, κ-carrageenan and nanoencapsulated vitamin K measured at various pH. Below its isoelectric point (pI) whey protein (WP) carries positive charges. At neutral pI both positive and negative charges exist in the structure of whey protein. At very high pH>10, whey protein carries negative charges. Therefore, repulsions between whey protein and κ-carrageenan are likely to occur.

The isoelectric pH (pI) of whey protein is 5.2. At a pH level below this value, whey protein is readily in a cationic form and partially soluble. The turbidity of the whey protein solution slightly increased from 33 to 54 NPU. Above the pI, the turbidity of the solution decreased gradually and falls below 10 NPU at pH 10. This decrease in turbidity indicated that whey protein is soluble in a relatively basic medium. In the 2-5.5 pH range, κ-carrageenan exhibited low turbidity due to attractions between the negatively charged sulfate groups in its molecules and the cationic medium reflected in the value of the pH. Above a pH value of 9.0, the turbidity increased as a consequence of the absence of intermolecular attractions between negatively charged sulfate groups of κ-carrageenan and that of the anionic solution

When the solutions of whey-protein and κ-carrageenan were mixed, the turbidity of the mixture remained low around 30 NTU and started to increase around the pI of whey protein. Indeed, in the vicinity of pI, whey protein exhibits an amphoteric character. In this condition, positive moieties in the whey protein molecules are involved in electrostatic interactions with the negatively charged sulfate groups of κ-carrageenan. As the pH of the mixture is raised above the pI, whey protein molecules carry negative charges and experienced no apparent interaction with the SO_4^- groups in the κ-carrageenan molecules above pH 5.6. The turbidity decreased below 10 NTU. This sudden decrease in turbidity observed above pH 5.6 is the result of intermolecular attractions, eventually electrostatic interactions, between SO_4^- groups, the protein, and κ-carrageenan molecules. The protein polymer mixtures formed soluble complexes above pI and become turbid below pI. The interaction between protein and polymer was high at pI due to reactions between protonated amine in the whey protein and the SO_4^- of κ-carrageenan. The interactions were also high in the vicinity of pI due to the amphoteric character of the protein which allows electrostatic interactions between cations in the protein and sulfate groups in κ-carrageenan. The electrostatic interactions occurring between cationic and anionic substance around the core substance can be estimated using the Coulumb's law for point charge interaction.

$$F = \frac{q_1 q_2}{4 \pi \varepsilon_0 r^2}$$

(Eq.1)

The work associated with this force depends on r and is given as

$$w(r) = -\int_{\infty}^{r} F_{el} dr = \frac{q_1 q_2}{(4 \pi \varepsilon_0 \varepsilon r)}$$
(Eq.2)

where q_1 and q_2 are the charges on each particle, r is the distance between the charges, ε_0 the permitinity of the air, and ε the relative permittivity of the medium.

Size Characterization of Capsules

Figure 3 shows a 3D AFM image of nanoencapsulated Vitamin K. Nanosize particles were obtained by applying ultrasonication of frequency 10 MHz for 10 min. On the basis of the color scale in the top right corner of the image, the sizes of the capsules are around 100 nm.

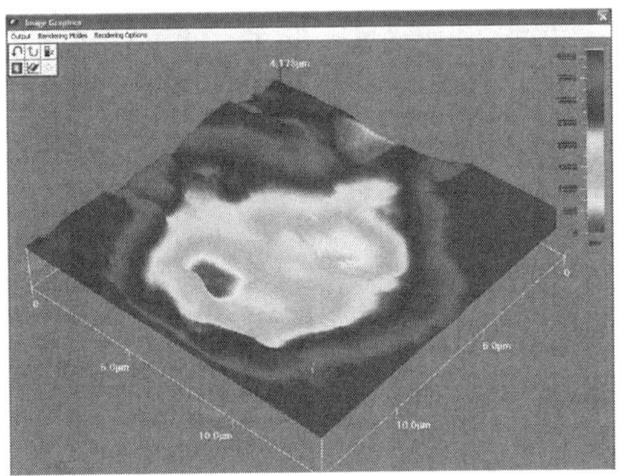

Figure 3. **3D** *AFM image of nanoencapsulated vitamin K. The core vitamin K droplets are surrounded by biopolymer layers.*

Figure 4 shows FlowCam images of microencapsulated Piroxicam. These images were obtained by setting the ultrasonication frequency to 7 MHz for 5 min after a coating biopolymer was added. The microcapsules have spherical shapes, and the layers surrounding the drug droplets were distinct.

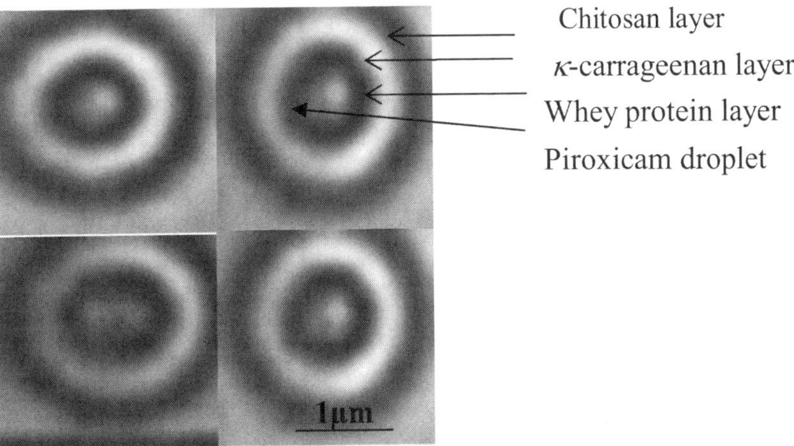

Figure 4. FlowCam screen shot images of microencapsulated Piroxicam, prepared using ultrasonication frequencies comprised between 5 and 7 MHz for 4-6 min.

Glass Transition, Water Sorption, and Controlled Release

DSC results revealed a broad transition indicating two successive glass transitions. The onset of the first transition was at 47.8°C and that of the second transition was at 59.10 °C. The presence of a broad T_g suggested that the layers surrounding the droplets coexist in separated phases. It can be assumed that whey protein is in one phase and κ-carrageenan and/or chitosan in another one. This is probably because proteins and polysaccharides exhibit poor miscibility (*10*). Figure 5 shows the glass transition of vitamin K encapsulated into whey protein and κ-carrageenan matrix.

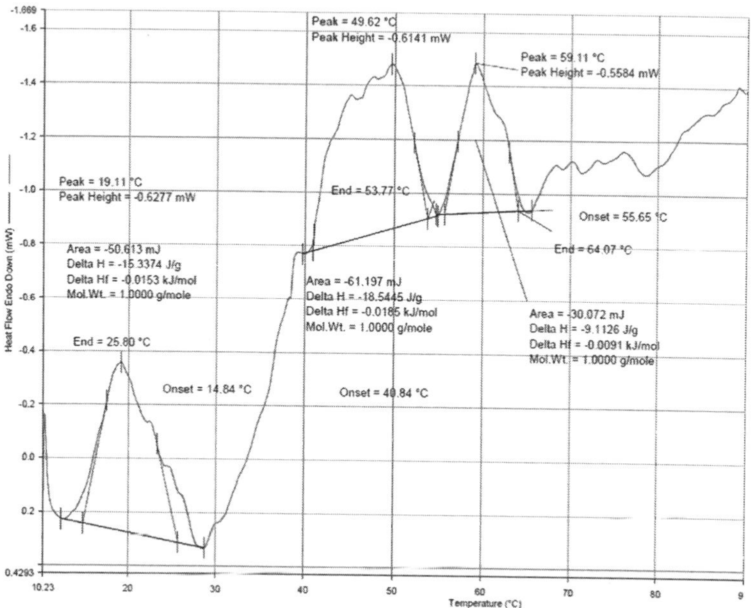

Figure 5. DSC thermogram of nanoparticles of vitamin K encapsulated into k-carrageenan and whey protein matrix; T_g of κ-carrageenan at 59.11°C; T_g of whey protein at 47.62°C.

The order in which the coating materials were added during emulsification and ultrasonication suggests that the whey protein forms the first layer around the core and κ-carragenan forms a second layer. In the flowcam image, chitosan forms the outmost layer. This broad transition indicated the presence of individual layers made of each polymer around the core droplet. The onset represents the starting point of the transition and the endset represents the end point.

Figure 6 and Figure 7 show HPLC chromatograms of samples of vitamin K and Pixoxicam, respectively. The retention time for vitamin K and Piroxicam were 7.3 min and 4.0 min, respectively.

The encapsulation efficiency (EE) was 92.1± 0.2% for vitamin K and 89.4 ± 0.3 for Piroxicam. This EE was much higher than the previously reported EE of 83% of linoleic acid in κ-carrageenan matrices (*1*). EE obtained in this study were also higher than 87.9% and 52.2% EE obtained by coating vitamin D3 with carbomethyl chitosan and zein nanoparticles, respectively (*14*).

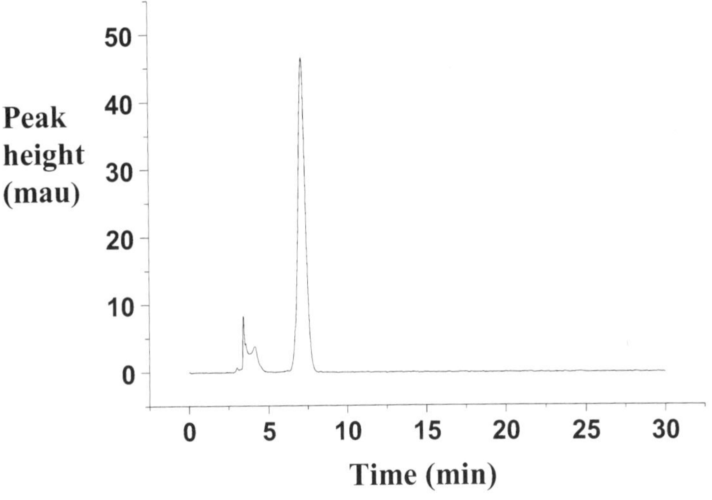

Figure 6. HPLC chromatogram of Vitamin K (1000 ppm).

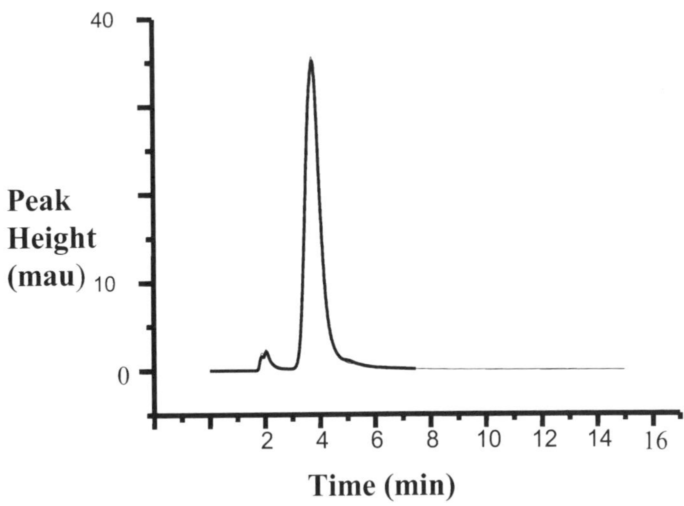

Figure 7. HPLC chromatogram of Piroxicam (1000 ppm).

231

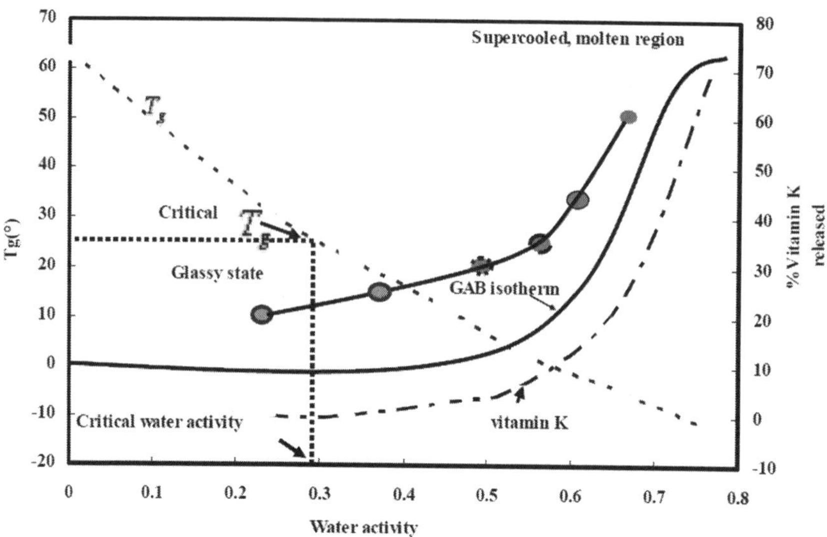

Figure 8. Water sorption properties and release profile of nanoencapsulated vitamin K. The water sorption isotherm of freeze-dried vitamin K was used to predict the Guggenheim-Anderson de Boer (GAB) sorption isotherm model. The glass transition temperature decreased as water activity increased. The supercooled molten region refers to the condition in which the coating matrices are plasticized, and the glassy state referes to the state in which vitrification occurs.

Since the temperature at which plasticization of the coating matrix occurred is useful to determine the conditions pertaining to the release of the core substance, the endset of the broad glass transition (T_g), can be considered as the temperature above which change in the physical state of the nano-microcapsules may lead to release of the core substance. Hence, below this reference temperature, vitrification is likely to occur and may lead to solidification of the coating layers (*10*). Above this temperature, according to the glass transition theory, plasticization of the coating matrices are likely to occur (*10*), while the core is gradually released from the polymeric coating matrix. Figure 8 illustrates the change in the glass transition temperature, T_g, of nanoencapsulated vitamin K. We assume that at temperatures above 60°C the whey protein layer around the lipophilic core is a super-cooled liquid state with liquid-like characteristics. Water sorption study of dried nano/microcapsules was conducted after storage of samples of about 100g at water activity conditions identical to those used for the DSC analysis. A sorption isotherm of nanoencapsulated vitamin K is shown in Figure 8. Water uptake started slightly around 0.444 a_w values and became significant at a_w values 0.662 and 0.764. The release of encapsulated vitamin K followed a similar trend with about 21% of encapsulated vitamin K released at aw value of 0.662, and 66% released at a_w value of 0.764. It appears that the release of encapsulated vitamin K occurred when the samples were in the supercooled molten state, and not in the glassy state. Figure 8 also shows a schematic representation of changes in the structure of the coating matrix in the range of water activity from 0 to 0.8. As a_w increased, T_g values decreased because the coating matrix underwent gradual plasticization.

In the glass state, the coating matrix is stable and no release of vitamin K occurred. This suggests that temperature and water activity, or water contents play important roles in the release of encapsulated vitamin K.

Conclusion

Vitamin K and Piroxicam were successively encapsulated into the whey protein-carrageenan-chitosan matrix. The sizes of the particles were in the nano-regime when the ultrasonic frequency was 10MHz for a treatment of 10 min, and around 2μm for a treatment of 5 min at 5MHz. The encapsulation was mainly dependent on coulumbic interactions between oppositely charged biopolymers. The release of vitamin K increased with increasing water activity due to plasticization of the coating matrix. The release of nano/microencapsulated bioactive and drugs is not likely when the capsules are in the glass state, but are significant in the ruberry, supercooled state. A good selection of temperature below the glass transition and water activity below 0.534 is essential to keep the encapsulated substances stable. The release profile obtained from HPLC analysis suggested that water activity values of 0.662 and higher were suitable for controlling the release of nano/microencapsulated substances. This study described efficient encapsulation methods of vitamin K and the complex drugs, Piroxicam. A future study, will address the effects of change in the physical state of the coating materials under conditions similar to the digestive system.

Acknowledgments

The authors thank the Western Illinois University Office of Sponsored Projects for supporting this research. The authors also thank Mr. Andrew Selinas at Flow Imaging Technology for conducting the flowCam imaging.

References

1. Kouassi, K. G.; Teriveedhi, V. K.; Milby, C. L.; Ahmad, T.; Boley, M. S.; Gowda, N. M.; Terry, J. R. *J. Encapsulation Adsorpt. Sci.* **2012**, *2*, 1–10.
2. Gharsallaoui, A.; Roudaut, G.; Chambin, O.; Voilley, A.; Saurel, R. *Food Res. Int.* **2007**, *40*, 1107–1121.
3. Gonnet, M; Lethuaut, L; Boury, F. *J. Controlled Release* **2010**, *146*, 276–290.
4. Luo, Y; Teng, Z.; Wang, Q. *J. Agric. Food Chem.* **2011**, *60*, 836–843.
5. Huang, Q. R; Yu, H. L; Ru, Q. M. *J. Food Sci.* **2010**, *75*, R50–R57.
6. Wilson, N; Shah, N. P. *ASEAN Food J.* **2007**, *14*, 1–14.
7. Liu, H.; Finn, N.; Yates, Z. M. *Langmuir* **2005**, *21*, 379–385.
8. Khan, T.; Park, J. K.; Kwon, J. H. *Korean J. Chem. Eng.* **2007**, *24*, 816–826.

9. Gbassi, K. G.; Vandammne, T. *Pharmaceutics* **2012**, *4*, 149–163.

10. Roos, H. Y. *Annual Rev. Food. Sci.* **2010**, *1*, 469–496.

11. Gbassi, G. K.; Vandamme, T. *Pharmaceutics* **2012**, *4*, 149–163.

12. Kouassi, K.; Joupilla, K.; Roos, Y. *J. Food. Sci.* **2002**, *67*, 2190–2195.

13. Thongngam, M.; McClements, J. *Langmuir* **2005**, *21*, 79–86.

14. Luo, Y; Zhang, B.; Whent, M.; Yu, L.; Wang, Q. *Colloids Surf., B* **2011**, *85*, 145–152.

15. Chen, L.; Remendetto, G. E.; Subirade, M. *Trends Food Sci. Technol.* **2006**, *17*, 272–2832006.

16. Kouassi, K.; Roos, H, Y. *J. Food Sci.* **2002**, *67*, 3402–3407.

17. Adibkia, K.; Siahi, S. M. R.; Nokhodchi, A.; Javadzedeh, A.; Barzegar–Jalali, M.; Barar, J.; Mohammadi, G.; Omidi, Y. *J. Drug Target.* **2007**, *15*, 407–416.

18. Rishavy, M. A.; Berkner, K. L. *J. Biochem.* **2008**, *47*, 9836–9846.

19. Van Hasselt, P. M; Jansens, G. E. P. J.; Slot, T. K.; van der Ham; Minderhoud, T. C; Talelli, M.; Akkermans, L. M; Ricken, C. J. F.; Nostrum, Van. *J. Controlled Release* **2009**, *133*, 161–168.

20. Dixit, M.; Kini, A. G.; Kulkarni, P. K. *Res. Pharm. Sci.* **2010**, *5* (2), 89–97.

Chapter 14

Encapsulation of Flavor Compounds as Helical Inclusion Complexes of Starch

K. Kasemwong[1] and T. Itthisoponkul[*,2]

[1]Nano Delivery System Laboratory, National Nanotechnology Center, National Science and Technology Development Agency, Klong-Luang, Pathumthani 12120, Thailand
[2]Faculty of Agricultural Product Innovation and Technology, Srinakharinwirot University, Bangkok 10110, Thailand
[*]E-mail: teerarat@swu.ac.th.

Inclusion complexes between starch and flavor compounds are of great interest in food science as they influence the retention and release of flavor in food systems. Starch is a mixture of the glucose polymers, amylose and amylopectin. Amylose, the linear chain is mainly responsible for the complex formation, while amylopectin, the highly branched component of starch, can also form complexes with certain types of guest molecules. In the presence of flavor compounds, amylose changes from a double helix to a single helix, forming a helical structure that has a hydrophobic cavity and a hydrophilic exterior enabling it to form inclusion complexes. The flavor molecules are included within the cavity, in between the helices, or in both locations, depending on the structure of the molecules. It has been suggested that amylose inclusion complexes can be used in the food industry to prevent the loss of volatile or labile flavoring materials during processing and storage because the complexes are markedly resistant to high temperature and oxidation. Furthermore, the release of the complexes can be controlled by α-amylase enzyme hydrolysis and changes in the moisture content and temperature.

Introduction

Encapsulation is widely used in the food industry for providing the stability and controlled release of flavor ingredients. The advantages of flavor encapsulation include protecting compounds against loss through evaporation, chemical degradation, or reaction with other components during processing and storage. The controlled release of flavor during the consumption of foods is also achieved by using this technology. It is well known that cyclodextrins, the cyclic oligosaccharides, are playing the important role of being a host to stabilize a number of flavor compounds. Cyclodextrins have a hydrophobic cavity and a hydrophilic exterior which enables them to accommodate a range of guest molecule shapes and sizes (*1*). Starch can be used as an alternative host for molecular encapsulation because it offers a unique advantage in this regard and the material cost would be less than for cyclodextrin. Starch, particularly amylose, can also form a helical structure with a hydrophobic core that can accommodate hydrophobic flavor molecules. The starch-flavor inclusion complexes are of interest in connection with flavor retention and release in foods as well as the texture of foods (*2–4*). The linear amylose fraction of starch has the ability to form molecular inclusion complexes, termed V_h-amylose, with small molecules. The molecular dimensions of the inclusion complexes are varying with external diameter of 1.35-1.62 nm and inner diameter of 5.4-8.5 Å (*5, 6*), and thus, V_h-amylose is considered as a possible platform for encapsulation.

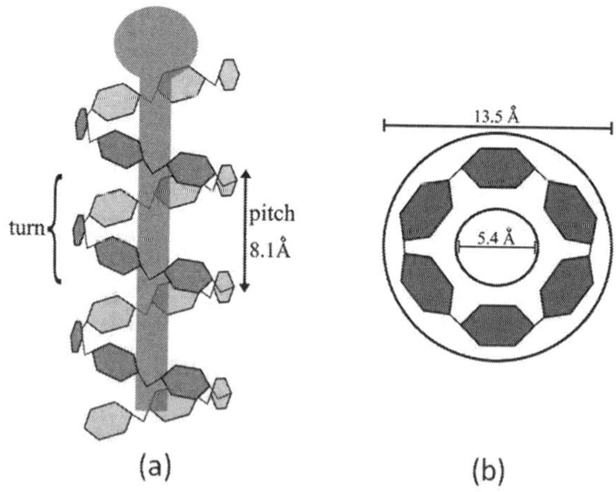

Figure 1. Left-handed single helix of V_h-amylose. (a) side view, (b) front view.

Amylose is a linear polymer consisting of $(1\rightarrow4)$-α-D-glucose units with very minimal α-$(1\rightarrow6)$ linked branching (*7*). Native amylose usually occurs as a double helical assembly. However, a simpler single helix is favored during the inclusion complexation with guest molecules. In such complexes, the hydrophobic parts of the ligand are entrapped in the central helical cavity of amylose. This type of complex, a so-called V_h arrangement (*8, 9*), normally features six glucose residues per turn in a left-handed single helix (Figure 1a). V_h-amylose has an overall helix diameter of approximately 13.5 Å, a channel of 5.4 Å width, and an axial pitch of 8.1 Å per turn (*9*) (Figure 1b). It is most often observed with small guest molecules or those of linear shape such as iodine, monoglycerides, fatty acids and long-chain alcohols (*9, 10*). Most flavor compounds are small molecules with short carbon chains and generally present a ring structure. It seems that lipids and most flavor compounds have a common property in term of the hydrophobic attribute and flavor compounds generally have poor water dispersibility. For this reason, it is possible to encapsulate flavor compounds by forming the inclusion complex with amylose.

Formation of Starch Inclusion Complexes

The important factors in forming starch inclusion complexes include starch or amylose, guest compounds and condition during preparation. The amylose form double helices in water through intramolecular hydrogen bonding. However, in the presence of suitable guest molecules, amylose undergoes conformational rearrangement to form a single helix. The other main component of starch, amylopectin, also has the ability to form inclusion complex but only amongst its longer, external branches (*2*). Inclusion complexes have been shown to be stabilized by a combination of hydrogen bonds, hydrophobic, dipolar and charge interactions (*11*). Typically, compounds that can form complexes with amylose posses both polar and nonpolar parts in their molecular structure (*12*).

Complex formation of pure amylose in aqueous system requires temperatures above 100 °C to avoid the spontaneous crystallization of amylose in the double helices form which impedes the formation of complexes due to the absence of a central channel (*2*). To obtain amylose from native starch it is first necessary to break up the starch granular structure by heating it in excess water. The method of preparing starch-flavor inclusion complexes can be explained as follows (*3*). First, the native starch or lipid free starch need to be suspended with water to make a proper concentration. The starch suspension is then placed in a seal containers and heating to 85°C with gently stirred. Second, just before cooking, the excess amount of flavor compound is added to starch paste. However, a concentration in the final products of 0.4 mmol of flavor compound per kg of starch paste is suggested. The starch-flavor mixtures are held at about 85°C for 15 min, then cool down to room temperature and held at room temperature over night to allow precipitation of complexes. The starch inclusion complexes are collected by centrifugation and rinse with ethanol to eliminate free flavor compounds. The formation of complexes can be indicated by the turbidity of solution and the formation of a precipitate during the cooling of starch mixtures (*13*). In the solid form, the single helix structure and its crystalline packing is revealed by wide-angle X-ray diffraction (*9, 10, 14*). Electron microscopy and electron diffraction has also be used to investigate the inclusion complexes of amylose (*15*). The quantitative determination of inclusion complexes can be carried out by measuring the iodine binding capacity (*12*). Moreover, calorimetric methods are appropriate to follow the formation, melting and characterization of complexes (*16*).

Molecular Structure of Amylose Inclusion Complexes

The nature of the guest molecules has a considerable influence on the helical inclusion complexes of amylose. Many different types of V_h-amylose complex are known, featuring a wide variety of guest molecules as shown in Table I.

Table I. Characteristic of amylose-flavor inclusion complexes

Type of guest compounds	Type of crystal	Characteristic of complex	Ref.
Decanal	V_h amylose or V_{6I}	Sixfold left-handed amylose helices in which the guest is included in the cavity.	(9)
Hexanal			(17)
Lactones			(18)
Butanol	V_{6II}	Six monomers of D-glycosyl per turn.	(9)
Isopropanol	V_{6III}	V_{6II}, V_{6III} : ligand could also be entrapped between helices	(19)
Menthone			(15)
Fenchone		II, III represent varying volume between helices in the crystalline stacking which are larger than Vh	(15, 20)
Geraniol			(15)
Thymol			(15)
Linalool			(3)
α–Naphthol	V_8	Amylose complex with largest helix diameter. The ligand is included in the helix and between the helices.	(9, 15)

Amylose complexes can be classified by the size of helix, which corresponds to the number of glucose monomers per helical turn, or by the packing of V_h-amylose in the crystalline structure. Linear flavor compounds such as decanal, hexanal and lactones induce the formation of amylose complexes with six glucose units per turn, and yield a V_h amylose or V_{6I} crystalline form (9, 17, 18). Crystalline amylose complexes features a larger size of cavity than V_h amylose have been observed with compounds such as menthone, fenchone, geraniol, thymol and linalool (3, 15, 20). It has been suggested that the ligands of V_{6III} complexes are entrapped between the helices in the crystal. For bulky ligands such as α–naphthol, a larger helical diameter involving eight glucose units per helical turn (V_8-amylose) has been reported (9, 15).

Flavor compounds are postulated to be accommodated within amylose helices (V_h amylose or V_{6I}) or may be entrapped outside the crystalline amylose helices (V_{6II}, V_{6III}), depending on the size of the ligands (Figure 2). In addition, as the exposed surface of the helices is more hydrophilic than the interior, the hydrophobic flavor compounds preferred to stay inside the helices while the hydrophilic flavor compounds preferentially lie entrapped between the helices (21).

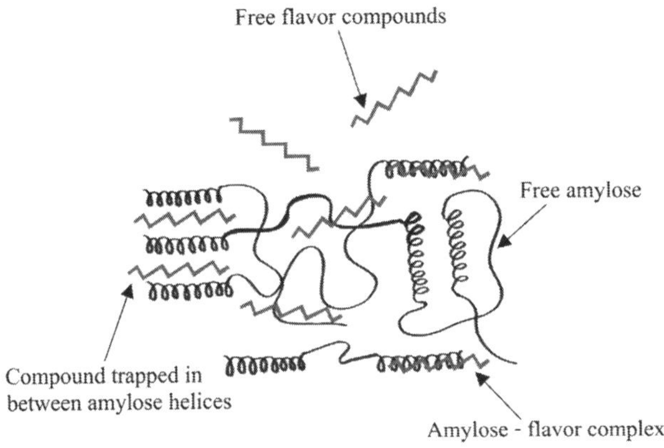

Figure 2. Schematic showing modes of complexation between amylose and ligand flavor compounds.

Inclusion Complexes between Starch and Flavor

Different starches exhibit different tendencies to form complexes, largely on account of their differing amylose content. As expected, amylose-rich starch tends to bind greater amount of compounds (*4, 6*). Furthermore, use of commercial native starch appeals because of its low cost compared to that of pure amylose. Native starch that contains little or no internal lipid is preferred because internal lipids can compete with flavor compounds and interfere with amylose complex formation (*22*). In the system that contains native lipid and low solubility flavor compounds, the structure may consist of starch-lipids and starch-lipid-flavor compounds. While in the presence of high solubility flavor compounds, the system may consist of starch-flavor compounds, starch-lipids and starch-lipid-flavor compounds (*21*). The mixed helical structure of complexes would result in compromised stability and an altered release profile of flavor compounds. For this reason, potato starch (internal lipid (<0.1 %) (*23*) has been used more often than other similar starches in studies on amylose inclusion complexes. It binds (-)fenchone, menthone and geraniol, resulting in a crystalline structure of V_{6III} (V-isopropanol pattern). Moreover, the complex with decanal showed a V_h amylose pattern (*9*). In addition, γ- and δ-lactones with a linear chain length $\geq$ C5 formed inclusion complexes with potato starch, whereas lactones with a short linear chain showed poor complexing ability. Most lactones induced the formation of V_h helices, except δ-decalactone (*18*).

Inclusion complexes of other native starches and flavors have previously been investigated including corn starch (*24*) and pea starch (*25*) with linear alcohols, tapioca starch (*26*) with alcohol and ketone compounds, high-amylose maize starch with limonene, cymene, thymol, menthone (*21*) and α–naphthol (*4*).

Several works have reported evidence of pure amylose complexed with flavor compounds such as *n*-butanol, isopropanol, thymol, linalool, α–naphthol, limonene, menthol and menthone (*6, 9, 12, 15, 27*). Here the amylose chain length significantly affects complex formation and the properties of the complexes themselves. In general, the longer the amylose chains, the more stable are the complexes that are formed (*28*). However, conformational disorder within the crystal structure may occur if the amylose chains are too long or too short. An amylose chain length of containing 8–16 glucose units per bound flavor molecule was suggested by Wulff *et al.* (*29*) to maximize stability.

The ratio of ligand to amylose also affects the characteristics of the amylose inclusion complex. With an excess of ligand some of the uncomplexed ligands become trapped between the amylose helices. At low flavor concentration, uncomplexed amylose remaining after ligand is utilized forms double helices conformation instead of single helix, thereby diminution of the formation of inclusion complexes. The optimum guest molecule-to-amylose ratio varies depending on the structure and solubility of guest molecules in the amylose solution. For example, 10% lipids is sufficient for complexing all amylose molecules (*30*), however for some flavor compounds, a loading of 4-10% flavor compounds is attainable maximum (*21, 29*). It is likely that, amylose or starch-inclusion complexes can exist in different polymorphs of V-amylose depending on the factors such as amylose fraction, flavor compound characteristics and concentration.

Thermostability and Melting Behavior of the Amylose-Flavor Complexes

Regardless of flavor compounds types, amylose complexes can form either amorphous or crystalline structures, depending on the temperature at which they are synthesised (*31*). Amorphous (type I) complexes are formed below 60 °C and exhibit a low order of crystallinity, while the type II complexes are obtained at temperature of at least 90 °C and are typically characterized by a well defined crystalline structure. Because of their lack of crystallinity, type I complexes usually melts 10-30 degrees below those of type II, depending on the ligands and the experimental conditions (*24, 25*). The nature of the bound flavor compound markedly influences the distribution of complex types formed and its melting temperature as shown in Table II.

Several studies have shown that the melting temperature of amylose complexes increases with the chain length for ligand compounds such as fatty acids and alcohols. Whittam *et al.* (*24*) reported that complexes formed between amylose and C4-C8 alcohols showed the endothermic transition increased with hydrocarbon chain length. The enthalpy of complex dissociation did not vary significantly with alcohol chain length. The same correlation between chain length of linear alcohols (C7-C10) and melting temperatures and enthalpies has also been reported (*26*).

In general, it is only possible to dissociate an inclusion complex at, or above its melting temperature. Since amylose-flavor complexes melt at high temperature, it is suggested that they offer a practical method for stabilizing flavor against high temperature and degradation during prolonged storage times. There have been only a limited number of studies on the stability of flavor-inclusion complexes against oxidation. It has been reported that tapioca starch-linalool complex had a great ability to protect linalool against oxidation across a wide range of relative humidity levels (11-75%RH) (*32*).

Table II. Type of amylose-flavor complexes and their melting temperature

Guest compounds	*Type of complexes and Melting temperature (°C)*	*Ref.*
Decanal	typeI (90) + typeII (100)	(*9*)
Butanol	typeI (66.5) + typeII (80)	(*9*)
Menthone	typeII (114)	(*15*)
1-Naphthol	typeII (121)	(*9*)
Carvone	typeII (91)	(*18*)
Geraniol	typeII (91)	(*15*)
Campher	typeI (76)	(*18*)
Thymol	typeII (105)	(*15*)
Fenchone	typeII (114)	(*15*)
Geraniol	typeII (92)	(*15*)
γ–Heptalactone	typeI (90) + typeII (117)	(*18*)
γ-Nonalactone	typeII (90)	(*18*)
γ-Decalactone	typeI (95) + typeII (127)	(*18*)
γ–Dodecalactone	typeI (72) + typeII (105)	(*18*)
δ–Decalactone	typeII (104)	(*18*)
δ –Dodecalactone	typeI (89) + typeII (115)	(*18*)

Release of Flavor from Amylose Inclusion Complexes

The release of bound flavor molecules from the encapsulated starch complexes is of importance in the flavor perception in food. The release of flavour compounds can occur by various methods including thermal, mechanical and hydrodynamic treatment (*33*). It has previously been stipulated that high temperatures are required to dissociate amylose-flavor complexes (*17, 26*). In addition, α-amylase (*34*) and moisture (*32*) are alternative options. The release of amylose-hexanal complex was investigated and found that at 30°C the guest molecules were stable but at high storage temperature (80°C), 9 % of hexanal had been released after 21 h. However, when the complexes were heated in the excess of water at 30°C, a sudden release of the hexanal occurred and the release increased with temperature. Almost no evaporation or oxidative degradation was observed after one year storage under dry conditions at 20°C (*29*).

The stability of complexes across a range of different water activities (a$_w$) was also investigated at 30°C, 50°C and 80°C. At 30°C and up to a$_w$=0.5 no hexanal was released; but at 50°C and 80°C the release of hexanal increased with a$_w$. These results are compatible with those of recent work (*6*) which found that the release of menthone and menthol increased with temperature. There was no aroma loss at low temperature (4°C and 25°C), but at 80°C only 65 % of aroma retained in the complex. It has also reported that volatile release of volatile compouds depended on relative humidity. The release of volatiles from tapioca starch-alcohol complexes was greater at high relative humidity (*32*).

As the dissociation of the amylose-flavor complexes is possible upon complex melting at high temperature, it is possible to apply this system to food products that are processed at high temperature such as baking, drying, extrusion and pasteurization in which a high retention of flavor is maintained. For example, in hard candy making, the flavor compounds are added to the molten candy at 120°C. It is expected that the good retention of flavor compounds in the hard candy was achieved if starch-flavor inclusion complexes were used. It should be noted that the food products such as instant soup and noodles, tea and other hot drinks often need to be consumed at higher temperatures as well to help fully release their flavors.

Studies involving α–amylase indicate that amylose-lipid complexes display a substantially reduced susceptibility to this hydrolytic enzyme in comparison to uncomplexed amylose. The enzymatic breakdown of partly crystalline amylose-lipid structures is initiated in the amorphous areas on the surface of lamellar regions of crystalline amylose helices. The remaining crystalline fragments are hydrolysed rather more slowly and the final extent of starch hydrolysis depends on the α–amylase activity (*35*). The digestibility of amylose complexes depends not only on the incubation time but also on the enzyme concentration and the specific structure of inclusion complex. It was showed that complexes of potato starch and either geraniol or γ-nonalactone were very slowly attacked at low enzyme concentrations which is similar to human saliva (α–amylase 25-200 U/g dry starch). Rapid breakdown of the complexes became possible in the gastrointestinal tract where the concentration of enzyme was high (2000 U/g dry starch) (*36*). The release of menthone and menthol from inclusion complexes in the mouth condition has been investigated and reported that at first 40 min only about half of the aroma was released and that full release was only achieved after 4h of hydrolysis (*6*). Release of flavor from amylose complexes is unlikely during the chewing process that lasts mere seconds. However, inclusion complexes can be used as flavor carriers in food systems where a slow flavor release is desired, such as in chewing gum. Furthermore, such complexes can be used to mask the bitter taste or off-flavor of certain compounds.

The starch-flavor complexes can considerably affect the texture and hence the flavor release properties of foods (*36*). For examples, starch-naphthol inclusion complex affected the rheological properties of wheat starch by enhancing the stronger shear-thinning properties of wheat starch gel (*4*). Potato starch-lactone inclusion complexes resulted in lower gelation times and higher gel strength (*18*). Starch inclusion complexes with decanal and fenchone would readily induce the gelation of starch dispersion (*20*). Cayot et al. (*37*) studied the release of isoamyl acetate from a starch-based food matrix and showed that potato starch cream had a better aroma retention than corn starch cream because potato starch has no internal lipid and may thus form stronger complexes with isoamylacetate. However, the researchers found that the release of the aroma compounds was not correlated only to the rheology of the starch gel because the experiments was conducted in a complex matrix and that aroma compounds can also be bound to other ingredients in the system.

It would be of great interest to find a way to prove if starch-inclusion complexes affect the texture and release of flavor in real food systems. In addition, more work is necessary to evaluate the effect of different types of starch-inclusion complexes on the flavor retention and release and on the rheploghy of the food matrix. To our knowledge, few works have been reported that probe flavor release from inclusion complexes during oral digestion. Interesting future studies could involve the in-vivo testing of both the release of required flavor and masking of off-flavor compounds.

Conclusions

Amylose inclusion complexes can be used as a platform to provide thermal and oxidative stability and a controlled release of flavor compounds. Under the right conditions, either pure amylose or raw starch form amylose single helices, when mixed with the desired flavor. These entrap the small-molecule flavor compounds either inside or between the amylose helices with a binding favorability that appears to depend on properties of the ligand such as its steric hindrance and solubility. Release from starch-flavor complexes increases with temperature and moisture content, while the release by α-amylase is possible but is time consuming. Retaining, stabilizing and controlling the release of foodstuff flavors via amylose inclusion complexes continues to be show exciting potential for the food industry for which more research is needed to improve this technology and its valuable applications.

References

1. Reineccius, T. A.; Reineccius, G. A.; Peppard, T. L. *J. Food Sci.* **2004**, *69*, FCT58–FCT62.
2. Conde-Petit, B.; Escher, F.; Nuessli, J. *Trends Food Sci. Technol.* **2006**, *17*, 227–235.
3. Arvisenet, G.; Le Bail, P.; Voilley, A.; Cayot, N. *J. Agric. Food Chem.* **2002**, *50*, 7088–7093.
4. Zhu, F; Wang, Y. J. *Food Chem.* **2013**, *138*, 256–262.
5. Immel, S.; Lichtenthaler, F. W. *Starch/Staerke* **2000**, *52*, 1–8.

6. Ades, H.; Kesselman, E.; Ungar, Y.; Shimoni, E. *LWT−Food Sci. Technol.* **2012**, *45*, 277–288.

7. Bertoft, E. In *Starch in Food:Structure, Function and Applications*; Eliassion, A. C., Ed.; Woodhead Publishing Limited and CRC Press: NY, 2004; pp. 57–96

8. Zobel, H. F.; French, A. D.; Hinkle, M. E. *Biopolymers* **1967**, *5*, 837–845.

9. Le Bail, P.; Rondeau, C.; Buleon, A. *Int. J. Biol. Macromol.* **2005**, *35*, 1–7.

10. Putseys, J. A.; Lamberts, L.; Delcour, J. A. *J. Cereal Sci.* **2010**, *51*, 238–247.

11. Kuge, T.; Takeo, K. *J. Agric. Biol. Chem.* **1968**, *32*, 753–758.

12. Rutschmann, M. A.; Solms, J. *LWT−Food Sci. Technol* **1990**, *23*, 84–87.

13. Osman-Ismail, F.; Solms, J. *LWT−Food Sci. Technol* **1973**, *6*, 147–150.

14. Kawada, J.; Marchessault, R. H. *Starch/Starke* **2004**, *56*, 13–19.

15. Nuessli, J.; Putaux, J. L.; Bail, P. L.; Buleon, A. *Int. J. Biol. Macromol.* **2003**, *33*, 227–234.

16. Kubik, S.; Wulff, G. *Starch/Starke* **1993**, *45*, 220–225.

17. Nuessli, J.; Sigg, B.; Conde-Petit, B.; Escher, F. *Food Hydrocolloids* **1997**, *11*, 27–34.

18. Heinemann, C.; Conde-Petit, B.; Nuessli, J.; Escher, F. *J. Agric. Food Chem.* **2001**, *49*, 1370–1376.

19. Buleon, A.; Delage, M. M.; Brisson, J.; Chanzy, H. *Int. J. Biol. Macromol.* **1990**, *12*, 25–33.

20. Nuessli, J.; Conde-Petit, B.; Trommsdorff, U. R.; Escher, F. *Carbohydr. Polym.* **1995**, *28*, 167–170.

21. Tapanapunnitikul, O.; Chaiseri, S.; Peterson, D. G.; Thomson, D. B. *J. Agric. Food Chem.* **2008**, *56*, 220–226.

22. Jouquand, C.; Ducruet, V.; Le Bail, P. *Food Chem.* **2006**, *96*, 461–470.

23. Moorthy, S. N. *Starch/Starke* **2002**, *54*, 559–592.

24. Whittam, M. A.; Orford, P. D.; Ring, S. G.; Clark, S. A.; Parker, M. L.; Cairns, P.; Miles, M. J. *Int. J. Biol. Macromol.* **1989**, *11*, 339–344.

25. Kowblansky, M. *Macromolecules* **1985**, *18*, 1776–1779.

26. Itthisoponkul, T.; Mitchell, J. R.; Taylor, A. J.; Farhat, I. A. *Carbohydr. Polym.* **2007**, *69*, 106–115.

27. Rondeau-Mouro, C.; Le Bail, P.; Buleon, A. *Int. J. Biol. Macromol.* **2004**, *34*, 309–315.

28. Gelders, G. G.; Vanderstukken, T. C.; Goesaert, H.; Delcour, J. A. *Carbohydr. Polym.* **2004**, *56*, 447–458.

29. Wulff, G.; Avgenaki, G.; Guzmann, G. M. S. P. *J. Cereal Sci.* **2005**, *41*, 239–249.

30. Lebail, P.; Buleon, A.; Shiftan, D.; Marchessault, R. H. *Carbohydr. Polym.* **2000**, *43*, 317–326.

31. Biliaderis, C. G.; Page, C. M.; Slade, L.; Sirett, R. R. *Carbohydr. Polym.* **1985**, *5*, 367–389.

32. Itthisoponkul, T. Ph.D. Thesis, University of Nottingham, Nottingham, U.K., 2009.

33. Whorton, C. In *Encapsulation and Controlled Release of Food Ingredients*; Risch, S. J., Reineccius, G. A., Eds.; ACS Symposium Series 590, American Chemical Society: Washington, DC, 1995; pp. 134-142

34. Nebesny, E.; Rosicka, J.; Tkaczyk, M. *Starch/Stärke* **2005**, *57*, 325–331.

35. Seneviratne, H. D.; Biliaderis, C. G. *J. Cereal Sci.* **1991**, *13*, 129–143.

36. Heinemann, C.; Zinsli, M.; Renggli, A.; Escher, F.; Conde-Petit, B. *LWT-Food Sci. Technol.* **2005**, *38*, 885–894.

37. Cayot, N.; Taisant, C.; Voilley, A. *J. Agric. Biol. Chem.* **1998**, *46*, 3201–3206.

Editors' Biographies

Michael Appell

Michael Appell is Research Chemist at the Agricultural Research Service, United States Department of Agriculture.

Bosoon Park

Bosoon Park is Research Agricultural Engineer at the Agricultural Research Service, United States Department of Agriculture.

Indexes

Author Index

Subject Index